continuum theory
and dynamical systems

LECTURE NOTES IN PURE AND APPLIED MATHEMATICS

1. *N. Jacobson,* Exceptional Lie Algebras
2. *L.-Å. Lindahl and F. Poulsen,* Thin Sets in Harmonic Analysis
3. *I. Satake,* Classification Theory of Semi-Simple Algebraic Groups
4. *F. Hirzebruch, W. D. Newmann, and S. S. Koh,* Differentiable Manifolds and Quadratic Forms
5. *I. Chavel,* Riemannian Symmetric Spaces of Rank One
6. *R. B. Burckel,* Characterization of C(X) Among Its Subalgebras
7. *B. R. McDonald, A. R. Magid, and K. C. Smith,* Ring Theory: Proceedings of the Oklahoma Conference
8. *Y.-T. Siu,* Techniques of Extension on Analytic Objects
9. *S. R. Caradus, W. E. Pfaffenberger, and B. Yood,* Calkin Algebras and Algebras of Operators on Banach Spaces
10. *E. O. Roxin, P.-T. Liu, and R. L. Sternberg,* Differential Games and Control Theory
11. *M. Orzech and C. Small,* The Brauer Group of Commutative Rings
12. *S. Thomier,* Topology and Its Applications
13. *J. M. Lopez and K. A. Ross,* Sidon Sets
14. *W. W. Comfort and S. Negrepontis,* Continuous Pseudometrics
15. *K. McKennon and J. M. Robertson,* Locally Convex Spaces
16. *M. Carmeli and S. Malin,* Representations of the Rotation and Lorentz Groups: An Introduction
17. *G. B. Seligman,* Rational Methods in Lie Algebras
18. *D. G. de Figueiredo,* Functional Analysis: Proceedings of the Brazilian Mathematical Society Symposium
19. *L. Cesari, R. Kannan, and J. D. Schuur,* Nonlinear Functional Analysis and Differential Equations: Proceedings of the Michigan State University Conference
20. *J. J. Schäffer,* Geometry of Spheres in Normed Spaces
21. *K. Yano and M. Kon,* Anti-Invariant Submanifolds
22. *W. V. Vasconcelos,* The Rings of Dimension Two
23. *R. E. Chandler,* Hausdorff Compactifications
24. *S. P. Franklin and B. V. S. Thomas,* Topology: Proceedings of the Memphis State University Conference
25. *S. K. Jain,* Ring Theory: Proceedings of the Ohio University Conference
26. *B. R. McDonald and R. A. Morris,* Ring Theory II: Proceedings of the Second Oklahoma Conference
27. *R. B. Mura and A. Rhemtulla,* Orderable Groups
28. *J. R. Graef,* Stability of Dynamical Systems: Theory and Applications
29. *H.-C. Wang,* Homogeneous Branch Algebras
30. *E. O. Roxin, P.-T. Liu, and R. L. Sternberg,* Differential Games and Control Theory II
31. *R. D. Porter,* Introduction to Fibre Bundles
32. *M. Altman,* Contractors and Contractor Directions Theory and Applications
33. *J. S. Golan,* Decomposition and Dimension in Module Categories
34. *G. Fairweather,* Finite Element Galerkin Methods for Differential Equations
35. *J. D. Sally,* Numbers of Generators of Ideals in Local Rings
36. *S. S. Miller,* Complex Analysis: Proceedings of the S.U.N.Y. Brockport Conference
37. *R. Gordon,* Representation Theory of Algebras: Proceedings of the Philadelphia Conference
38. *M. Goto and F. D. Grosshans,* Semisimple Lie Algebras
39. *A. I. Arruda, N. C. A. da Costa, and R. Chuaqui,* Mathematical Logic: Proceedings of the First Brazilian Conference
40. *F. Van Oystaeyen,* Ring Theory: Proceedings of the 1977 Antwerp Conference
41. *F. Van Oystaeyen and A. Verschoren,* Reflectors and Localization: Application to Sheaf Theory
42. *M. Satyanarayana,* Positively Ordered Semigroups
43. *D. L Russell,* Mathematics of Finite-Dimensional Control Systems
44. *P.-T. Liu and E. Roxin,* Differential Games and Control Theory III: Proceedings of the Third Kingston Conference, Part A
45. *A. Geramita and J. Seberry,* Orthogonal Designs: Quadratic Forms and Hadamard Matrices
46. *J. Cigler, V. Losert, and P. Michor,* Banach Modules and Functors on Categories of Banach Spaces

47. *P.-T. Liu and J. G. Sutinen*, Control Theory in Mathematical Economics: Proceedings of the Third Kingston Conference, Part B
48. *C. Byrnes*, Partial Differential Equations and Geometry
49. *G. Klambauer*, Problems and Propositions in Analysis
50. *J. Knopfmacher*, Analytic Arithmetic of Algebraic Function Fields
51. *F. Van Oystaeyen*, Ring Theory: Proceedings of the 1978 Antwerp Conference
52. *B. Kadem*, Binary Time Series
53. *J. Barros-Neto and R. A. Artino*, Hypoelliptic Boundary-Value Problems
54. *R. L. Sternberg, A. J. Kalinowski, and J. S. Papadakis*, Nonlinear Partial Differential Equations in Engineering and Applied Science
55. *B. R. McDonald*, Ring Theory and Algebra III: Proceedngs of the Third Oklahoma Conference
56. *J. S. Golan*, Structure Sheaves Over a Noncommutative Ring
57. *T. V. Narayana, J. G. Williams, and R. M. Mathsen*, Combinatorics, Representation Theory and Statistical Methods in Groups: YOUNG DAY Proceedings
58. *T. A. Burton*, Modeling and Differential Equations in Biology
59. *K. H. Kim and F. W. Roush*, Introduction to Mathematical Consensus Theory
60. *J. Banas and K. Goebel*, Measures of Noncompactness in Banach Spaces
61. *O. A. Nielson*, Direct Integral Theory
62. *J. E. Smith, G. O. Kenny, and R. N. Ball*, Ordered Groups: Proceedings of the Boise State Conference
63. *J. Cronin*, Mathematics of Cell Electrophysiology
64. *J. W. Brewer*, Power Series Over Commutative Rings
65. *P. K. Kamthan and M. Gupta*, Sequence Spaces and Series
66. *T. G. McLaughlin*, Regressive Sets and the Theory of Isols
67. *T. L. Herdman, S. M. Rankin III, and H. W. Stech*, Integral and Functional Differential Equations
68. *R. Draper*, Commutative Algebra: Analytic Methods
69. *W. G. McKay and J. Patera*, Tables of Dimensions, Indices, and Branching Rules for Representations of Simple Lie Algebras
70. *R. L. Devaney and Z. H. Nitecki*, Classical Mechanics and Dynamical Systems
71. *J. Van Geel*, Places and Valuations in Noncommutative Ring Theory
72. *C. Faith*, Injective Modules and Injective Quotient Rings
73. *A. Fiacco*, Mathematical Programming with Data Perturbations I
74. *P. Schultz, C. Praeger, and R. Sullivan*, Algebraic Structures and Applications: Proceedings of the First Western Australian Conference on Algebra
75. *L Bican, T. Kepka, and P. Nemec*, Rings, Modules, and Preradicals
76. *D. C. Kay and M. Breen*, Convexity and Related Combinatorial Geometry: Proceedings of the Second University of Oklahoma Conference
77. *P. Fletcher and W. F. Lindgren*, Quasi-Uniform Spaces
78. *C.-C. Yang*, Factorization Theory of Meromorphic Functions
79. *O. Taussky*, Ternary Quadratic Forms and Norms
80. *S. P. Singh and J. H. Burry*, Nonlinear Analysis and Applications
81. *K. B. Hannsgen, T. L. Herdman, H. W. Stech, and R. L. Wheeler*, Volterra and Functional Differential Equations
82. *N. L. Johnson, M. J. Kallaher, and C. T. Long*, Finite Geometries: Proceedings of a Conference in Honor of T. G. Ostrom
83. *G. I. Zapata*, Functional Analysis, Holomorphy, and Approximation Theory
84. *S. Greco and G. Valla*, Commutative Algebra: Proceedings of the Trento Conference
85. *A. V. Fiacco*, Mathematical Programming with Data Perturbations II
86. *J.-B. Hiriart-Urruty, W. Oettli, and J. Stoer*, Optimization: Theory and Algorithms
87. *A. Figa Talamanca and M. A. Picardello*, Harmonic Analysis on Free Groups
88. *M. Harada*, Factor Categories with Applications to Direct Decomposition of Modules
89. *V. I. Istrățescu*, Strict Convexity and Complex Strict Convexity
90. *V. Lakshmikantham*, Trends in Theory and Practice of Nonlinear Differential Equations
91. *H. L. Manocha and J. B. Srivastava*, Algebra and Its Applications
92. *D. V. Chudnovsky and G. V. Chudnovsky*, Classical and Quantum Models and Arithmetic Problems
93. *J. W. Longley*, Least Squares Computations Using Orthogonalization Methods
94. *L. P. de Alcantara*, Mathematical Logic and Formal Systems
95. *C. E. Aull*, Rings of Continuous Functions
96. *R. Chuaqui*, Analysis, Geometry, and Probability
97. *L. Fuchs and L. Salce*, Modules Over Valuation Domains

98. *P. Fischer and W. R. Smith,* Chaos, Fractals, and Dynamics

99. *W. B. Powell and C. Tsinakis,* Ordered Algebraic Structures

100. *G. M. Rassias and T. M. Rassias,* Differential Geometry, Calculus of Variations, and Their Applications

101. *R.-E. Hoffmann and K. H. Hofmann,* Continuous Lattices and Their Applications

102. *J. H. Lightbourne III and S. M. Rankin III,* Physical Mathematics and Nonlinear Partial Differential Equations

103. *C. A. Baker and L, M. Batten,* Finite Geometrics

104. *J. W. Brewer, J. W. Bunce, and F. S. Van Vleck,* Linear Systems Over Commutative Rings

105. *C. McCrory and T. Shifrin,* Geometry and Topology: Manifolds, Varieties, and Knots

106. *D. W. Kueker, E. G. K. Lopez-Escobar, and C. H. Smith,* Mathematical Logic and Theoretical Computer Science

107. *B.-L. Lin and S. Simons,* Nonlinear and Convex Analysis: Proceedings in Honor of Ky Fan

108. *S. J. Lee,* Operator Methods for Optimal Control Problems

109. *V. Lakshmikantham,* Nonlinear Analysis and Applications

110. *S. F. McCormick,* Multigrid Methods: Theory, Applications, and Supercomputing

111. *M. C. Tangora,* Computers in Algebra

112. *D. V. Chudnovsky and G. V. Chudnovsky,* Search Theory: Some Recent Developments

113. *D. V. Chudnovsky and R. D. Jenks,* Computer Algebra

114. *M. C. Tangora,* Computers in Geometry and Topology

115. *P. Nelson, V. Faber, T. A. Manteuffel, D. L. Seth, and A. B. White, Jr.,* Transport Theory, Invariant Imbedding, and Integral Equations: Proceedings in Honor of G. M. Wing's 65th Birthday

116. *P. Clément, S. Invernizzi, E. Mitidieri, and I. I. Vrabie,* Semigroup Theory and Applications

117. *J. Vinuesa,* Orthogonal Polynomials and Their Applications: Proceedings of the International Congress

118. *C. M. Dafermos, G. Ladas, and G. Papanicolaou,* Differential Equations: Proceedings of the EQUADIFF Conference

119. *E. O. Roxin,* Modern Optimal Control: A Conference in Honor of Solomon Lefschetz and Joseph P. Lasalle

120. *J. C. Díaz,* Mathematics for Large Scale Computing

121. *P. S. Milojević,* Nonlinear Functional Analysis

122. *C. Sadosky,* Analysis and Partial Differential Equations: A Collection of Papers Dedicated to Mischa Cotlar

123. *R. M. Shortt,* General Topology and Applications: Proceedings of the 1988 Northeast Conference

124. *R. Wong,* Asymptotic and Computational Analysis: Conference in Honor of Frank W. J. Olver's 65th Birthday

125. *D. V. Chudnovsky and R. D. Jenks,* Computers in Mathematics

126. *W. D. Wallis, H. Shen, W. Wei, and L. Zhu,* Combinatorial Designs and Applications

127. *S. Elaydi,* Differential Equations: Stability and Control

128. *G. Chen, E. B. Lee, W. Littman, and L. Markus,* Distributed Parameter Control Systems: New Trends and Applications

129. *W. N. Everitt,* Inequalities: Fifty Years On from Hardy, Littlewood and Pólya

130. *H. G. Kaper and M. Garbey,* Asymptotic Analysis and the Numerical Solution of Partial Differential Equations

131. *O. Arino, D. E. Axelrod, and M. Kimmel,* Mathematical Population Dynamics: Proceedings of the Second International Conference

132. *S. Coen,* Geometry and Complex Variables

133. *J. A. Goldstein, F. Kappel, and W. Schappacher,* Differential Equations with Applications in Biology, Physics, and Engineering

134. *S. J. Andima, R. Kopperman, P. R. Misra, J. Z. Reichman, and A. R. Todd,* General Topology and Applications

135. *P Clément, E. Mitidieri, B. de Pagter,* Semigroup Theory and Evolution Equations: The Second International Conference

136. *K. Jarosz,* Function Spaces

137. *J. M. Bayod, N. De Grande-De Kimpe, and J. Martínez-Maurica,* p-adic Functional Analysis

138. *G. A. Anastassiou,* Approximation Theory: Proceedings of the Sixth Southeastern Approximation Theorists Annual Conference

139. *R. S. Rees,* Graphs, Matrices, and Designs: Festschrift in Honor of Norman J. Pullman

140. *G. Abrams, J. Haefner, and K. M. Rangaswamy,* Methods in Module Theory

141. *G. L. Mullen and P. J.-S. Shiue*, Finite Fields, Coding Theory, and Advances in Communications and Computing
142. *M. C. Joshi and A. V. Balakrishnan*, Mathematical Theory of Control: Proceedings of the International Conference
143. *G. Komatsu and Y. Sakane*, Complex Geometry: Proceedings of the Osaka International Conference
144. *I. J. Bakelman*, Geometric Analysis and Nonlinear Partial Differential Equations
145. *T. Mabuchi and S. Mukai*, Einstein Metrics and Yang–Mills Connections: Proceedings of the 27th Taniguchi International Symposium
146. *L. Fuchs and R. Göbel*, Abelian Groups: Proceedings of the 1991 Curaçao Conference
147. *A. D. Pollington and W. Moran*, Number Theory with an Emphasis on the Markoff Spectrum
148. *G. Dore, A. Favini, E. Obrecht, and A. Venni*, Differential Equations in Banach Spaces
149. *T. West*, Continuum Theory and Dynamical Systems
150. *K. D. Bierstedt, A. Pietsch, W. Ruess, and D. Vogt*, Functional Analysis

Additional Volumes in Preparation

continuum theory and dynamical systems

edited by
Thelma West
The University of Southwestern Louisiana
Lafayette, Louisiana

Marcel Dekker, Inc.   **New York • Basel • Hong Kong**

Library of Congress Cataloging-in-Publication Data

Continuum theory and dynamical systems / edited by Thelma West.
    p. cm. -- (Lecture notes in pure and applied mathematics ; 149)
    Includes bibliographical references and index.
    ISBN 0-8247-9072-3 (acid-free)
    1. Continuum (Mathematics) 2. Differentiable dynamical systems I. West, Thelma.
    II. Series: Lecture notes in pure and applied mathematics : v. 149.
    QA611.28.C66   1993
    515'.32--dc20

                                                                93-14090
                                                                CIP

The publisher offers discounts on this book when ordered in bulk quantities. For more information, write to Special Sales/Professional Marketing at the address below.

This book is printed on acid-free paper.

Marcel Dekker, Inc.
270 Madison Avenue, New York, New York 10016

Current printing (last digit):
10 9 8 7 6 5 4 3 2 1

PRINTED IN THE UNITED STATES OF AMERICA

Preface

Expansion of knowledge of the relationship between continuum theory and dynamical systems and an increased interest among mathematicians has created a need for a forum where researchers in each of these fields could learn from one another. This need led to the Conference/Workshop on Continuum Theory and Dynamical Systems, sponsored by the Department of the Navy, the Naval Surface Warfare Center, and the University of Southwestern Louisiana, Office of Research and Sponsored Programs, and held in Lafayette, Louisiana.

The first part of this meeting consisted of a series of lectures on basic dynamical systems by Marcy Barge and a series of lectures on basic continuum theory by Sam B. Nadler, Jr. The second part consisted of various talks on continuum theory, dynamical systems, and the relationships between the two areas. The papers in this volume are representative of the topics covered at the conference; some come directly from talks given, while others are on related topics. This volume includes research in final form, preliminary results, and expository papers. Robert Cawley of the Naval Surface Warfare Center was largely responsible for making the Conference/Workshop on Continuum Theory and Dynamical Systems possible; I thank him and the Naval Surface Warfare Center for their support. I would like to thank Marcy Barge and Sam B. Nadler, Jr., who both gave a series of lectures at the conference. Thanks are due the invited speakers who each gave one-hour talks: Kathleen Alligood, Morton Brown, Robert Devaney, John Franks, Jenny Harrison, Judy Kennedy, and James T. Rogers, Jr. I would also like to acknowledge the efforts of the other speakers at the conference.

Many people at the University of Southwestern Louisiana helped to make this conference a success. Thanks are due to Wayne Denton, who as Vice President of the Office of Research and Sponsored Programs was supportive of this conference from its inception. Many faculty members and graduate students in the Mathematics Department helped to make this conference a success; but I particularly want to recognize Bradd Clark, Vic Schneider, Lloyd Roeling, Roger Waggoner, Kathleen Lopez, Donna Thompson, Judith Covington, Brad Davis, Jennifer Prejean, and Kevin Reeves for their contributions.

Many people helped to make this volume possible. I appreciate the authors who contributed papers and those who refereed articles. I also thank Maria Allegra of Marcel Dekker, Inc., who worked with me to complete these proceedings.

Thelma West

Contents

Preface   iii

Contributors   vii

Accessible Rotation Numbers for Chaotic States   1
Kathleen T. Alligood and Timothy Sauer

Prime End Rotation Numbers Associated with the Hénon Maps   15
Marcy Barge

On the Rotation Shadowing Property for Annulus Maps   35
Fernanda Botelho and Liang Chen

A Nielsen–Type Theorem for Area–Preserving Homeomorphisms
of the Two Disc   43
Kenneth Boucher, Morton Brown, and Edward Slaminka

Irrational Rotations on Simply Connected Domains   51
Beverly L. Brechner

The Rotational Dynamics of Cofrontiers   59
Beverly L. Brechner, Merle D. Guay, and John C. Mayer

A Periodic Homeomorphism of the Plane   83
Morton Brown

Dynamical Conections Between a Continuous Map and Its
Inverse Limit Space   89
Liang Chen and Shihai Li

Horseshoelike Mappings and Chainability   99
James F. Davis

Iterated Function Systems, Compact Semigroups, and Topological
Contractions   113
P. F. Duvall, Jr., John Wesley Emert, and Laurence S. Husch

Contents

The Forced Damped Pendulum and the Wada Property   157
Judy Kennedy and James A. Yorke

Denjoy Meets Rotation on an Indecomposable Cofrontier   183
John C. Mayer and Lex G. Oversteegen

New Problems in Continuum Theory   201
Sam B. Nadler, Jr.

A Continuum Separated by Each of Its Nondegenerate Proper
Subcontinua   231
Sam B. Nadler, Jr., and Gary A. Seldomridge

Dense Embeddings into Cubes and Manifolds   243
J. Nikiel, H. M. Tuncali, and E. D. Tymchatyn

An Example Concerning Disconnection Numbers   261
Robert Pierce

Indecomposable Continua, Prime Ends, and Julia Sets   263
James T. Rogers, Jr.

Homeomorphisms on Cofrontiers with Unique Rotation Number   277
Mark H. Turpin

Self-Homeomorphic Star Figures   283
Wlodzimierz J. Charatonik, Anne Dilks Dye, and James F. Reed

Index   291

Contributors

KATHLEEN T. ALLIGOOD   George Mason University, Fairfax, Virginia

MARCY BARGE   Montana State University, Bozeman, Montana

FERNANDA BOTELHO   Memphis State University, Memphis, Tennessee

KENNETH BOUCHER   University of Utah, Salt Lake City, Utah

BEVERLY L. BRECHNER   University of Florida, Gainesville, Florida

MORTON BROWN   University of Michigan, Ann Arbor, Michigan

WLODZIMIERZ J. CHARATONIK   University of Wroclaw, Wroclaw, Poland, and McNeese State University, Lake Charles, Louisiana

LIANG CHEN   Memphis State University, Memphis, Tennessee

JAMES F. DAVIS   West Virginia University, Morgantown, West Virginia

P. F. DUVALL, JR.   University of North Carolina at Greensboro, Greensboro, North Carolina

ANNE DILKS DYE   McNeese State University, Lake Charles, Louisiana

JOHN WESLEY EMERT   Ball State University, Muncie, Indiana

MERLE D. GUAY   University of Southern Maine, Portland, Maine

LAURENCE S. HUSCH   University of Tennessee, Knoxville, Tennessee

JUDY KENNEDY   University of Delaware, Newark, Delaware

SHIHAI LI   University of Florida, Gainesville, Florida, and National University of Singapore, Singapore

JOHN C. MAYER   University of Alabama at Birmingham, Birmingham, Alabama

SAM B. NADLER, JR.   West Virginia University, Morgantown, West Virginia

J. NIKIEL   Texas A&M University, College Station, Texas

LEX G. OVERSTEEGEN   University of Alabama at Birmingham, Birmingham, Alabama

ROBERT PIERCE   West Virginia University, Morgantown, West Virginia

JAMES F. REED   McNeese State University, Lake Charles, Louisiana

JAMES T. ROGERS, JR.   Tulane University, New Orleans, Louisiana

TIMOTHY SAUER   George Mason University, Fairfax, Virginia

GARY A. SELDOMRIDGE   Potomac State College, Keyser, West Virginia

EDWARD E. SLAMINKA   Auburn University, Auburn University, Alabama

H. M. TUNCALI   Nipissing University, North Bay, Ontario, Canada

MARK H. TURPIN   Boston University, Boston, Massachusetts

E. D. TYMCHATYN   University of Saskatchewan, Saskatoon, Saskatchewan, Canada

JAMES A. YORKE   University of Maryland, College Park, Maryland

Accessible Rotation Numbers for Chaotic States

KATHLEEN T. ALLIGOOD and TIMOTHY SAUER   George Mason University, Fairfax, Virginia

1. Introduction.

When a homeomorphism f of the plane has a fixed point attractor, the basin of attraction W is topologically equivalent to a disk. If, in addition, the map is orientation preserving, information about the dynamics of certain points on the basin boundary can be obtained from the prime end rotation number, i.e., from the action induced by the homeomorphism on the Carathéodory prime end compactification of the disk.

A basin boundary point is called *accessible* from a basin if it is the first boundary point hit by a path emanating from the basin. In a study of fractal basin boundaries by Grebogi et al. [14], it was found repeatedly in numerical examples that the set of accessible points consisted of a periodic saddle and its stable manifold, as illustrated in Fig. 1. In [4] it is shown that if the prime end rotation number is rational, say p/q, and if the fixed points of f^q are isolated, then there are accessible periodic orbits. (See [8], [9], and [4] for the relevant theory of prime ends.) Furthermore, if f is a diffeomorphism and if the accessible periodic orbits are hyperbolic, then the set of points accessible from the basin consists entirely of periodic orbits (of minimum period q, unless $q = 2$) and their stable manifolds. For this reason, we refer to the prime end rotation number as the *accessible rotation number* in [2] and here.

One-piece chaotic attractors, like fractal basin boundaries, are typically observed to have accessible periodic saddles. A *chaotic attractor* is a compact invariant set which is (1) chaotic, i.e., the closure of an orbit with a positive Lyapunov exponent; and (2) attracting, in the sense that it attracts a positive measure set of initial conditions. (A review of elementary concepts can be found in [12].) For an accessible point on a chaotic attractor, we mean accessible by a path emanating from the complement of the attractor. The attractor can be thought of as the boundary of its complement. When viewed as a subset of the sphere $R^2 \cup \{\infty\}$ instead of the plane R^2, each component of the complement of the attractor is a simply-connected open set. If the original planar homeomorphism f is area–contracting, then the complement is also connected, and again we can assign a rotation number to the accessible orbits. In [4], it is shown that if the accessible rotation number of the attractor is rational, then there are accessible periodic orbits on the attractor. See Fig. 2.

For dissipative maps of the annulus, the study of rotation numbers of orbits was begun by Birkhoff [7]. Le Calvez [18] and Casdagli [10] studied rotation intervals of Birkhoff attractors for dissipative monotone twist maps of the annulus. They give conditions for the existence of an interval I of rotation numbers for such maps; i.e., for each number r in I, there is a point

1

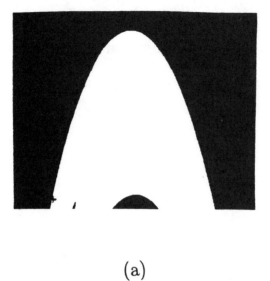

(a)

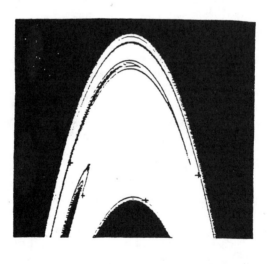

(b)

Figure 1

A portion of the basin of infinity of the Hénon map (2.1) is shown in black for B fixed at 0.3 and each of two values of the parameter t. The x and y values shown are in the rectangle $[-3, 3] \times [-4, 12]$. (a) At $t = 1.31$, the set of accessible points consists of a saddle fixed point and its stable manifold. (b) At $t = 1.39$, the set of accessible points consists of a period–four saddle and its stable manifold.

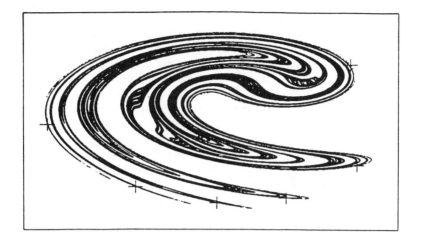

(a)

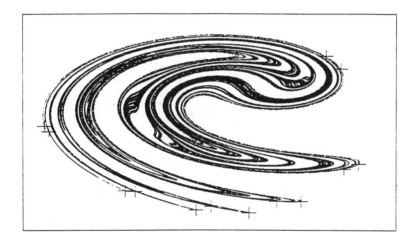

(b)

Figure 2

A chaotic attractor of the Ikeda map

$$f(z) = (t + 0.9z \exp\{i(0.4 - \frac{6.0}{1 + |z|^2})\},$$

where z is a number in the complex plane and t is a real parameter, is shown. In (a), there is an accessible period–six orbit at $t = .97$. In (b), there is an accessible period–thirteen orbit (with accessible rotation number 2/13) at $t = .975$.

p in the attractor with rotation number r. Of course, this interval may degenerate to a single number. Aronson et al. [5] and Hockett and Holmes [17] prove an analogous result for maps of the annulus which contain transverse homoclinic points such that the intersecting stable and unstable manifolds encircle the annulus. Assigning a rotation number to an attractor as the boundary of its complement enables us to study the rotation behavior of a more general class of attractors than those dealt with in the references above. That is, the accessible rotation number is defined for attractors which do not necessarily come from maps of the annulus or bifurcate from invariant circles.

In this regard, Barge and Gillette [6] studied rotations on sets that can be viewed as co–basin boundaries (i.e., all boundary points are on the boundary of two basins). Their techniques extend to some prototypical attractors, such as the closure of the unstable manifold of a fixed point. They show that such sets contain periodic orbits of all rational rotation numbers in an interval.

Chaotic attractors are observed to undergo discontinuous changes in size and shape as a parameter is varied. These global bifurcations have been called *crises* [15]. In particular, a chaotic attractor can suddenly disappear. This last type of crisis occurs when an attractor collides with its basin boundary. As we will illustrate, the accessible orbits of a chaotic attractor go through a predictable sequence of rotation numbers as the attractor approaches a crisis. In fact, the accessible rotation number of the attractor appears to converge to the accessible rotation number of the basin boundary, as the parameter approaches the crisis value.

With the existence of accessible rotation numbers for basin boundaries and one-piece chaotic attractors, the question remains as to whether this rotation number can be irrational for area-contracting maps. While the answer appears to be yes, there is a difference between the basin boundary case and the attractor case. One can design an example of an area-contracting map with a fixed point attractor whose basin boundary has an irrational accessible rotation number (see Sec. 3). However, such behavior does not appear to be typical. Consider, for example, the qualitative change in a basin boundary called a *metamorphosis* (see [14] and [3]). A metamorphosis causes the rotation number to jump discontinuously from one rational number to another rational number, bypassing the irrationals (and rationals) in between. In Sec. 2, we describe this phenomenon and review some aspects of a numerical study of metamorphoses of the Henon map which appears in [2].

In Sec. 3, we present the results of numerical studies of accessible rotation numbers of some numerically observed connected chaotic attractors, and interpret the results theoretically. The studies show how crises differ from metamorphoses and provide evidence supporting the existence of irrational accessible rotation numbers for typical attractors.

Figures 1 and 2 in this article were drawn using [22].

2. Accessible Rotation Numbers for Basin Boundaries.

Let f be an orientation-preserving, area-contracting C^∞ map of the plane. Let S be a saddle fixed point for f, i.e. one of the eigenvalues of the derivative $Df(S)$ is less than one in magnitude, and the other is greater than one. In particular, the eigenvalues are real. Assume that the stable manifold of S forms the entire boundary of a connected, simply–connected basin of an attractor A. This situation occurs, for example, just after a saddle–node bifurcation, when a saddle fixed point S and attracting fixed point A are created, but also in much more general contexts. The accessible rotation number ρ of the basin boundary is then 0.

Suppose next that f is a member of a one–parameter family f_t of plane maps, and that as the parameter t varies, a homoclinic intersection of the stable and unstable manifolds of the saddle S is created at the parameter $t = t_*$. This causes the qualitative change in the basin boundary called a metamorphosis. In particular, the basin boundary changes from a smooth curve to a fractal set.

This is illustrated by the Henon family of maps, defined by

$$f_t(x, y) = (t - x^2 - By, x). \qquad (2.1)$$

The parameter B is the Jacobian determinant of the map; thus the map contracts area by a factor of B. When $B = 0.3$ and t is varied, a homoclinic intersection of the stable and unstable manifolds of the fixed point occurs for approximately $t = t_* \approx 1.315$. Computer–generated pictures of the boundary of the basin of the fixed point attractor for values of t slightly smaller and larger than t_* are shown in Fig. 1. The accessible rotation number of the basin boundary jumps from a value of $\rho = 0$ for $t \leq t_*$ to a value of $\rho = 1/4$ for t immediately larger than t_*.

This change in rotation number is a discontinuous one. At $t = t_*$ the boundary jumps so that for $t > t_*$ the set of points accessible from the basin consists of the four points of a period four saddle and their stable manifolds. Each of these new accessible points is a non-zero distance from the boundary for $t \leq t_*$.

The first result linking homoclinic tangencies and jumping stable manifolds is in [16]. The theory underlying the phenomenon of metamorphosis is introduced in [14], with proofs given in [3]. We describe the major ideas of this analysis in the following.

For $t < t_*$, the stable and unstable manifolds of the fixed saddle S do not intersect except at S, and at $t = t_*$ the stable and unstable manifolds become tangent at a point p. For t near t_*, the map f_t has a large number of periodic orbits in the region near p. In particular, there is a number $K_1 > 0$ and domains B_n near p such that f^n restricted to B_n is a horseshoe map, for each $n \geq K_1$. These horseshoe maps were originally studied by Newhouse [19] and Gavrilov and Silnikov [13], and there is a large body of literature (see also [21] and [20]) concerning the complicated dynamics at and near a homoclinic point. We mention several facts which are important for our analysis:

(H1) The horseshoe map f^n, $n \geq K_1$, has a saddle fixed point. We call the corresponding orbits of f the *principal saddles* associated with the homoclinic tangency. There is a sequence p_n of periodic points (one from each principal saddle orbit) converging to each point of tangency (and, hence, to p). In the following, we use the term "principal saddle" to refer to both the periodic orbit and to individual points in the orbit, when there is no confusion.

(H2) There exists an increasing sequence t_n of parameter values converging to t_* such that the principal saddle p_n is "born" in a saddle node bifurcation at t_n, for each $n \geq K_1$.

(H3) There exists a number K_2, $K_2 \geq K_1 > 0$, such that $W^u(p_n)$, the unstable manifold of p_n, crosses $W^s(p_{n+1})$, the stable manifold of p_{n+1}, for all $n \geq K_2$ and t in the interval $[t_*, t_* + \delta)$, δ small.

It is this last point which accounts for the metamorphosis phenomenon which occurs at homoclinic tangencies. The complicated web of crossings given by (H3) implies that $W^u(p)$ is contained in the set of limit points (the closure) of $W^u(p_n)$, for all $n \geq K_2$. Thus when $W^u(p)$ crosses $W^s(p)$ for each $t \in [t_*, t_* + \delta)$, so does $W^u(p_n)$. At all parameter values of the Henon map where the accessible rotation number is 0, $W^s(p)$ forms the boundary between the basin of infinity and its complement. After the homoclinic intersection is created, (i.e., for all $t > t_*$), points on $W^u(p_n)$ go to infinity, and p_n is no longer in the interior of the basin. For $t \in [t_*, t_* + \delta)$, we observe that the new accessible orbits is p_{K_2} (where the linking of stable and unstable manifolds of principal saddles described in (H3) begins) and $W^s(p_{K_2})$. Hence there is a jump in the accessible rotation number from 0 to $1/K_2$ at $t = t_*$.

In the case of the Hénon map (2.1), for $B = 0.3$ and t growing beyond $t_0 \approx 1.315$, $K_2 = 4$. When the homoclinic intersection is created, the rotation number ρ jumps from 0 to 1/4. For larger values of the Jacobian B, K_2 tends to increase. This is due to the fact that the less extreme the area contraction, the further separated the unstable manifolds of the principal saddles are relative to the stable manifolds. Thus the values of n at which $W^u(p_n)$ and $W^s(p_{n+1})$ will begin to link are larger. (We refer the reader to [2] for details.) For example, when the area contraction is $B = 0.9$, the rotation number first jumps from 0 to 1/8 at $t = t_* \approx -0.244$; the period 4 saddle does not become accessible until $t = 0.203$, after several intermediate jumps (see Fig. 3).

For $B = 0.9$ and $t \approx .815$, the stable and unstable manifolds of adjacent points in the orbit p_4 become tangent (as shown in Fig. 4) and then cross. We call this type of tangency a *rotary homoclinic tangency*, after Holmes and Hockett [17]. Because portions of the intersecting invariant manifolds form a closed curve, horseshoe maps develop just as in the case of a homoclinic tangency. However, the principal saddles from these horseshoes do not have rotation numbers of the form $1/n$. The principal saddle which becomes accessible beyond the homoclinic tangency has rotation number 3/11.

In fact, there are infinitely many levels of rotary homoclinic tangencies and associated principal saddles. In [2], we give a bifurcation analysis of these principal saddles in terms of the continued fractions representation of their rotation numbers.

The sequence $\{S_n\}$ of saddles which exists when p_K (the simple orbit of period K) forms a rotary homoclinic tangency have rotation numbers $\{(n + 1)/(nK + K - 1)\}$, for n sufficiently large. (Details are given in [2].) Fig. 4 illustrates this formula for $K = 4$. The rotation number of the principal saddle S_n is given with respect to $\Gamma(S_n)$, the simple closed curve formed by portions of the intersecting stable and unstable manifolds at a rotary homoclinic tangency of the saddle S_n. As a continued fraction, $(n + 1)/(nK + K - 1)$ is

$$\frac{1}{K - 1/(n + 1)}.$$

In general, for integers $a_i \geq 2$ let $[a_1] = 1/a_1$ and define inductively

$$[a_1, \ldots, a_j] = \frac{1}{a_1 - [a_2, \ldots, a_j]}.$$

Each rational number strictly between 0 and 1 can be written in a unique way as $[a_1, \ldots, a_j]$ for some $j \geq 1$ and $a_i \geq 2$. The following theorem is proved in [2]:

Theorem 2.1 *Let f_t be a C^3-family of real analytic, area-contracting, orientation-preserving maps of the plane. Suppose that a periodic saddle p forms a rotary homoclinic tangency at a point q when $t = t_*$. Then there exists a sequence $\{t_n\}$ of parameter values with limit t_* and a sequence $\{S_n\}$ of principal saddles with limit q such that for n sufficiently large, each principal saddle S_n has a rotary homoclinic tangency at t_n.*

Moreover, if the rotation number of p with respect to $\Gamma(p)$ is $[a_1, \ldots, a_j]$, then the rotation number of S_n with respect to $\Gamma(S_n)$ is $[a_1, \ldots, a_j, n + 1]$.

In the case $j = 1$, this is essentially the Gavrilov-Silnikov result. For larger j, the theorem implies that at any rotary homoclinic tangency there are infinitely many levels of rotary homoclinic tangencies and principal saddles described by the continued fraction representations of their rotation numbers. A basin boundary metamorphosis can occur at any level, depending of course on the level of the accessible saddle.

As an illustration, we interpret the jumps which occur when B is fixed at 0.9 and t is varied. See Fig. 3. For $t = .8$, there is an an accessible period 4 saddle orbit, and so the rotation number is 1/4. As t is increased, this accessible saddle has a rotary homoclinic tangency. The sequence of principal saddles converging to the point of tangency has rotation numbers $1/(4 - 1/n)$ for

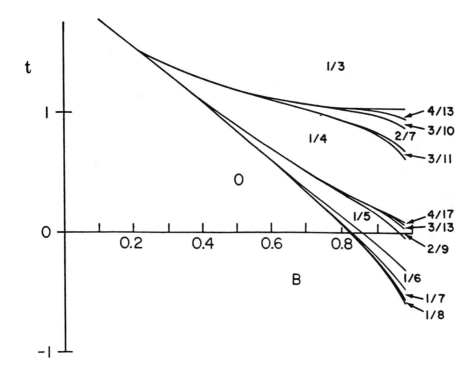

Figure 3

Rotation numbers of accessible boundary orbits of the two–parameter Hénon map are shown. The fractions represent rotation numbers of orbits accessible from the region U, i.e., the set of points not in the closure of the basin of infinity. In each of Fig. 1a,b, U is the region shown in white.

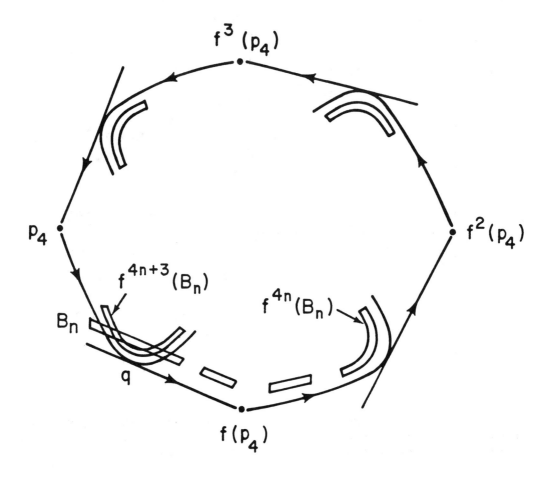

Figure 4

The domain B_n and the range $f^{4n+3}(B_n)$ of a horseshoe map formed near a rotary homoclinic point q are shown. The point p_4 is a periodic saddle of period four.

$n \geq 2$. For values of t slightly greater than the tangency value (approximately $t = .815$), the accessible orbit has rotation number $[4, 3] = 1/(4 - 1/3) = 3/11$. The other orbits created by the tangency, of rotation number $[4, n]$ for $n \geq 4$, are buried in the basin boundary and not accessible. At approximately $t = .831$, the $3/11$ orbit is buried by a new accessible orbit with rotation number $[4, 2] = 2/7$. As t increases beyond $.972$, the $2/7$ orbit has a rotary tangency, resulting in a sequence of principal saddles of periods $[4, 2, n]$ for $n \geq 2$. It turns out that the $n = 2$ orbit is accessible for t immediately beyond $t = .972$, so the accessible orbit has rotation number $[4,2,2]=3/10$. At $t = 1.013$, the $3/10$ orbit has a rotary tangency, resulting in a new accessible orbit with rotation number $[4,2,2,2]=4/13$. When the $4/13$ orbit has a rotary tangency (at approximately $t = 1.036$), the entire sequence of orbits converging to the tangency is buried by the stable manifold of the period 3 saddle orbit remaining from the original period 1 homoclinic tangency, and the rotation number jumps to $[3]=1/3$.

In summary, for $B = 0.9$ and t between 0.8 and 1.04, the only accessible rotation numbers which occur on the basin boundary are the rational numbers $1/4, 3/11, 2/7, 3/10, 4/13$, and $1/3$. For positive values of B closer to 0 there are larger and fewer jumps in the rotation number as t is varied.

3. Accessible Rotation Numbers for Chaotic Attractors.

The analysis begun for basin boundaries in [4] and [2] and described in the previous section can be applied to connected chaotic attractors. It has been suggested by D. Rand and others that one-piece chaotic attractors can be viewed as basin boundaries under the inverse of the map. It was shown in [4], under mild assumptions on the attractor, that if the accessible rotation number of the attractor is rational, then there are accessible periodic orbits. In this section we exhibit numerical evidence of how accessible rotation numbers change at a typical crisis and discuss the existence of irrational accessible rotation numbers.

One can design an example of an area-contracting map with a fixed point attractor whose basin has an irrational accessible rotation number, which we briefly describe. Start with a circle with pure irrational rotation. Such a set cannot be the boundary of a basin for an area-contracting map, due to the fact that since basins are invariant sets and have positive area, they must have infinite area. But we can modify this set using a Denjoy construction (see, e.g., [11]). Replace the points in one orbit by intervals, the sum of whose lengths is finite. Then replace each interval by a "tunnel" which extends to infinity. The interior of the original circle together with the interior of the tunnels form the basin of an attracting fixed point. On the boundary, tunnels map to tunnels, and the remaining points map as before under the irrational rotation. Thus the accessible rotation number remains irrational.

As we have demonstrated, however, such behavior is not expected for typical maps, due to the phenomenon of metamorphosis. Evidence provided by the Hénon map basin boundary study supports this idea. As described in Sec. 1 and 2, the horseshoes which are already linked at tangency determine the new accessible orbits. For example, if at a simple homoclinic tangency the horseshoes of f^n are linked for all $n > K$, but not for $n < K$, then the rotation number after tangency is $1/K$.

The situation for chaotic attractors appears to be different. It has been shown that there are area–contracting diffeomorphisms of the plane for which the closure of an unstable manifold, viewed as a prototype of a chaotic attractor, has an irrational accessible rotation number [1]. Figure 2 shows some of the accessible rotation numbers for the chaotic attractor of the Ikeda map

$$f_t(z) = t + Bz \exp[i(.4 - \frac{6}{1 + |z|^2})], \tag{3.1}$$

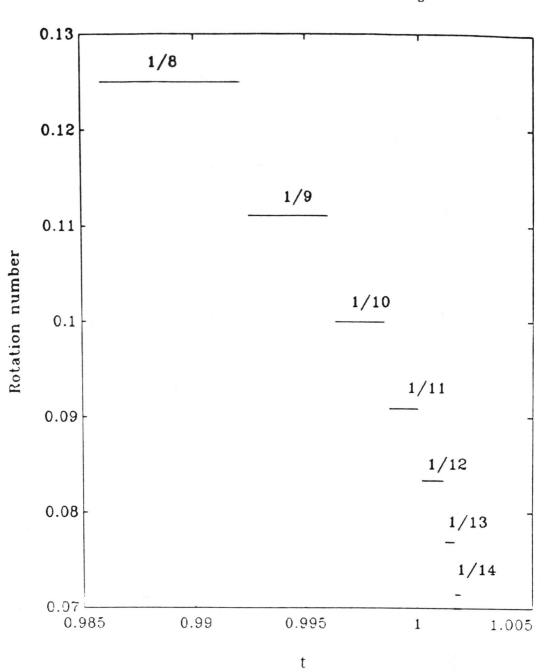

Figure 5

The accessible rotation number of the Ikeda attractor is graphed as a function of the parameter t.

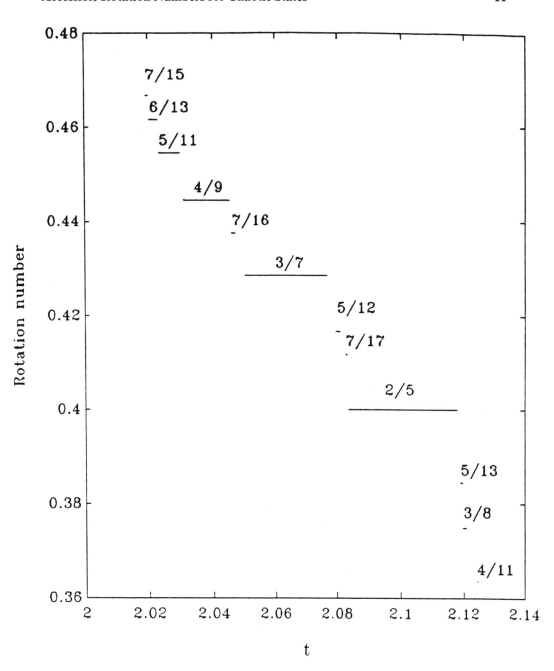

Figure 6

The accessible rotation number of the Hénon attractor is graphed as a function of the parameter *t*.

where z is a complex number. The Ikeda map has what is numerically observed to be a one-piece (i.e., connected) attractor Θ_t for $B = 0.9$ and various values of t in the range 0.814 to 1.004. Fig. 2 shows accessible periodic orbits on Θ_t for two different values of t. This map is area contracting and orientation preserving. (The Jacobian is B^2.) The basin of attraction of Θ_t is bounded by the stable manifold of a fixed point saddle S (outside the attractor). As t approaches 1.004, Θ_t is very close to $W^u(S)$. At $t = t_0 \approx 1.004$ there is a homoclinic tangency of $W^u(S)$ and $W^s(S)$, and for larger t the attractor Θ_t no longer exists. At $t = t_0$, there is a crisis in which the attractor collides with its basin boundary $W^s(S)$. As $W^u(S)$ approaches $W^s(S)$, the principal saddles described in Sec. 2 are formed.

In Fig. 5 we have shown all accessible rotation numbers which exist over an interval of at least 0.0002 for t in the range 0.986 to 1.004. All rotation numbers shown are of the form $[n] = 1/n$. However, there are values of t for which the accessible rotation number is not of this form. For example, at $t = 0.9962$, the accessible rotation number is 2/19. It appears that as the attractor approaches a crisis at $t = t_0$, the accessible rotation numbers converge to 0, the rotation number of accessible points on the basin boundary.

We have not mentioned the problem of whether Θ_t is actually a chaotic attractor throughout this parameter range. In fact, Θ_t is observed to disappear for brief ranges of parameter values. For example, when $t = .98$ (which is outside the parameter range shown in Fig. 5), the attractor has temporarily disappeared, and the points that had formed the basin of the attractor fall into the basin of a period 5 sink. This occurs because the attractor has crossed the stable manifold basin boundary of the basin of the periodic sink before the sink has lost its stability. At all the parameter values sampled in the parameter range $[0.986, 1.004]$, however, a chaotic attractor was observed.

The Henon map (2.1) provides another illustration of the evolution of accessible rotation numbers of an attractor as a parameter is varied. For $B = 0.3$ and most t values in the interval $[2.02, 2.12]$, there appears to be a connected chaotic attractor Φ_t. The rational accessible rotation numbers for the attractor in this range are displayed in Fig. 6. Rational numbers with denominator ≤ 17 are listed. There is a crisis at $t = t_0 \approx 2.124$ when the attractor collides with the boundary of its basin and disappears. The basin boundary in this case consists of three points of a period 3 saddle orbit and their stable manifolds. Thus the rotation number of the accessible points on the basin boundary is 1/3, and the accessible rotation number of Φ_t appears to converge to 1/3 as the attractor approaches a crisis at $t = t_0$.

These numerical studies indicate that the accessible rotation number of an attractor changes in a much different manner than that of a basin boundary. In Fig. 6, in particular, the graph of the accessible rotation number looks like the graph of a Cantor function or "devil's staircase". In fact, it may be possible for the accessible rotation number of an attractor to change continuously with the parameter, even at a crisis, whereas the accessible rotation number of a basin boundary has a jump discontinuity whenever it changes.

References.

[1] K. Alligood, Is the accessible rotation number of a homoclinic tangle always rational? In preparation.

[2] K. Alligood, T. Sauer, Rotation numbers of periodic orbits in the Hénon map. Commun. Math. Phys. **120** (1988), 105–119.

[3] K. Alligood, L. Tedeschini-Lalli, and J. Yorke, Metamorphoses: Sudden jumps in basin boundaries. Commun. Math. Phys. **141** (1991), 1–8.

[4] K. Alligood and J. Yorke, Accessible saddles on fractal basin boundaries. Ergod. Th. Dynam. Syst. **12** (1992), 377–400.

[5] D. Aronson, M. Chory, G. Hall, R. McGehee, Bifurcations from an invariant circle for two-parameter families of maps of the plane: a computer assisted study. Commun. Math. Phys.**83** (1982) 303–354.

[6] M. Barge and R. Gillette, Rotation and periodicity in plane separating continua. Ergod. Th. Dynam. Syst. **11** (1991), 619–631.

[7] G.D. Birkhoff, Sur quelques courbes fermees remarquables. Bull. Soc. Math. France **60** (1932) 1–26.

[8] C. Caratheodory, Uber die Begrenzung einfach zusammenhangender Gebiete. Math. Ann. **73** (1913) 323–370.

[9] M. Cartwright and J. Littlewood, Some fixed point theorems. Ann. of Math. **54** (1951), 1–37.

[10] M. Casdagli, Periodic orbits for dissipative twist maps. Ergod. Th. Dynam. Syst. **7** (1987) 165–173.

[11] R. Devaney, An Introduction to Chaotic Dynamical Systems. Benjamin Cummings (1985).

[12] J.–P. Eckmann and D. Ruelle, Ergodic theory of chaos and strange attractors. Rev. Mod. Phys. **57** (1985), 617–656.

[13] N. Gavrilov, L. Silnikov, On the three dimensional dynamical systems close to a system with a structurally unstable homoclinic curve. Math. USSR Sbornik **17** (1972) 467–485. Math. USSR Sbornik **19** (1973) 139–156.

[14] C. Grebogi, E. Ott, and J. Yorke, Basin boundary metamorphoses: changes in accessible boundary orbits. Physica **24D** (1987) 243–262.

[15] C. Grebogi, E. Ott, and J. Yorke, Crises, sudden changes in chaotic attractors, and transient chaos, Physica **7D** (1983) 181.

[16] S. Hammel, C. Jones, Jumping stable manifolds for dissipative maps of the plane, Physica **35D** (1989), 87–106.

[17] K. Hockett, P. Holmes, Josephson's junction, annulus maps, Birkhoff attractors, horseshoes and rotation sets. Ergod. Th. Dynam. Syst. **6** (1986) 205–239.

[18] P. Le Calvez, Propriétés des attracteurs de Birkhoff, Ergod. Th. Dynam. Syst. **8** (1987), 241–310.

[19] S. Newhouse, Diffeomorphisms with infinitely many sinks. Topology **13** (1974) 9–18.

[20] C. Robinson, Bifurcation to infinitely many sinks. Commun. Math. Phys. **90** (1983) 433–459.

[21] S. Smale, Differentiable Dynamical Systems, Bull. Amer. Math. Soc. **73** (1967) 747–817.

[22] J. Yorke, Dynamics: A program for IBM PC clones.

Prime End Rotation Numbers Associated with the Hénon Maps

MARCY BARGE   Montana State University, Bozeman, Montana

Introduction

The two-parameter family of maps of the plane,

$$H_{a,b}(x,y) = (1 + y - ax^2, -by),$$

called the Hénon family ([He]), has received considerable attention as an algebraically simple system with very complex dynamical behavior. It has recently been proved that for $b < 0$ (the orientation reversing case), $|b|$ small, there is an abundance of values of the parameter a at which $H_{a,b}$ possesses a strange attractor—a compact connected attracting set, containing more than a single point, on which $H_{a,b}$ is topologically transitive ([BC]). The situation when b is positive (the orientation preserving case) is less well understood.

For $1 < a < 2$ and $b > 0$, b small, $H_{a,b}$ has an attracting set that is a non-separating, nowhere dense, plane continuum. Associated with $H_{a,b}$ and its attracting set there is a circle of prime ends and an induced orientation preserving homeomorphism (the prime end homeomorphism) on this circle of prime ends. We prove in Section 2 that if the prime end homeomorphism is transitive then $H_{a,b}$ is chain-transitive on its attracting set; in particular, the attracting set contains no stable periodic orbits.

It is difficult at present to determine much about the prime end homeomorphism for $b > 0$. We will show in Section 3 that the limit as $b \to 0^+$ of the rotation number of the prime end homeomorphism (the prime end rotation number) varies continuously from $1/2$ to 1 as a varies from 1 to 2. This

15

will be established by "blowing up" the one-dimensional ($b = 0$) limit, using inverse limits, to obtain a family of orientation preserving homeomorphisms $F_{a,b}$, $1 < a < 2$, $b \geq 0$ and small, with the property that for $b > 0$, $F_{a,b}$ and $H_{a,b}$ are topologically conjugate on neighborhoods of their attracting sets. It will be shown that the prime end rotation number associated with $F_{a,b}$ (and hence that of $H_{a,b}$) limits on the prime end rotation number associated with $F_{a,0}$ as $b \to 0^+$ and that the latter prime end rotation number varies continuously from $1/2$ to 1 as a varies from 1 to 2. Moreover, we will show that $F_{a,0}$, restricted to its attracting set, is chain-transitive when the prime end rotation number is irrational (even through the prime end homeomorphism is not transitive).

We don't know whether the prime end rotation number associated with $H_{a,b}$, $b > 0$, varies continuously with a and b or even whether this rotation number is ever irrational. We will see in the next section that if this rotation number is discontinuous, then there must be a discontinuity not only in the attracting set but also in the local basin of attraction.

1 Definitions and Notation

Associated with a monotone degree one map $g : S^1 \to S^1$ of the circle S^1 there is a <u>rotation number</u>, $\rho(g)$, defined as follows. Let $\pi : \mathbb{R} \to S^1$ be the (period one) universal cover and let $\tilde{g} : \mathbb{R} \to \mathbb{R}$ be a lift of g. Then

$$\rho(g) = \lim_{n \to \infty} \frac{\tilde{g}^n(t)}{n} (mod\, 1).$$

This number exists and is independent of the choice of lift $\tilde{g}$ and point $t \in \mathbb{R}$.

Suppose now that Λ is a continuum (compact and connected) in the plane that consists of more than one point and that is invariant under the orientation preserving homeomorphism $F : \mathbb{R} \to \mathbb{R}$. We will define the (exterior) <u>prime end rotation number</u> of F associated with Λ in two equivalent ways. If U is the unbounded component of $\mathbb{R}^2 \setminus \Lambda$ the equivalence classes $[\{C_n\}]$ of prime chains of cross-cuts $C_n \subset \mathcal{U}$ associated with Λ form a circle $\mathcal{S}$ (see, for example, [CoL] or [M]), called the <u>circle of prime ends</u>, and F induces a degree one homeomorphism $\mathcal{F} : \mathcal{S} \to \mathcal{S}$ by $\mathcal{F}([\{C_n\}]) = [\{F(C_n)\}]$.

The (exterior) prime end rotation number of F associated with Λ is then $\rho_e(F, \Lambda) = \rho(\mathcal{F})$.

Alternatively, this rotation number can be calculated in terms of the action of F on the points of Λ accessible from $\mathcal{U}$. If $x \in \Lambda$ and $\gamma : [0, 1] \to \mathbf{R}^2$ is a simple arc with $\gamma(0) = x$ and $\gamma((0, 1]) \subset \mathcal{U}$ then x is said to be accessible (from U) and γ is called an accessing arc. If γ_0 and γ_1 are two accessing arcs then γ_0 and γ_1 are equivalent if $\gamma_0(0) = x = \gamma_1(0)$ and there is an isotopy H_t from γ_0 to γ_1, with values in $\mathcal{U} \cup \{x\}$, such that $H_t(0) = x$ for all t. There is then a natural cyclic order on the equivalence classes of accessing arcs that is preserved by the transformation $[\gamma] \to [F \circ \gamma]$. This transformation then has a rotation number which agrees with $\rho_e(F, \Lambda)$ as previously defined (see [B], [LeC], and [AY]).

Given a metric space (X, d), d bounded, and a continuous map $f : X \to X$, the inverse limit space with bonding map f is

$$\varprojlim(X, f) = \{\underline{x} = (x_0, x_1, \cdots) | \; f(x_{n+1}) = x_n, \; x_n \in X, n = 0, 1, 2, \cdots\}.$$

with metric

$$\underline{d}(\underline{x}, \underline{y}) = \sum_{n=0}^{\infty} \frac{d(x_n, y_n)}{2^n}.$$

The shift homeomorphism, $\hat{f}$, on $\varprojlim(X, f)$ is defined by $\hat{f}((x_0, x_1, \ldots)) = (f(x_0), x_0, x_1, \ldots)$. If X is a continuum, so is $\varprojlim(X, f)$.

If x, $y \in X$ and $\epsilon > 0$, an $\epsilon-$chain from x to y is a sequence $\{x_i\}_{i=0}^n$, $n > 0$, in X such that $x_0 = x$, $x_n = y$, and $d(f(x_i), x_{i+1}) < \epsilon$ for $i = 0, 1, \ldots, n-1$. The point $x \in X$ is said to be chain recurrent if for each $\epsilon > 0$ there is an $\epsilon - chain$ from x to x and f is said to be chain transitive if for each $\epsilon > 0$ and $x, y \in X$ there is an $\epsilon - chain$ from x to y. If X is a continuum and each point of X is a chain recurrent, then f is chain transitive ([F]).

Finally, we will call the continuum $\Lambda \subset \mathbf{R}^2$ an attracting set for the map $F : \mathbf{R}^2 \to \mathbf{R}^2$ provided there is a compact neighborhood B of Λ such that $F(B) \subset \overset{\circ}{B}$ and $\Lambda = \cap_{n \geq 0} F^n(B)$.

2   General Properties of the Prime End Rotation Number

There are two distinct behaviors for degree one circle homeomorphisms g with irrational rotation number: g may be transitive (have a dense orbit), in which case it is topologically conjugate with an irrational rotation; or g may be "Denjoy", in which case it has a wandering interval and is only semi-conjugate to irrational rotation. In either case, if $\rho(g)$ is irrational, g is chain transitive. Several authors have investigated the dynamical implications of a rational prime end rotation number ([AY, CL, BG1, BG2]) but little is known in the irrational case. Our first proposition provides some information when the prime end homeomorphism is transitive.

Proposition 2.1: Suppose that Λ is a nontrivial and nowhere dense (in $\mathbb{R}^2$) attracting set for the orientation perserving homeomorphism $F : \mathbb{R}^2 \to \mathbb{R}^2$. If the associated prime end homeomorphism is transitive then $F|_\Lambda$ is chain transitive.

Proof: Let x and y be two points of Λ and let $\epsilon > 0$ be given. Since Λ is nowhere dense and consists of more than one point there is a cross-cut C (an arc lying in $\mathcal{U} = \mathbb{R}^2 \setminus \Lambda$ except for its endpoints, which lie in Λ) with distinct endpoints and contained in the disk of radius ϵ centered at y. Similarly, there is a cross-cut C' with distinct endpoints contained in the ball of radius ϵ centered at $F(x)$. Now C and C' each determine two distinct (accessible) prime ends—those corresponding to the two sub-arcs of each of C and C' accessing the endpoints—call these prime ends ξ_0 and ξ_1 (corresponding to C) and ξ_0' and ξ_1' (coresponding to C').

Now since the prime end homeomorphism $\mathcal{F} : \mathcal{S} \to \mathcal{S}$ associated with F and Λ is transitive, it is conjugate to rigid rotation of S^1 through an irrational angle. It follows that there is a positive integer m such that the points $\mathcal{F}^m(\xi_0')$ and $\mathcal{F}^m(\xi_1')$ are in opposite components of $\mathcal{S} \setminus \{\xi_0, \xi_1\}$ and from this it follows that $F^m(C') \cap C \neq 0$.

Let $x_1 \epsilon C' \cap F^{-m}(C)$. Then

$$x_0 = x, \; x_1, \; x_2 = F(x_1), \ldots, \; x_m = F^{m-1}(x_1), \; x_{m+1} = y$$

is an ϵ-chain, in $\mathbb{R}^2$, from x to y. Thus Λ is contained in a chain transitive component of F (see, for example, [F]). Since Λ is an attracting set, no other points of $\mathbf{R}^2$ are in the chain transitive component containing Λ so that Λ itself is a chain transitive component of F and $F|_\Lambda$ is chain transitive (again, see [F]).   ∎

Proposition 2.2: Let $J \subset \mathbf{R}$ be an interval and suppose that $F_a, a \in J$, is a continuous one parameter family of orientation preserving homeomorphisms of the plane. Suppose further that for each $a \in J, \Lambda(a)$ is a nondegenerate, non-separating plane continuum invariant under F_a and that the $\Lambda(a)$ vary continuously (in the Hausdorff topology) with a. Then the prime end rotation numbers $\rho_e(F_a, \Lambda(a))$ vary continuously with a.

Proof: Fix $a' \in J$ and let I be a compact interval with $a' \in I \subset J$. Since the $\Lambda(a)$ vary continuously there is a compact disk $D \subset \mathbf{R}^2$ with $\Lambda(a) \subset \overset{\circ}{D}$ for all $a \in I$. We may modify the F_a, $a \in I$, to a continuous family G_a having the properties: $G_a : \mathbf{R}^2 \to \mathbf{R}^2$ is an orientation preserving homeomorphism for $a \in I$; $G_a = F_a$ in a neighborhood of $\Lambda(a)$ for each $a \in I$; and $G_a = id$ on ∂D for each $a \in I$. Clearly $\rho_e(G_a, \Lambda(a)) = \rho_e(F_a, \Lambda(a)) \equiv \rho_e(a)$.

We will say that $\eta : [0,1] \to D$ is an arc from ∂D to $\Lambda(a)$ provided: η is continuous and one-to-one; $\eta(0) \in \partial D$; $\eta(1) \in \Lambda(a)$; and $\eta((0,1)) \subset \overset{\circ}{D} \backslash \Lambda(a)$. Given two arcs η and γ from ∂D to $\Lambda(a)$ we define an intersection number as follows. First, if $\gamma(t_0), t_0 \in (0,1)$, is a point of topologically transverse intersection of $\eta([0,1])$ with $\gamma([0,1])$ we will say that the intersection of γ with η at $\gamma(t_0)$ is positive if $\gamma(t)$ crosses $\eta([0,1])$ from the (locally) clockwise side of $\eta([0,1])$ to the counterclockwise side of $\eta([0,1])$; otherwise the intersection is negative. If γ and η are two arcs from ∂D to $\Lambda(a)$ there is an isotopy from γ to an arc γ', rel. $\{0,1\}$ and with values in $(\overset{\circ}{D} \backslash \Lambda(a)) \cup \{\gamma(0), \gamma(1)\}$, such that the intersections of $\gamma'((0,1))$ with $\eta((0,1))$ are finite in number, topologically transverse, and are either all positive or all negative. Note that if $\gamma(1) = \eta(1)$ or $\gamma(0) = \eta(0)$ then the number of intersections of two such γ' (as above) with η may differ by one. Let $n(\gamma, \eta) = \min\{\#(\gamma'((0,1)) \cap \eta((0,1)))| \ \gamma'$ is as above$\}$ and let $i(\gamma, \eta) = \pm n(\gamma, \eta)$, with the plus sign taken if the intersections of γ' with η are positive and the negative sign if the intersections are negative.

Now given an arc γ from ∂D to $\Lambda(a)$,

$$\lim_{n \to \infty} \frac{i(G_a^n \circ \gamma, \gamma)}{n} = \rho_e(G_a, \Lambda(a)) = \rho_e(a).$$

In fact

$$\left| \frac{i(G_a{}^n \circ \gamma, \, \gamma)}{n} - \rho_e(a) \right| \leq \frac{2}{n}$$

for $n = 1, 2, \ldots$. Let $a_n \in I$ be a sequence with $a_n \to a'$. Then $G_{a_n} \to G_{a'}$ on D and $\Lambda(a_n) \to \Lambda(a')$ in the Hausdorff topology. There is then a sequence of arcs γ_k from ∂D to $\Lambda(a_k)$ such that $\{\gamma_k\}$ converges uniformly to an arc γ from ∂D to $\Lambda(a')$.

Let $\epsilon > 0$ be given and let n be large enough so that $5/n < \epsilon$. Let K be large enough so that

$$\left| i(G_{a_k}^n \circ \gamma_k, \gamma_k) - i(G_{a'}^n \circ \gamma, \gamma) \right| \leq 1$$

for all $k \geq K$ (This is possible since diam $(\Lambda(a))$ is bounded above zero for $a \in I$). Now, for $k \geq K$,

$$
\begin{aligned}
|\rho_e(a_k) - \rho_e(a')| \; &\leq \; \left| \rho_e(a_k) - \frac{i(G_{a_k}^n \circ \gamma_k, \gamma_k)}{n} \right| \\
&\quad + \left| \frac{i(G_{a_k}^n \circ \gamma_k, \gamma_k)}{n} - \frac{i(G_{a'}^n \circ \gamma, \gamma)}{n} \right| \\
&\quad + \left| \frac{i(G_{a'}^n \circ \gamma, \gamma)}{n} - \rho_e(a') \right| \\
&< \frac{2}{n} + \frac{1}{n} + \frac{2}{n} = \frac{5}{n} \\
&< \epsilon.
\end{aligned}
$$

∎

The point p is a fixed hyperbolic saddle of the diffeomorphism $F\colon \mathbf{R}^2 \to \mathbf{R}^2$ provided $F(p) = p$ and the derivative matrix, $DF(p)$, has an eigenvalue of absolute value less than 1 and an eigenvalue of absolute value greater than 1. The unstable manifold of the fixed hyperbolic saddle p, $W^u(p) = \{x \in \mathbf{R}^2 \mid F^{-n}(x) \to p$ as $n \to \infty\}$, is a continuous one-to-one image of $\mathbf{R}$.

<u>Proposition 2.3:</u> Suppose that F_a, $a \in J$, is a C^1 family of C^1 diffeomorphisms of the plane, that $p(a)$ is a fixed hyperbolic saddle of F_a for each $a \in J$, and that $p(a)$ varies continuously with a. For each $a \in J$ let $\Lambda(a) = cl(W^u(p(a)))$ and suppose that there is a continuous (in the Hausdorff topology) one parameter family of compact sets $B(a) \subset \mathbf{R}^2$ such that

$F^a(B(a)) \subset \overset{\circ}{B}(a)$ and $\Lambda(a) = \cap_{n \geq 0} F_a^n(B(a))$ for each $a \in J$. Then the $\Lambda(a)$ vary continuously with a in the Hausdorff topology.

<u>Proof:</u> Suppose that a_n is a sequence in J with $a_n \to a \in J$. Let $\mathcal{U} \subset \mathbb{R}^2$ be an open set with $\Lambda(a) \subset \mathcal{U}$. There is then an $m \geq 0$ such that $F_a^m(B(a)) \subset \mathcal{U}$. Since $F_{a_n}^m \to F_a^m$ and $B(a_n) \to B(a)$ as $n \to \infty$, there is an N so that $F_{a_n}^m(B(a_n)) \subset \mathcal{U}$ for all $n \geq N$. Then $\Lambda(a_n) \subset \mathcal{U}$ for all $n \geq N$ and we see that $a \to \Lambda(a)$ is upper semi-continuous.

On the other hand, the local unstable manifolds $W_{loc}^u(p(a))$ vary continuously (see, for example, [H]) and it follows that $a \to \Lambda(a) = cl(W^u(p(a)))$ is lower semi-continuous. ∎

<u>Corollary 2.4:</u> If the F_a and $\Lambda(a)$ are as in Proposition 2.4 and the F_a are orientation preserving, then the prime end rotation numbers $\rho_e(F_a, \Lambda(a))$ vary continuously with $a \in J$.

3 Prime End Rotation Numbers for a Model of the Hénon Attractor

The Hénon map ([He]), $H = H_{a,b} : \mathbb{R}^2 \to \mathbb{R}^2$ and given by

$$H(x,y) = (1 + y - ax^2, -bx),$$

is a two-parameter family of maps of the plane. H is an orientation preserving homeomorphism if $b > 0$ and an orientation reversing homeomorphism if $b < 0$. If $b = 0$, the range of H is contained in $\mathbb{R} \times \{0\}$ and H restricted to $\mathbb{R} \times \{0\}$ is, in the first coordinate, the quadratic family $f_a : \mathbb{R} \to \mathbb{R}$ given by $f_a(x) = 1 - ax^2$.

Let J_a be the interval $J_a = [1 - a, 1]$. Then, for $1 \leq a \leq 2$, $f_a(J_a) = J_a$. Now, with $\alpha(a) = -\beta(a) = -\sqrt{(2/a)}$ and $I_a = [\alpha(a), \beta(a)]$, it is easy to check that for $1 < a < 2$: $J_a \subset \overset{\circ}{I}_a$; $f_a(I_a) \subset \overset{\circ}{I}_a$; and $f_a^2(I_a) = J_a$. Let $B(a) = \{(x,y)| -2 + ax^2 \leq y \leq 1/2(\sqrt{(2/a)} - 1)\} \subset \mathbb{R}^2$. Then $H_{a,0}(B(a)) \subset \overset{\circ}{B}(a)$ for $1 < a < 2$ so there is a continuous $\gamma : (1,2) \to \mathbb{R}^+$ such that $H_{a,b}(B(a)) \subset \overset{\circ}{B}(a)$ whenever $1 < a < 2$ and $0 \leq b \leq \gamma(a)$.

For (a, b) satisfying $1 < a < 2$ and $0 \leq b \leq \gamma(a)$, the "Hénon attractor" is the invariant set

$$\Lambda(a, b) = \bigcap_{n \geq 0} H_{a,b}^n(B(a)).$$

For this range of parameters, there is a fixed hyperbolic saddle $p = p(a, b) \in \Lambda(a, b)$.

<u>Question</u>: For $1 < a < 2$ and $0 \leq b \leq \gamma(a)$, is

$$\Lambda(a, b) = cl(W^u(p(a, b)))\,?$$

<u>Note</u>: For $1 < a < 2$ there is a $\gamma' = \gamma'(a) < 0$ so that if $\gamma'(a) < b \leq 0$ then $\Lambda(a, b) = cl(W^u(p(a, b)))$, see [BC], but the geometry of $W^u(p(a, b))$ is not so clear in the orientation preserving case.

We are interested in the behavior of the prime end rotation number $\rho_e(a, b) \equiv \rho_e(\Lambda(a, b), H_{a,b})$. If the answer to the above question is "yes", then, by corollary 2.4, ρ_e is continuous for $1 < a < 2$ and $0 < b < \gamma(a)$.

For $b = 0$ and $1 < a < 2$, $\Lambda(a, b)$ degenerates to $J_a \times \{0\}$. We will "blow up" this one-dimensional limit of the Hénon family by means of inverse limits in order to understand something of the behavior of $\rho_e(a, b)$ for b near zero.

To this end, let $T_{a,b} : D \to D$, $1 < a < 2$, $0 \leq b \leq \gamma(a)$, be a continuous family of maps of a compact disk D, with $B(a) \subset \overset{\circ}{D}$, onto itself that satisfies:

i) $T_{a,b}(D) = D$;

ii) $T_{a,b}|_{\partial D} = id$;

iii) $T_{a,b}|_{B(a)} = H_{a,b}|_{B(a)}$;

iv) $T_{a,0}^{-1}(J_a \times \{0\}) \subset B(a)$;

v) $T_{a,0}$ is one-to-one off $B(a)$, and

vi) $T_{a,b}$ is a homeomorphism for $b > 0$.

Now let $\varprojlim(D, T_{a,b})$ be the inverse limit space with bonding map $T_{a,b}$ and let $\hat{T}_{a,b} : \varprojlim(D, T_{a,b}) \to \varprojlim(D, T_{a,b})$ be the shift homeomorphism.

<u>Proposition 3.1:</u> Fix $1 < a_1 < a_2 < 2$ and let $R = \{(a,b)|a_1 \leq a \leq a_2$ and $0 \leq b \leq \gamma(a)\}$. There is then a family of homeomorphisms $\psi_{a,b} : \varprojlim(D, T_{a,b}) \to D$ such that the homeomorphisms $F_{a,b} = \psi_{a,b} \circ \hat{T}_{a,b} \circ \psi_{a,b}^{-1} : D \to D$ vary continuously with $(a,b) \in R$.

<u>Proof:</u> Let $T : D \times R \to D \times R$ by $T(x,y,a,b) = (T_{a,b}(x,y), a, b)$. Given ϵ, $0 < \epsilon \leq \inf_{a_1 \leq a \leq a_2} \gamma(a)$, let $T_\epsilon : D \times R \to D \times R$ be defined by

$$T_\epsilon(x,y,a,b) = \begin{cases} T(x,y,a,b) & \text{if } b \geq \epsilon \\ (T_{a,\epsilon}(x,y), a, b) & \text{if } b < \epsilon. \end{cases}$$

T_ϵ is then a homeomorphism and $T_\epsilon \to T$ uniformly as $\epsilon \to 0$. It follows from a theorem of M. Brown ([Br]) that $\varprojlim(D \times R, T)$ is homeomorphic with $D \times R$. In fact, if $\{\epsilon_n\}$ is a sequence that goes to zero quickly enough, then

$$\psi : \varprojlim(D \times R, T) \to \varprojlim(D \times R, T)$$

by

$$\psi(z_0, z_1, \cdots) = \lim_{n \to \infty} T_{\epsilon_1} \circ \cdots \circ T_{\epsilon_n}(z_n)$$

is a homeomorphism. Let $F : D \times R \to D \times R$ be given by $F = \psi \circ \hat{T} \circ \psi^{-1}$. We see from the form of ψ and $\hat{T}$ that $F(x,y,a,b) = (F_{a,b}(x,y), a, b)$ where $F_{a,b} : D \to D$ is a homeomorphism of the form $F_{a,b} = \psi_{a,b}^{-1} \circ \hat{T}_{a,b} \circ \psi_{a,b}$, for $(a,b) \in R$, where $\psi_{a,b} : \varprojlim(D, T_{a,b}) \to D$ is a homeomorphism. ∎

The $F_{a,b}$, $(a,b) \in R$, described by Proposition 3.1 are only specified up to topological conjugacy ($F_{a,b}$ depends on the choice of a_1, a_2, and the $\epsilon_n \to 0$ in the proof). But this is enough to guarantee that the prime end rotation numbers $\rho_e(a,b)$, to be defined below, are independent of the $\psi_{a,b}$.

Since for $b > 0$ $T_{a,b}$ is a homeomorphism, the projection

$$\pi_0 : \varprojlim(D, T_{a,b}) \to D$$

onto the first coordinate is also a homeomorphism. Moreover $\pi_0 \circ \hat{T}_{a,b} = T_{a,b} \circ \pi_0$ so, for $b > 0$, $F_{a,b}$ is topologically conjugate with $T_{a,b}$ and hence with $H_{a,b}$ on a neighborhood of $\Lambda(a,b)$. For $(a,b) \in R, b > 0$, let

$$\begin{aligned}\Gamma(a,b) &= \psi_{a,b} \circ \pi_0^{-1}(\Lambda(a,b)) \\ &= \psi_{a,b}(\{\underline{z} \in \varprojlim(D, T_{a,b}) |\ z_n \in B(a) \text{ for } n = 0,1,2,\ldots\}\end{aligned}$$

and let

$$\begin{aligned}\Gamma(a,0) &= \psi_{a,0}(\{\underline{z} \in \varprojlim(D, T_{a,0}) | z_n \in B(a) \text{ for } n = 0,1,2,\ldots\} \\ &= \psi_{a,0}(\{\underline{z} \in \varprojlim(D, T_{a,0}) | z_n \in J_a \times \{0\} \text{for } n = 0,1,2,\ldots\}.\end{aligned}$$

Finally, let $\rho_e(a,b)$ be the prime end rotation number of $F_{a,b}$ associated with $\Gamma(a,b)$. Since, for $b > 0, F_{a,b}$ near $\Gamma(a,b)$ is topologically conjugate with $H_{a,b}$ near $\Lambda(a,b)$, $\rho_e(a,b)$ equals the prime end rotation number of $H_{a,b}$ associated with $\Lambda(a,b)$ for $(a,b) \in R, b > 0$.

Theorem 3.2: The function $a \mapsto \rho_e(a,0)$ is continuous on $(1,2)$ and has the following properties:

i) $\displaystyle\lim_{a \to 1^+} \rho_e(a,0) = 1/2$

ii) $\displaystyle\lim_{a \to 2^-} \rho_e(a,0) = 1$

iii) for $1 < a < 2$, $\displaystyle\lim_{b \to 0^+} \rho_e(a,b) = \rho_e(a,0)$;

iv) if $\rho_e(a,0)$ is irrational then the prime end homeomorphism $\mathcal{F}_a$ associated with $F_{a,0}$ and $\Gamma(a,0)$ is Denjoy and $F_{a,0}|_{\Gamma(a,0)}$ is chain transitive.

Proof: We will calculate $\rho_e(a,0)$ by opening up the interval $J_a \times \{0\}$ into a circle $S(a)$. A degree one map g_a of the circle $S(a)$ will be constructed and we will find that $\rho_e(a,0)$ is equal to the maximum rotation number of g_a (this is the rotation number of the g_a^+ that appears later). The conclusions of the Theorem will then follow from a study of g_a.

Recall that

$$T_{a,0}(B(a)) = H_{a,0}(B(a)) \subset I_a \times \{0\}$$

and that

$$\cap_{n\geq 0} T_{a,0}^n(B(a)) = (\cap_{n\geq 0} f_a^n(I_a)) \times \{0\} = J_a \times \{0\}$$

where $f_a(x) = 1 - ax^2$ and $J_a = [1 - a, 1]$. Fix $a \in (1, 2)$ and let $D(a) \subset \mathbf{R}^2$ be the closed disk of radius $a/2$ centered at $(\frac{2-a}{2}, 0)$. Let $S(a) = \partial D(a)$ and let $D_1 \subset \mathbf{R}^2$ be a compact disk with $D(a) \subset \overset{\circ}{D}_1$. There is then a continuous surjection $P_a : D_1 \to D$ such that $P_a(x, y) = (x, 0)$ if $(x, y) \in D(a)$ and P_a maps $D_1 \backslash D(a)$ homeomorphically onto $D \backslash (J_a \times \{0\})$.

Let $i_\pm : J(a) \to S(a) = \partial D(a)$ be given by $i_\pm(x) = (x, y) \in S(a)$ where $\pm y \geq 0$. Note that $P_a \circ i_\pm(x) = (x, 0)$ for $x \in J_a$. Now let $g_a : S(a) \to S(a)$ be defined by

$$g_a(x, y) = \begin{cases} i_+(f_a(x)) & \text{if } y \leq 0 \\ i_-(f_a(x)) & \text{if } y \geq 0 \text{ and } 0 \leq x \leq 1 \\ i_+(f_a(x)) & \text{if } y \geq 0 \text{ and } 1 - a \leq x \leq 0 . \end{cases}$$

Then g_a is a degree one circle map that is semi-conjugate, by way of P_a, to f_a.

Let $G_a : D_1 \to D_1$ be a map extending g_a and having the properties:

i) $G_a|_{S(a)} = g_a$;

ii) G_a is a near homeomorphism (that is, G_a is a uniform limit of homeomorphisms of D_1); and

iii) $P_a \circ G_a = T_{a,0} \circ P_a$.

Since G_a is a near homeomorphism, the inverse limit space $\varprojlim(D_1, G_a)$ is homeomorphic with D_1([Br]). Also, the map

$$\hat{P}_a : \varprojlim(D_1, G_a) \to \varprojlim(D, T_{a,0})$$

defined by

$$\hat{P}_a(z_0, z_1, \ldots) = (P_a(z_0), P_a(z_1), \ldots)$$

is a continuous surjection that semi-conjugates $\hat{G}_a$ and $\hat{T}_{a,0}$; that is, $\hat{P}_a \circ \hat{G}_a = \hat{T}_{a,o} \circ \hat{P}_a$.

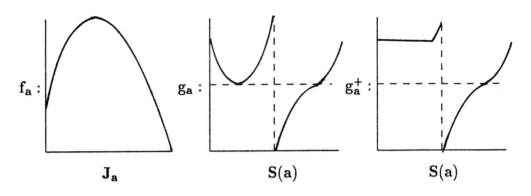

Figure 1

The "attractor" for $\hat{T}_{a,0}$ is

$$\Psi_{a,0}^{-1}\left(\Gamma(a,0)\right) = \{(z_0, z_1, \ldots) \in \varprojlim(D, T_{a,0})|z_n \in J_a \times \{0\} \text{ for all n}\}$$

and the attractor for $\hat{G}_a$ is

$$\hat{P}_a^{-1} \circ \Psi_{a,0}^{-1}\left(\Gamma(a,0)\right) = \{(z_0, z_1, \ldots) \in \varprojlim(D_1, G_a)|z_n \in D(a) \text{ for all n}\}.$$

Since P_a maps $D_1 \backslash D(a)$ homeomorphically onto $D \backslash (J_a \times \{0\})$, $\hat{P}_a$ maps

$$(\varprojlim(D_1, G_a)) \backslash \hat{P}_a^{-1} \circ \Psi_{a,0}^{-1}\left(\Gamma(a,0)\right) \equiv \mathcal{U}$$

homeomorphically onto

$$\varprojlim(D, T_{a,0}) \backslash \Psi_{a,0}^{-1}\left(\Gamma(a,0)\right) \equiv \mathcal{V}.$$

Thus $\hat{G}_a|_{\mathcal{U}}$ is topologically conjugate with $\hat{T}_{a,0}|_{\mathcal{V}}$ and it follows that the prime end rotation numbers

$$\begin{aligned}\rho_e(\hat{G}_a, \hat{P}_a^{-1} \circ \Psi_{a,0}^{-1}(\Gamma(a,0))) &= \rho_e(\hat{T}_{a,0}, \Psi_{a,0}^{-1}(\Gamma(a,0))) \\ &= \rho_e(F_{a,0}, \Gamma(a,0))\end{aligned}$$

are equal.

Let $g_a^+ : S(a) \to S(a)$ be the upper truncation of g_a defined by

$$g_a^+(x,y) = \begin{cases} g_a(1-a,0) & \text{if } 1-a \leq x \leq a-1 \text{ and } y \leq 0 \\ g_a(x,y) & \text{otherwise.} \end{cases}$$

g_a^+ is then a monotone degree one map of the circle $S(a)$ that varies continuously with a (in the sense that there is a one-parameter family of homeomorphisms $\Phi_a : S(a) \to S^1$ onto the unit circle such that the $\Phi_a \circ g_a^+ \circ \Phi_a^{-1}$ vary continuously with a). Thus the rotation number $\rho(g_a^+)$ varies continuously with a.

We wish to show that $\rho_e(\hat{G}_a, \hat{P}_a^{-1} \circ \Psi_{a,0}^{-1}(\Gamma(a,0))) = \rho(g_a^+)$. To do this we first construct a near homeomorphism $G_a^+ : D_1 \to D_1$ that extends g_a^+ and that agrees, as much as possible, with G_a. Let Δ be the disk in D_1 bounded by the arc $A = P_a^{-1}(\{(x,y)| \ y = ax^2 - a(1-a)^2, \ y < 0\})$ and the arc $L(a) = \{(x,y)|y = -\sqrt{(a/2)^2 - (x - (\frac{2-a}{2}))^2} \ , \ 1-a \le x \le a-1\} \subset S(a)$. (Note that $L(a)$ is the arc of $S(a)$ on which g_a^+ is constant.) Then $G_a(\Delta) = G_a(L(a))$ and $G_a(A) = G_a(1-a,0) = g_a^+(1-a,0)$. Define $G_a^+ : D_1 \to D_1$ so that:

i) $G_a^+ = G_a$ on $\mathcal{D}_1\backslash(\mathcal{D}(a) \cup \Delta)$;

ii) $G_a^+ = g_a^+$ on $S(a)$;

iii) $G_a^+(D(a)) \subset D(a)$; and

iv) G_a^+ is a near homeomorphism.

We have that $\varprojlim(D_1, G_a^+)$ is homeomorphic with D_1 and the attractor for G_a^+ is $\varprojlim(D(a), G_a^+|_{D(a)})$, the boundary of which is $\varprojlim(S(a), g_a^+)$. Since $g_a^+ : S(a) \to S(a)$ is a near homeomorphism (its monotone), $\varprojlim(S(a), g_a^+)$ is homeomorphic with $S(a)$. Thus the prime end rotation number of $\hat{G}_a^+$ associated with $\varprojlim(D(a), G_a^+|_{D(a)})$ is just the rotation number $\rho(\hat{g}_a^+)$ of the circle homeomorphism

$$\hat{g}_a^+ = \hat{G}_a^+|_{\varprojlim(S(a), G_a^+|S(a))}.$$

Furthermore, since $\hat{g}_a^+$ is semi-conjugate to g_a^+ (by way of the projection π_0), $\rho(\hat{g}_a^+) = \rho(g_a^+)$.

To show that

$$\rho_e(\hat{G}_a^+, \hat{P}_a^{-1} \circ \Psi_{a,0}^{-1}(\Gamma(a,0))) = \rho(g_a^+),$$

it remains to show that

$$\rho_e(\hat{G}_a, \hat{P}_a^{-1} \circ \Psi_{a,0}^{-1}(\Gamma(a,0))) = \rho_e(\hat{G}_a^+, \varprojlim(D(a), G_a^+|_{D(a)})).$$

We will accomplish this by considering the action of $\hat{G}_a$ and $\hat{G}_a^+$ on certain accessing arcs.

If w is a point on $(J_a \backslash [1-a, a-1]) \times \{0\}$, then $T_{a,0}^{-1}\,(T_{a,0}(w))$ is a parabolic arc that meets $J_a \times \{0\}$ only at w and $(T_{a,0}^{-1}\,(T_{a,0}(w)))\backslash\{w\}$ is a pair of (half open) arcs, each meeting $\partial B(a)$, that access w from opposite sides of $J_a \times \{0\}$. It follows that if z is a point on $S(a)\backslash L(a)$ (recall that $L(a) = P_a^{-1}([1-a, a-1] \times \{0\}) \cap S(a) \cap \{(x,y)|y \leq 0\}))$, then there is an arc α that lies in $P_a^{-1}(B(a))\backslash \overset{\circ}{D}(a)$, that has one endpoint at z and the other on $\partial P_a^{-1}(B(a))$, and has the property that $G_a(\alpha)$ is the single point $G_a(z) = g_a(z)$. Also, since $T_{a,0}$ is one-to-one off $P_a^{-1}(B(a))$, G_a is one-to-one off $P_a^{-1}(B(a))$ and, moreover, $G_a(D_1 \backslash P_a^{-1}(B(a))) \supset D_1 \backslash P_a^{-1}(B(a))$.

It follows from the above that if

$$z_n \in S(a), \ z_{n+1} \in S(a)\backslash L(a), \ g_a(z_{n+1}) = z_n,$$

and if $\alpha_n : [0,1] \to D_1 \backslash \overset{\circ}{D}(a)$ is an arc with $\alpha_n(0) \in \partial D_1, \alpha_n([0,1]) \subset D_1 \backslash D(a)$, and $\alpha_n(1) = z_n$, then there is an arc $\alpha_{n+1} : [0,1] \to D_1 \backslash \overset{\circ}{D}(a)$ such that

$$a_{n+1}(0) \in \partial D_1, \alpha_{n+1}([0,1]) \subset D_1 \backslash D(a), \alpha_{n+1}(1) = z_{n+1},$$

and $G_a(\alpha_{n+1}([0,1])) = \alpha_n([0,1])$. It follows now that if $\underline{z} = (z_0, z_1, \ldots) \in \varprojlim(S(a), g_a) \subset \varprojlim(D(a), G_a)$ is such that $z_n \in S(a)\backslash L(a)$ for all n then $\underline{z}$ is accessible from $\mathcal{U} = (\varprojlim(D_1, G_a))\backslash(\varprojlim(D(a), G_a|_{D(a)}))$.

Now let

$$\underline{z} = (z_0, z_1, \ldots) \in \varprojlim(S(a), g_a)$$

be such that

$$z_0 \in S(a)\backslash \cup_{n \geq 0} (g_a^+)^n(L(a)).$$

Then $\underline{z}$ is accessible from $\mathcal{U}$ and we may choose an accessing arc $\underline{\alpha} = (\alpha_0, \alpha_1, \ldots)$ such that $\alpha_n(t) \notin \Delta$ for all $n \geq 1$ and all $t \in [0,1]$.

Then $\underline{\alpha}$ is also an arc in $\varprojlim(D_1, G_a^+)$ (since $G_a^+ = G_a$ off $D(a) \cup \Delta$) accessing $\underline{z} \in \varprojlim(S(a), g_a^+)$ from $(\varprojlim(D_1, G_a^+))\backslash(\varprojlim(D(a), G_a^+|_{D(a)})$. Clearly $\hat{G}_a^{-n} \circ \underline{\alpha} = (\hat{G}_a^+)^{-n} \circ \underline{\alpha}$ for all $n \geq 0$ and it follows that

$$\rho_e(\hat{G}_a, \hat{P}_a^{-1} \circ \Psi_{a,0}^{-1}(\Gamma(a,0))) = \rho_e(\varprojlim(D(a), \hat{G}_a^+), G_a^+|_{D(a)}).$$

We now have established that $\rho_e(a,0) = \rho_e(F_{a,0}, \Gamma(a,0)) = \rho(g_a^+)$. Thus $\rho_e(a,0)$ varies continuously with $a \in (1,2)$ as asserted in the theorem.

It is an easy matter to check that $\rho(g_1^+) = 1/2$ and $\rho(g_2^+) = 1$ (g_1^+ has a periodic orbit of period 2, g_2^+ has fixed point, and, for the appropriate continuous family of lifts $\tilde{g}_a^+ : \mathbf{R} \to \mathbf{R}, \tilde{g}_2^+ \geq \tilde{g}_1^+$). Thus $\lim_{t \to 1+} \rho_e(a,0) = \lim_{t \to 1+} \rho(g_a^+) = 1/2$ and $\lim_{t \to 2-} \rho_e(a,0) = \lim_{t \to 2-} \rho(g_a^+) = 1$, proving i) and ii).

The proof of iii), that $\lim_{b \to 0+} \rho_e(a,b) = \rho_e(a,0)$ for $1 < a < 2$, is along the lines of the proofs of Propositions 2.2 and 2.3. We will argue that the attractors $\Gamma(a,b)$ converge to $\Gamma(a,0)$ in the Hausdorff topology as $b \to 0^+$ so that (as in the proof of Proposition 2.2) $\lim_{b \to 0+} \rho_e(a,b) = \rho_e(a,0)$. For fixed $a \in (1,2)$, the basin $B(a)$ of the attractor $A(a,b) = \cap_{n \geq 0} T_{a,b}^n(B(a))$ is constant in b so that $\limsup_{b \to 0+} A(a,b) \subset A(a,0)$ (see the proof of Proposition 2.3). Also, for each (a,b), $b \geq 0$ and small, $A(a,b)$ contains a hyperbolic saddle $p(a,b)$. These saddles, and their local unstable manifolds, vary continuously so that (again as in the proof of Proposition 2.3) $\liminf_{b \to 0+} A(a,b) \supset cl(W^u(p(a,0)))$. But it is easy to check that $cl(W^u(p(a,0)) = J_a \times \{0\} = A(a,0)$. Thus $A(a,b) \to A(a,0)$ in the Hausdorff topology as $b \to 0^+$. It follows that the attractors

$$\{(z_0, z_1, \ldots) \in \varprojlim(D, T_{a,b}) | z_n \in A(a,b) \text{ for all n}\} \text{ of } \hat{T}_{a,b}$$

limit on

$$\{(z_0, z_1, \ldots) \in \varprojlim(D, T_{a,0}) | z_n \in A(a,0) \text{ for all n}\}$$

in the Hausdorff topology (in D^N) and from this we find that the attractors

$$\Gamma(a,b) = \Psi_{a,b}(\{(z_0, z_1, \ldots) \in \varprojlim(D, T_{a,b}) | z_n \in A(a,b) \text{ for all n}\})$$

of $F_{a,b}$ converge to $\Gamma(a,0)$ as $b \to 0^+$, as desired, proving iii).

We now prove that if $1 < a < 2$ and $\rho_e(a,0)$ is irrational, then $F_{a,0}|_{\Gamma(a,0)}$ is transitive. Let a be such a value. It was established above that $\rho_e(a,0) = \rho(g_a^+)$. Thus $\rho_e(a,0)$ irrational implies the g_a^+ has no periodic orbits so $((g_a^+)^n(L(a))) \cap L(a) = \emptyset$ for all $n \geq 1$ (recall that $L(a)$ is the interval on which g_a^+ is constant). It follows that the intervals $(g_a^+)^{-n} (\overset{\circ}{L}(a)), n = 0, 1, 2, \ldots$, are pairwise disjoint and hence that

$$\text{diam}((g_a^+)^{-n}(\overset{\circ}{L}(a))) \to 0 \text{ as n} \to \infty.$$

Let

$$\mathcal{U} = \cup_{n \geq 0} (g_a^+)^{-n} (\overset{\circ}{L}(a)).$$

It is straightforward to check that $g_a^+(S(a) \backslash \mathcal{U}) = S(a) \backslash \mathcal{U}$ and, since $g_a = g_a^+$ off $\overset{\circ}{L}(a)$ and hence off $\mathcal{U}$, $g_a(S(a) \backslash \mathcal{U}) = S(a) \backslash \mathcal{U}$.

Since g_a^+ is monotone and has irrational rotation number,

$$\hat{g}_a^+ : \varprojlim(S(a), g_a^+) \rightarrow \varprojlim(S(a), g_a^+)$$

is a circle homeomorphism with irrational rotation number. Thus $\hat{g}_a^+$ is chain transitive and g_a^+ is also chain transitive. It follows that $g_a^+|_{S(a) \backslash \mathcal{U}}$ is chain transitive so $g_a|_{S(a) \backslash \mathcal{U}}$ is also chain transitive.

We will now prove that $f_a : J_a \rightarrow J_a$ is chain transitive. Since J_a is connected, it suffices to show that each point in J_a is chain recurrent under f_a ([F]).

Recall that $P_a(S(a)) = J_a \times \{0\}$, that P_a semi-conjugates g_a on $S(a)$ with $T_{a,0}$ on $J_a \times \{0\}$, and that $T_{a,0}(x, 0) = (f_a(x), 0)$ for $x \in J_a$. Abusing notation, we will consider the range of $P_a|_{S(a)}$ to be J_a so that P_a semi-conjugates g_a with f_a.

We claim that f_a has no attracting periodic orbits in J_a. For suppose, to the contrary, that $\mathcal{O}(x)$ is an attracting periodic orbit in J_a. Then $f_a^n(0)$ approaches $\mathcal{O}(x)$ as $n \rightarrow \infty$ ([S]). There is then an attracting periodic orbit $\mathcal{O}(z)$ in $S(a)$ under g_a and $g_a^n(w)$ approaches $\mathcal{O}(z)$ as $n \rightarrow \infty$ for $w \in P_a^{-1}(0)$. Then $g_a^n((1 - a, 0)) = g_a^n(g_a^2(w))$ approaches $\mathcal{O}(z)$ as $n \rightarrow \infty$. But $(1 - a, 0) \in \partial \overset{\circ}{L}(a) \subset S(a) \backslash \mathcal{U}$ and $\mathcal{O}(z) \subset \mathcal{U}$ since $\rho(g_a^+)$ is irrational. Thus $g_a^n((1 - a, 0))$ can't approach $\mathcal{O}(z)$ ($S(a) \backslash \mathcal{U}$ is invariant under g_a) and we conclude that f_a can't have an attracting periodic orbit, as claimed.

Now since f_a has no attracting periodic orbits it is the case that for each nondegenerate interval $K \subset J_a$, either $f_a^n(K) \cap K \neq \emptyset$ for some $n \geq 1$ or $0 \in f_a^n(K)$ for some $n \geq 1$ (this is the "no homterval" property, see [G] and [Mi]). Now let $x \in J_a$ and let $\epsilon > 0$ be given. Let K be an interval containing x such that $\text{diam}(K) < \epsilon/2$ and $\text{diam}(f_a(K)) < \epsilon/2$. If $f_a^n(K) \cap K \neq \emptyset$ for some $n \geq 1$ then there is clearly an ϵ-chain from x to x. Otherwise, $0 \in f_a^n(K)$ for some $n \geq 1$ so there is an ϵ-chain from x to 0. Now let $w \in P_a^{-1}(0)$ and let $z \in P_a^{-1}(x)$. Then $w \in S(a) \backslash \mathcal{U}$ and if $z \in S(a) \backslash \mathcal{U}$ there is an ϵ-chain from w to z under g_a (since $g_a|_{S(a) \backslash U}$ is chain transitive). If $z \in \mathcal{U}$ then $z \in (g_a^+)^{-m}(\overset{\circ}{L}(a))$ for some $m \geq 0$. Let $k \geq m$ be large enough so that

$\text{diam}((g_a^+)^{-k}(\overset{\circ}{L}(a))) < \epsilon$. Since $\partial(g_a^+)^{-k-1}(\overset{\circ}{L}(a)) \subset S(a)\backslash\mathcal{U}$, there is an ϵ-chain under g_a from w to a point of $\partial(g_a^+)^{-k-1}(\overset{\circ}{L}(a))$. This ϵ-chain may be extended to $(g_a^+)^{m-k}(z) \in (g_a^+)^{-k}(\overset{\circ}{L}(a))$ and thence to z.

Thus there is an ϵ-chain under g_a from $w \in P^{-1}(0)$ to $z \in P^{-1}(x)$. Since P_a shrinks distances and semi-conjugates g_a with f_a, such a chain pushes down by P_a to an ϵ-chain under f_a from 0 to x. Concatenating this with the ϵ-chain from x to 0 we see that x is ϵ-chain recurrent and hence chain recurrent. Thus f_a is chain transitive on J_a.

It is simple to check that $\hat{f}_a$ must be chain transitive on $\varprojlim(J_a, f_a)$. Since $\hat{f}_a$ is conjugate to $\hat{T}_{a,0}$ on $A(a,0)$ and $\hat{T}_{a,0}|_{A(a,0)}$ is conjugate to $F_{a,0}$ on $\Gamma(a,0)$, we have that $F_{a,0}$ is chain transitive on its attractor $\Gamma(a,0)$, as was to be proved.

It remains to show that when $\rho_e(a,0)$ is irrational the prime end homeomorphism $\mathcal{F}_a$ associated with $F_{a,0}$ and $\Gamma(a,0)$ is Denjoy. Since, as proved earlier, $\mathcal{F}_a$ is topologically conjugate with the prime end homeomorphism $\mathcal{G}_a$ associated with $\hat{G}_a$ and $\varprojlim(D(a), G_a|_{D(a)})$, the question will be settled if we show that $\mathcal{G}_a$ is not transitive. Were $\mathcal{G}_a$ transitive it would be the case that for each neighborhood $\mathcal{W}$ of a point of $\partial(\varprojlim(D(a), G_a|_{D(a)}))$ and each arc $\underline{\alpha}$ accessing a point of $\partial(\varprojlim(D(a), G_a|_{D(a)}))$ from the complement, there would be an $n > 0$ such that $\hat{G}_a^{-n}(\underline{\alpha}) \cap \mathcal{W} \neq \emptyset$. But let Δ be the disk defined earlier in the construction of G_a^+ and let $\mathcal{W} = \{\underline{z} \in \varprojlim(D_1, G_a)|z_0 \in \Delta \cup \overset{\circ}{D}(a)\}$. Then $\mathcal{W}$ is a neighborhood of a point $\underline{z}$ on $\partial(\varprojlim(D(a), G_a|_{D(a)}))$ with $z_0 \in \overset{\circ}{L}(a)$. Let $\underline{\alpha}$ be the accessing arc constructed earlier with $\alpha_n(t) \notin \Delta$ for all $n \geq 0$ and all t. Since $\alpha_n(t)$ also lies outside $\overset{\circ}{D}(a)$ for all n and t we see that

$$\hat{G}_a^{-n} \circ \underline{\alpha}(t) = (\alpha_n(t), \alpha_{n+1}(t), \ldots) \notin \mathcal{W}$$

for all $n \geq 0$ and t. That is $\hat{G}_a^{-n}(\underline{\alpha})$ is disjoint from $\mathcal{W}$ for all $n \geq 0$ so $\mathcal{G}_a$, and hence $\mathcal{F}_a$, is not transitive. ■

<u>Corollary:</u> There are uncountably many $a, 1 < a < 2$, such that the interval map $f_a : J_a \to J_a$, $J_a = [1-a, 1]$, $f_a(x) = 1 - ax^2$, is chain transitive.

<u>Proof:</u> It follows from the continuity of $\rho_e(a,0)$ and i) and ii) of Theorem 3.2 that $\rho_e(a,0)$ is irrational for infinitely many a in $(1,2)$. For each such

$a, F_{a,0}$ restricted to $\Gamma(a,0)$ is chain transitive. Moreover $F_{a,0}|_{\Gamma(a,0)}$ is conjugate to $\hat{T}_{a,0}|_{A(a,0)}$ which is semiconjugate to f_a.   ■

Acknowledgement: The author thanks Jason Holt for numerical work and helpful discussions during the preparation of this paper.

References

[AY] K. Alligood and J. Yorke. Accessible saddles on fractal basin boundaries. *Ergodic Th. and Dynam. Sys.* To appear.

[B]   G. D. Birkhoff. Sur quelques courbes fermés remarquables. *Bull. Soc. Math. France* 60 (1932), 1-26.

[BC] M. Benedicts and L. Carleson. The dynamics of the Hénon maps. preprint (1988).

[Br] M. Brown. Some applications of an approximation theorem for inverse limits. *Proc. Amer. Math. Soc.* 11 (1960), 478-483.

[BG1] M. Barge and R. Gillette. Rotation and periodicity in plane separating continua. *Ergod. Th. and Dynam. Sys.* To appear.

[BG2] M. Barge and R. Gillette. A fixed point theorem for plane separating continua. *Top. and its Applications.* To appear

[CL] M. L. Cartwright and J. E. Littlewood. Some fixed point theorems. *Ann. Math.* 54 (1951), 1-37.

[CoL] E. F. Collingwood and A. J. Lohwater. *Theory of Cluster Sets, Cambridge Tracts in Mathematics and Mathematical Physics*, No. 56. Cambridge University Press, Cambridge, 1966.

[F]   J. Franks. Recurrence and fixed points of surface homeomorphisms. *Ergod. Th. and Dynam. Sys.* 8 (1988), 99-107.

[G]   J. Guckenheimer. Sensitive dependence on initial conditions for one-dimensional maps. *Comm. Math. Phys.* 70 (1979), 133-160.

[H] P. Hartman. *Ordinary Differential Equations.* Wiley, 2nd edition (1973).

[He] M. Hénon. A two-dimensional mapping with a strange attractor. *Comm. Math. Phys.* 50 (1976), 69-78.

[LeC] P. LeCalvez. Propriétés des attracteurs de Birkhoff. *Ergod. Th. and Dynam. Sys.* 8 (1988), 241-310.

[M] J. Mather. Topological proofs of some purely topological consequences of Caratheodory's Theory of prime ends. Th. M. Rassias. G. M. Rassias, eds., *Selected Studies*, North-Holland (1982), 225-255.

[Mi] M. Misiurewicz. Entropy of piecewise monotone mappings. *Studia Math.* 67 (1980).

[S] D. Singer. Stable orbits and bifurcations of maps of the interval. *SIAM J. Math.* 35, 260, (1978)

On the Rotation Shadowing Property for Annulus Maps

FERNANDA BOTELHO and LIANG CHEN   Memphis State University, Memphis, Tennessee

Abstract. *The relationship between the rotation shadowing property and the structure of the rotation sets for annulus homeomorphisms is discussed.*

Let A denote a compact annulus with the Euclidean metric d, $\tilde{A}$ its *universal covering space* $R \times [0,1]$ with the covering map $p : \tilde{A} \to A$. Points in A are represented by small roman letters, i.e. x; points in $\tilde{A}$ are represented by small letters with a tilde on top, i.e. $\tilde{x}$ satisfying $p(\tilde{x}) = x$.

Let $f : A \to A$ be a homeomorphism of A isotopic to the identity and F be a *lift* of f to $\tilde{A}$ (i.e. it satisfies $p \circ F = f \circ p$.)

Let $\tilde{x}_1 = \pi_1(\tilde{x})$ and $F_1(\tilde{x}) = \pi_1[F(\tilde{x})]$, where π_1 is the natural projection onto the first coordinate.

A sequence $\{z_n\}$ of points in A is called a *δ-pseudo-orbit* of f if for all $n \geq 0$

$$d(z_{n+1}, f(z_n)) < \delta.$$

In particular, a finite δ-pseudo-orbit of f having x as its first term and y as its last term is called a *δ-chain of f from x to y*. A subset Z of A is *chain-transitive* if for $x, y \in Z$ and any $\delta > 0$ there exists a δ-chain from x to y.

The definition of pseudo-orbits can be extended to any endomorphism of a metric space. In particular, a pseudo-orbit in A can be lifted to a pseudo-orbit in $\tilde{A}$. (cf. Komuro [6])

Given $x \in A$ *the rotation sequence* $\{\theta_n(x)\}_{n \geq 1}$ of x is given by

$$\theta_n(x) = \frac{F_1^n(\tilde{x}) - \tilde{x}_1}{n}.$$

The rotation interval of x is defined to be

35

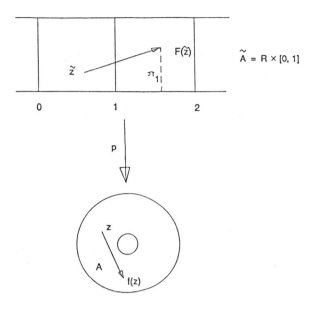

$$\rho(x) = [\liminf_{n\to\infty} \theta_n(x), \limsup_{n\to\infty} \theta_n(x)].$$

When the rotation interval of x collapses to a single point set, we say that x has a *rotation number* $\rho(x)$. The union of all rotation intervals is called *the rotation set* of f, denoted by $\rho(f)$ (cf. Botelho [2]). For an invariant subset Z of A, the rotation set of Z is defined by $\rho(Z) = \rho(f|Z)$. When $\rho(Z)$ consists of a singleton, we say that Z has a *trivial* rotation set.

An annulus map $f : A \to A$ has *the rotation shadowing property* if for any $\varepsilon > 0$, there exists $\delta > 0$ such that for every δ-pseudo-orbit $\{z_k\}$ in A, there exists $z \in A$ satisfying

$$\limsup_{k\to\infty} \frac{|\pi_1(\tilde{z}_k) - F_1^k(\tilde{z})|}{k} < \varepsilon.$$

The rotation shadowing property was introduced by Barge and Swanson [1] to simulate the longterm behavior of rotation sequences. They proved that any circle endomorphism has this property. In this note we study the rotation shadowing property for a large class of annulus homeomorphisms. The following examples show that there exists some connection between the rotation shadowing property and the structure of the rotation set of an annulus map.

Example 1. A generalized *shear map* is an annulus homeomorphism $f : A \rightarrow A$ whose lift is of the form

$$F(\tilde{x}, \tilde{y}) = (\tilde{x} + \phi(\tilde{y}), \tilde{y})$$

where $\phi : [0,1] \rightarrow [\alpha, \beta]$ is a continuous onto map. It is easy to see that, under f, every point of A has a rotation number and the rotation set $\rho(f) = [\alpha, \beta]$. But for any $\delta > 0$, we will always be able to construct a δ-pseudo-orbit which has "chaotic" rotation around the annulus.

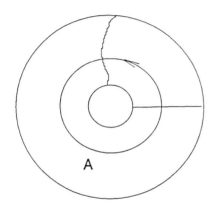
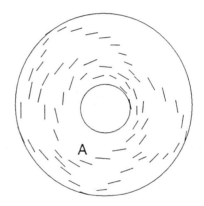
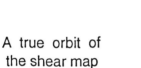

A true orbit of
the shear map

A pseudo orbit of
the shear map

Example 2. A *spiral map* $f : A \rightarrow A$ satisfies the condition: $\omega(f) \cup \alpha(f) \subset \partial A$, where $\omega(f)$ is the ω-limit set of f, $\alpha(f)$ is the α-limit set of f and ∂A is the boundary of A. The rotation set of f consists of two numbers, i.e. $\rho(f) = \{\alpha, \beta\}$ and α, β are the rotation numbers of f on the inner and the outer boundaries of A respectively. All the orbits of a sp iral map, even pseudo-orbits, in the interior of the annulus approach asymptotically the boundary of the annulus.

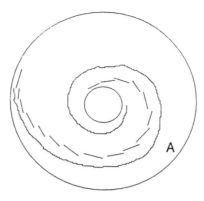

A pseudo orbit of a
spiral map

The next theorem applies to spiral maps and rigid rotations as well as to the class of all annulus homeomorphisms isotopic to the identity with finitely many periods. (see Franks [4])

Theorem 1.   *Let $f : A \to A$ be an annulus homeomorphism isotopic to the identity. If the complement of the rotation set of f is dense in the real line R, then f has the rotation shadowing property.*

In particular, for annulus homeomorphisms with finitely many periods we have:
Corollary 2.   *Let $f : A \to A$ be an annulus homeomorphism isotopic to the identity with finitely many periods. The rotation set of f contains at most finitely many rationals. Moreover, f has the rotation shadowing property.*

One may notice that shear maps do not satisfy the condition of Theorem 1. But the following theorem provides a criteria for the rotation shadowing property that can be applied to a much larger class of annulus maps (e.g. shear maps). We refer the reader to [3] for the complete proof.

Theorem 3.   *If f is an annulus map isotopic to the identity such that every point has a rotation number, then f has the rotation shadowing property if and only if every chain-transitive, invariant closed subset of the annulus has a trivial rotation set.*

The whole annulus is chain-transitive for any shear map, therefore it follows from Theorem 3 that if the rotation set is nontrivial then the map can not have the rotation shadowing property.

First we recall the definition of pseudo-rotation sets due to Barge and Swanson [1].

If $\{z_n\}$ is a δ-pseudo-orbit of f and $\{\tilde{z}_n\}$ is its lift, *the pseudo-rotation interval* is defined by

$$\rho(\{\tilde{z}_n\}) = [\liminf_{n\to\infty} \frac{\pi_1(\tilde{z}_n)}{n}, \limsup_{n\to\infty} \frac{\pi_1(\tilde{z}_n)}{n}].$$

Let $\rho(F,\delta)$ denote the union of all pseudo-rotation intervals of δ-pseudo-orbits for F, *the pseudo-rotation set of F* is given by

$$\rho_\psi(F) = \bigcap_{\delta>0} \rho(F,\delta)$$

and *the pseudo-rotation set of f* is given by $\rho_\psi(f) \equiv \rho_\psi(F)$ (mod 1).

Given a sequence of sets X_n, let

$$\limsup_{n\to\infty} X_n = \bigcap_{N\geq 0} cl(\bigcup_{n=N}^{\infty} X_n)$$

and for a given sequence $\{x_k\}$ of points let $\omega(\{x_k\}) = \limsup_{n\to\infty} X_n$ with $X_n = \{x_k : k \geq n\}$. A set Z in A is called α-*invariant* if $d(f(x),Z) < \alpha$ for all $x \in Z$ and Z is said to be α-*chain transitive* if for any two points $x, y \in Z$ there exists a α-chain from x to y. A set Z in A is said to be *chain transitive* if for any $\alpha > 0$, Z is α-chain transitive. For a closed invariant chain transitive set Z let $\rho(Z) = \rho(f|_Z)$ (resp. $\rho_\psi(Z) = \rho_\psi(f|_Z)$).

The following lemma is a minor modification of Lemma 3.5. and 3.6. in Barge and Swanson [1].

Lemma 4   *Let $f : A \to A$ be a homeomorphism and for each $n \in N$ let $z_n = \{z_k^n\}_k$ be a $\frac{1}{2n}$-pseudo-orbit of f and $S_n = \omega(\{z_k^n\}_{k=0}^{\infty})$, then the following holds:*

(i) S_n is $\frac{1}{n}$-invariant and $\frac{1}{n}$-chain transitive;

(ii) there exists a subsequence $\{S_{n_i}\}_i$ such that $\limsup_{i\to\infty} S_{n_i}$ is invariant, chain transitive and

$$\limsup_{i\to\infty} \rho(S_{n_i}, \frac{1}{n_i}) \subseteq \rho_\psi(\limsup_{i\to\infty} S_{n_i}).$$

Handel [5] has proved that the rotation set of any annulus homeomorphism isotopic to the identity is closed, so a result due to Barge and Swanson [1] can be stated as follows:

Lemma 5.   *Let $f : A \to A$ be an annulus homeomorphism isotopic to the identity. Then $\rho_\psi(f) = \rho(f)$.*

If the complement of $\rho(f)$ is dense in R, then for any $\varepsilon > 0$, we can choose $\alpha_1, \ldots, \alpha_n \notin \rho(f)$ satisfying

$$\alpha_0 < \inf \rho(f) < \alpha_1 < \ldots < \alpha_{n-1} < \sup \rho(f) < \alpha_n$$

and

$$|\alpha_{i+1} - \alpha_i| < \varepsilon \qquad \text{for } j = 0, 1, \ldots, n - 1.$$

Let $J_i = [\alpha_i, \alpha_{i+1}]$, $\rho_i = J_i \cap \rho(f)$ for $j = 0, \ldots, n - 1$ and

$$\eta := \min_i \{ |\inf \rho_i - \alpha_i|, |\sup \rho_i - \alpha_{i+1}| \} > 0.$$

For each $i = 0, \ldots, n - 1$, define

$$V_i = \{ x \in R \; : \; d(x, y) < \eta \quad \text{for some } y \in \rho_i, \}$$

then $V_i \subseteq (\alpha_i, \alpha_{i+1})$ and

$$\rho(f) \subseteq \bigcup_{i=0}^{n-1} V_i = V.$$

Lemma 6.   *There exists $\delta > 0$ such that*

$$\rho(f, \delta) \subseteq V.$$

Proof. Assume that for each $n \in N$ there exists $l_n \in \rho(f, \frac{1}{2n}) \setminus V$. Without loss of generality we may assume that l_n converges to some $l \notin V$. There exist sequences $z_n = \{z_k^n\}_{k=0}^\infty$ of $\frac{1}{2n}$-pseudo-orbits in A with $l_n \in \rho(\{z_k^n\}_{k=0}^\infty)$ for each $n = 1, 2, \ldots$.

Let $S_n = \omega(\{z_k^n\}_{k=0}^\infty)$. For each $n \in N$, S_n is $\frac{1}{n}$-chain transitive and $\frac{1}{n}$-invariant. We select a subsequence $\{S_{n_i}\}_i$ as in Lemma 4, therefore by Lemma 5

$$l \in \limsup_{i \to \infty} \rho(S_{n_i}, \frac{1}{n_i}) \subseteq \rho_\psi(\limsup_{i \to \infty} S_{n_i}) \subseteq \rho_\psi(f) = \rho(f).$$

This is impossible since $l \notin V$ and $\rho(f) \subset V$. $\square$

The proof of Theorem 1 is based on the previous lemma.

Proof of Theorem 1. Given $\varepsilon > 0$, there exists $\delta > 0$ satisfying Lemma 6. For any δ-pseudo-orbit $\{z_n\}$

$$\rho(\{\tilde{z}_n\}) = [\liminf_{n \to \infty} \frac{\pi_1(\tilde{z}_n)}{n}, \limsup_{n \to \infty} \frac{\pi_1(\tilde{z}_n)}{n}]$$
$$\subseteq \rho(f, \delta) \cap J_i \subseteq V_i \quad \text{for some } i.$$

Therefore there exists $N > 0$ such that $\frac{\pi_1(\tilde{z}_n)}{n} \in V_i$ for every $n \geq N$. We have

$$|\frac{\pi_1(\tilde{z}_n)}{n} - \alpha| < \varepsilon \quad \text{for all } \alpha \in \rho_i \text{ and } n \geq N.$$

Finally, since the complement of the rotation set is dense, for any $\alpha \in \rho_i$, there exists an $x \in A$ such that $\rho(x) = \alpha$. Then

$$|\frac{\pi_1(\tilde{z}_n)}{n} - \frac{F_1^n(\tilde{x})}{n}| \leq \varepsilon \quad \text{for all } n \geq N_0, \quad \text{for some } N_0. \quad \square$$

Proof of Corollary 2. If α is a rational in $\rho(f)$, then it follows from a result in Franks [4] that α is realized by a periodic point $x \in A$. Let n be the period of x, then for some integer m

$$F^n(\tilde{x}) = \tilde{x} + (m, 0).$$

The rotation number of x is

$$\alpha = \rho(x) = \lim_{k \to \infty} \theta_{kn}(x) = \frac{m}{n}.$$

For a given $n > 0$ there exist at most $n[\sup \rho(f) - \inf \rho(f)] + 1$ rationals in $\rho(f)$ corresponding to periodic points of period n. Since f has finitely many periods, $\rho(f)$ contains at most finitely many rationals. $\square$

References

[1] M.Barge, R. Swanson, The rotation shadowing property of circle and annulus maps, *Ergod. Th. & Dynam. Sys.* (1988), 8, 509-521.

[2] F.Botelho, Rotation sets of maps of the annulus, *Paci. J. of Math.*, Vol. 133, No. 2, 1988, 251-266.

[3] F.Botelho, L.Chen, Chain transitivity and rotation shadowing for annulus endomorphisms, *preprint*, 1991.

[4] J.Franks, Recurrence and fixed points of surface homeomorphisms. *Ergod. Th. & Dynam. Sys.* (1988), 8*, 99-107.

[5] M.Handel, The rotation set of a homeomorphism of the annulus is closed. *Comm. Math. Phys.*, (1990).

[6] M.Komuro, The pseudo orbit tracing properties on the space of probability measures, *Tokyo J. Math.*, Vol. 7, No. 21, 1984, 461-468.

A Nielsen-Type Theorem for Area-Preserving Homeomorphisms of the Two Disc

KENNETH BOUCHER   University of Utah, Salt Lake City, Utah

MORTON BROWN   University of Michigan, Ann Arbor, Michigan

EDWARD E. SLAMINKA   Auburn University, Auburn University, Alabama

Abstract

We prove the following theorem concerning the mininum number of fixed points for an area preserving homeomorphism of the two disc.

Theorem *Let $h: D \rightarrow D$ be an area preserving, orientation preserving homeomorphism of the two disc to itself. Furthermore, suppose that the fixed point set of h restricted to the boundary of D is finite and consists of precisely n attracting and n repelling fixed points. Then there exists at least $n + 1$ fixed points in $int(D)$.*

The proof utilizes the Brouwer translation arc lemma combined with the notion of free modification for orientation preserving homeomorphisms of two manifolds, developed by M. Brown, and its extension to area preserving homeomorphisms by Pelikan and Slaminka.

1 Introduction.

Let $h: M^n \to M^n$ be a self map of a compact, connected n–manifold with boundary. What can one say about the number of fixed points for h on the interior of M^n? In general this is an extremely difficult question to answer. Following the lead of Nielsen (cf. Nielsen [N], Jiang [J] or Kiang [K]), we alter the question to one that may be more amenable to analysis: what is the minimum number of fixed points of any map homotopic to h relative to the boundary of M^n? We will further restrict our focus upon area preserving, orientation preserving homeomorphisms, which, for example, arise in Hamiltonian dynamical systems, and isotopy classes of such maps. The first case that naturally arises is the two–disc, D. Alexander's Isotopy Lemma assures us that if $h: D \to D$ is a homeomorphism which is the identity on the boundary, ∂D, then h is isotopic to the identity map relative to ∂D. Thus if $g, h: D \to D$ are any two homeomorphisms such that $g|_{\partial D} = h|_{\partial D}$, then g is isotopic to h relative to ∂D. In this case the Nielsen problem is transformed into an extension problem. That is, given a homeomorphism of ∂D to itself, find the extension of h to D with the least number of fixed points. If h has a finite number of fixed points on ∂D of which n are attractors and, consequently, n are repellors then it is quite easy to give an estimate of $n+1$ as the lower bound for the number of fixed points. Figure 1 shows the procedure for constructing this extension.

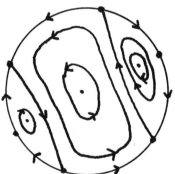

Figure 1

Our theorem proves that this estimate is indeed the minimum number of fixed points that is able to be achieved for area preserving, orientation preserving homeomorphisms of the two disc. This generalizes a theorem of

Montgomery [M] which states that every area preserving, orientation preserving homeomorphism of the open two disc must contain a fixed point.

The proof of our theorem relies upon the concept of free modification for homeomorphisms of two manifolds which was developed by Brown [Br1] and later extended to area preserving homeomorphism by Pelikan and Slaminka [PS]. In Section 2 we state the relevant definitions and theorems needed for the proof of the main result which is given in Section 3.

2   Preliminary Definitions and Facts.

We begin by defining attracting/repelling fixed points, free homeomorphisms, free modifications and stating the relevant theorems (though leaving the proofs to the references) which are needed for our main theorem.

Definition: Let $h: S^1 \to S^1$ be an orientation preserving homeomorphism of the one–sphere having an isolated fixed point p. Let I be an interval on S^1 with endpoints a and b such that $p \in I - \{a \cup b\}$ and such that $\mathrm{Fix}(h) \cap I = p$. We say that p is an *attractor* (resp. *repellor*) if $h(I) \subset I$ (resp. $h^{-1}(I) \subset I$). We also say that the interval I is an *attracting* (resp. *repelling*) *neighborhood* for the fixed point p. An isolated fixed point which is neither an attractor nor a repellor will be referred to as an *inessential fixed point*. An interval $I = I_a \cup I_b$ where I_a is the interval from a to p and I_b is the interval from b to p is called *an inessential neighborhood* if $h(I_a) \subset I_a$ and $h^{-1}(I_b) \subset I_b$ (or $h(I_b) \subset I_b$ and $h^{-1}(I_a) \subset I_a$.)

The technique of free modifications, which we have used to study the dynamics of homeomorphisms of surfaces, is based on the notion of free homeomorphism (cf. Brown [Br2]).

Definition: A homeomorphism $h: M^2 \to M^2$ of an orientable two manifold is said to be *free* if and only if given any disc $D \subset M^2$ such that $D \cap h(D) = \emptyset$, then $D \cap h^n(D) = \emptyset$ for all $n > 0$.

The Brouwer Translation Arc Lemma (cf. [B], [Br1] or [F] for a proof) states that if h is an orientation preserving homeomorphism of the plane which is fixed point free, then h is free. A simple application of the Brouwer Translation Arc Lemma proves Montgomery's Theorem (cf. Montgomery [M] or Andrea [A]).

Definition: A homeomorphism $g: M^2 \to M^2$ of an orientable two manifold is said to be a *free modification* of the homeomorphism $h: M^2 \to M^2$ if there is a homeomorphism $f: M^2 \to M^2$ such that

1. There exists a disc D such that $h(D) \cap D = \emptyset$;

2. The homeomorphism f is supported on D (i.e. $f(x) = x$ for all $x \notin D$); and,

3. $g = h \circ f$.

A finite number of free modifications will also be called a free modification.

Free modifications are useful in the study of the dynamics of homeomorphisms since they do not alter the fixed point set of the homeomorphism and they preserve the local fixed point index, which we now define.

Definition: If $h: M^2 \to M^2$ is a homeomorphism of an orientable two manifold and C is a simple closed curve in M^2 such that $\text{Fix}(h) \cap C = \emptyset$ then the *index of C with respect to h*, denoted by $\text{Ind}(h, C)$, is the degree of the map

$$g: C \to S^1 \quad \text{given by} \quad g(x) = \frac{h(x) - x}{\|h(x) - x\|}.$$

It can be shown (cf. Dold [D]) that if A is an annulus having as its boundary the simple closed curves C_1 and C_2 such that $\text{Fix}(h) \cap A = \emptyset$ then $\text{Ind}(h, C_1) = \text{Ind}(h, C_2)$. An isolated fixed point p for h then inherits an index in the following manner.

Definition: If $h: M^2 \to M^2$ is a homeomorphism of an orientable two manifold and p is an isolated fixed point for h and D is a disc in M^2 such that $p \in \text{int}(D)$ and $D \cap \text{Fix}(h) = p$ then $\text{Ind}(h, p)$ is defined to be the $\text{Ind}(h, C)$ where C is the boundary of D.

Free modifications are helpful in computing the local fixed point index. We modify the homeomorphism so that a simple closed curve meets its image in a certain canonical manner so that the index is easy to compute. The details of this construction is contained in Slaminka [Sl].

Our main theorem relies upon the following topological version of a result by Simon [Si] concerning the index of isolated fixed points for area preserving diffeomorphisms.

Theorem (Pelikan-Slaminka [PS]) *Let $h: M^2 \to M^2$ be an area preserving, orientation preserving homeomorphism of an orientable two manifold. If p is an isolated fixed point of h, then the $\text{Ind}(h, p) \leq +1$.*

The proof of this theorem amounts to showing that after employing suitable free modifications, a simple closed curve which surrounds an isolated fixed point can be represented in a canonical fashion as one which consists entirely of "hyperbolic arcs". We now define this concept for the special case required by our theorem.

Definition: Let H denote the upper half–plane including the x–axis (which will be denoted by R). Let $\alpha \subset H$ be an arc having endpoints a and b so that $R \cap \alpha = \{a \cup b\}$ and α and R bound a disc D. Give α the natural ordering so that $a < b$. Let $h: H \to H$ be an orientation preserving homeomorphism of the upper half plane. Assume that $h(\alpha) \cap \alpha$ is finite, $h(\alpha)$ intersects α transversely and that $h(D) \cap D$ is connected. Let $\{\beta_i\} \subset h(\alpha)$ be the finite set of arcs with endpoints $h(a_i), h(b_i)$ where $a_i < b_i$ such that $\beta_i \subset \overline{D^c}$. We say that α *consists entirely of hyperbolic arcs with respect to* h if whenever $a_i \neq a$ and $b_i \neq b$ then $a_i < h(a_i) < h(b_i) < b_i$ and either

1. $h(a)$ and $h(b) \subset D$ (See Figure 2); or,

2. $h(a)$ and $h(b) \subset D^c$ (here $a_1 = a$ and $b_n = b$) and $h(b_1) < b_1$, $a_n < h(a_n)$ (See Figure 3); or,

3. $h(a) \subset D$ and $h(b) \subset D^c$ and $a_n < h(a_n)$ (See Figure 4); or,

4. $h(a) \subset D^c$ and $h(b) \subset D$ and $h(b_1) < b_1$ (See Figure 5).

It is fairly routine to calculate the index of a curve consisting entirely of hyperbolic arcs. (See Brown-Neuman [BN] for a discussion of index along arcs.) If the curve has n hyperbolic arcs, $\{\beta_i\}$ where $a_i \neq a$ and $b_i \neq b$ then the index is: Case 1) $(n-1) + 1/2$, Case 2) $n + 1/2$, Case 3) n, and Case 4) n, where in Case 1) $n \geq 1$ and in Cases 2), 3) and 4) $n \geq 0$ due to area preservation. Thus in Figures 2 and 3 the index is $1\frac{1}{2}$ and in Figures 4 and 5 the index is 1.

3 Main Theorem.

We now state and prove the main theorem.

Theorem *Let $h: D \to D$ be an area preserving, orientation preserving, homeomorphism of the two disc to itself. Furthermore, suppose that the fixed point set of h restricted to the boundary of D consists of a finite number of fixed*

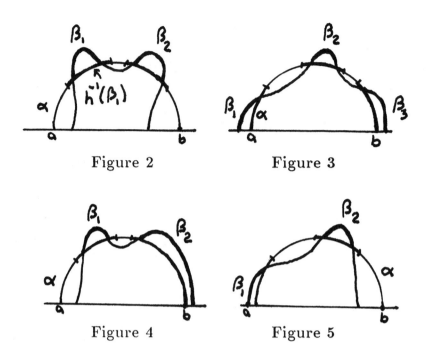

Figure 2   Figure 3

Figure 4   Figure 5

points, n of which are attracting (and consequently n are repelling). Then there exists at least $n+1$ fixed points in int(D).

Proof. The Brouwer Translation Arc Lemma implies, as noted above, that there exists at least one fixed point in int(D). We assume that there exist at most a finite number, N, of fixed points in int(D). Thus, the universal cover of $D - \text{Fix}(h|_{\text{int}(D)})$ is the upper half plane $H = \{(x,y) \mid y \geq 0\}$ where the boundary of D lifts to the x-axis. Let $\pi \colon H \to D$ be the projection and let $\hat{h} \colon H \to H$ be the covering homeomorphism which is fixed on $\pi^{-1}(\text{Fix}(h|_{\partial D}))$. Let the x-axis be denoted by R. Since $\hat{h}$ is fixed point free on $\{(x,y) \mid y > 0\}$, it follows easily that $\hat{h}$ is free on H. Let $N \subset H$ be a neighborhood of a fixed point $p \in R$ for $\hat{h}$ such that π restricted to N is a homeomorphism and such that $N \cap R$ is an attracting, repelling or inessential neighborhood for p on R. Let C be a simple arc in N with endpoints a and $b \in R$ such that $C \cap R = \{a, b\}$ and such that both $h(C)$ and $h^{-1}(C)$ are contained in N.

We modify $\hat{h}$ and C as in [Sl] to obtain a homeomorphism $\tilde{h}$ and an arc C' such that:

1. $C' \cap R = \{a, b\}$;

2. $\tilde{h}$ is a homeomorphism which preserves a quasi–Lebesgue measure (in the terminology of [Sl] a measure μ on a 2-manifold is *quasi-Lebesgue* if it is absolutely continuous with respect to Lebesgue measure, zero on points and there exist positive constants K_1, K_2 such that $K_1 m(A) \leq \mu(A) \leq K_2 m(A)$ where A is a Lebesgue measurable set contained in some compact set and m is Lebesgue measure);

3. $\mathrm{Fix}(\tilde{h}) = \mathrm{Fix}(\hat{h})$;

4. $\mathrm{Ind}(\hat{h}, C) = \mathrm{Ind}(\tilde{h}, C')$; and,

5. C' consists entirely of hyperbolic arcs.

Since C' consists entirely of hyperbolic arcs, we can compute the index using the formulas given in Section 2. Since $C \subset N$ and π is a local homeomorphism, we have thus computed the index of $\pi(C)$. Construct a simple closed curve $J \subset D$ which consists of the arcs $\pi(C)$ for each of the attracting/repelling/inessential fixed points plus the arcs on ∂D which connect the $\pi(C)$. Note that $J \cap \mathrm{Fix}(h) = \emptyset$. Computing the index of J we find that $\mathrm{ind}(h, J) = \sum((k_i - 1) + 1/2) + \sum(l_i + 1/2) + \sum m_i + 1 \geq n + 1$, where $k_i \geq 1$ is the number of hyperbolic arcs for Case 1), $l_i \geq 0$ is the number of hyperbolic arcs for Case 2) and $m_i \geq 0$ is the number of hyperbolic arcs for Cases 3) and 4) (excluding those for which the hyperbolic arcs β_i has either $a_i = a$ or $b_i = b$). The 1 is added for the full rotation about the simple closed curve C. The theorem of Pelikan and Slaminka implies that $\mathrm{ind}(h, J) = \sum_{i=1}^{N} \mathrm{ind}(h, x_i) \leq N$. Thus $n + 1 \leq \mathrm{ind}(h, J) \leq N$.

References

[A]   Andrea, S., *On homeomorphisms of the plane which have no fixed points*, Abh. Math. Sem. Univ. Hamburg **30** (1967), 61–74.

[B]   Brouwer, L. E. J., *Beweiss des ebenen Translationssatzes*, Math. Ann. **72** (1912), 37–54.

[Br1] Brown, M., *A new proof of Brouwer's lemma on translation arcs*, Houston J. Math. **10** (1984), 35–41.

[Br2] Brown, M., *Homeomorphisms of two-dimensional manifolds*, Houston J. of Math. **11** (1985), 455–469.

[BN] Brown, M. and Neumann, W., *Proof of the Poincaré–Birkhoff fixed point theorem*, Mich. Math. J. **24** (1977), 21–31.

[D] Dold, A., **Lectures on Algebraic Topology**, Springer–Verlag, Berlin, 1972.

[F] Fathi, A., *An orbit closing proof of Brouwer's lemma on translation arcs*, L'Ens. Math. **33** (1987), 315–322.

[J] Jiang, B., **Lectures on Nielsen Theory**, Cont. Math., Amer. Math. Soc. **14** (1983).

[K] Kiang, T., **The Theory of Fixed Point Classes**, Springer–Verlag, Berlin, 1989.

[M] Montgomery, D., *Measure preserving transformations at fixed points*, Bull. Amer. Math. Soc. **51** (1945), 949–953.

[N] Nielsen, J., *Untersuchungen zur Topologie der gescholossenen zweiseitigen Flächen,* I, II, III, Acta Math., **50** (1927), 189–358; **53** (1929), 1–76; **58** (1932), 87–167; IV, Det. Kgl. Danske Videnskabernes Salskab, Math. - fis. Medd. Band XXI/2, 1944.

[PS] Pelikan, P. and Slaminka, E. E., *A bound for the fixed point index of area preserving homeomorphisms of two manifolds*, Ergodic Th. and Dyn. Systems, 7, Part 3 (1987), 463–479.

[Si] Simon, C., *A bound for the fixed-point index of an area-preserving map with applications to mechanics*, Inventiones Math. **226** (1974), 187–200.

[Sl] Slaminka, E. E., *Removing index 0 fixed points for area preserving maps of two manifolds*, to appear in Trans. of the Amer. Math. Soc.

Irrational Rotations on Simply Connected Domains

BEVERLY L. BRECHNER   University of Florida, Gainesville, Florida

ABSTRACT. We construct uncountably many simply connected, planar, bounded domains, whose closures are topologically distinct, and such that each of these domains admits a homeomorphism on its closure which induces a Denjoy homeomorphism on the prime end disk. In particular, each of these homeomorphisms is topologically conjugate to the same irrational rotation on the interior of the unit disk, and is "Denjoy-like" on the boundary. The closed unit disk is one of the examples. By using uncountably many different irrational rotations, we see that the closed unit disk admits uncountably many inequivalent homeomorphisms which are Denjoy on the boundary, and are (distinct) irrational rotations on the interior.

1. INTRODUCTION

This work was motivated by the well known Siegel Disk Problem, which asks the following: Suppose f is a rational function of degree at least two in the complex plane, and M is a Siegel disk of this function; that is, $Int(M)$ is a bounded, simply connected domain on which f is conjugate to an irrational rotation, and further, $Int(M)$ is maximal with respect to this property. Must $Bd(M)$ be a simple closed curve?

Some partial results, after relaxing the hypotheses, have been obtained by Herman [He], by Moeckel [Moe], and by Pommerenke and Rodin [PR].

More recently, J.T. Rogers, Jr. in [R1,R2], and Mayer and Oversteegen in [MO], have obtained some very interesting related results. Rogers defines a "local Siegel disk" and proves the following: The boundary $Bd(G)$ of a bounded, irreducible, local Siegel disk satifies exactly one of the following two properties:

(1) If $\phi : Int(D) \longrightarrow G$ is the Riemann map from the interior of the unit disk to G, then $\phi^{-1} : G \longrightarrow Int(D)$ must extend continuously, taking $Bd(G)$ to $Bd(D) = S^1$, or
(2) $Bd(G)$ is an indecomposable continuum.

These results were presented at the American Mathematical Society's Special Session on Function Spaces, Muncie, Indiana, Oct. 1989 .

Mayer and Oversteegen construct an example of an indecomposable cofrontier invariant under a homeomorphism of the plane such that the prime end number from both sides is the same, but the induced homeomorphism on the prime end circle from one side is conjugate to a rigid rotation, while the induced homeomorphism on the other side is conjugate to a Denjoy homeomorphism. Both of these results are more topological in nature than the results mentioned in the previous paragraph.

In this paper, we modify the Siegel Disk Problem to a purely topological question, making the only requirement on f be that f is a homeomorphism. The following is proved:

For each irrational number β, there exist uncountably many pairs $\{(U_\alpha, h_{\beta_\alpha})\}_\alpha$, where U_α is a simply connected, bounded domain in the plane, and h_{β_α} is a homeomorphism of the plane onto itself keeping U_α invariant such that

(1) $h_{\beta_\alpha} \restriction U_\alpha$ is conjugate to the irrational, counter-clockwise rotation of angle β on the interior of the unit disk, D,

(2) For $\alpha \neq \beta, Bd(U_\alpha)$ is not homeomorphic to $Bd(U_\beta)$, and

(3) The induced prime end map on the boundary of the unit disk is conjugate to a Denjoy homeomorphism. (This shows that $Cl(U_\alpha)$ may be taken to be a disk, although we also construct a Denjoy homeomorphism on a disk precisely.)

See [B] to review the terminology of prime ends.

Note that the boundaries of our examples are decomposable, and thus, by the results of Rogers, none of these can be turned into an example of a Siegel disk with boundary which is not a simple closed curve.

The author is indebted to Timo Erkama for bringing the Siegel Disk Problem to her attention, and for pointing out the references [He], [Moe], and [PR]. She also wishes to express her gratitude to Jaime Zapata for the beautiful computer drawings which he prepared for this paper.

2. DENJOY HOMEOMORPHISMS OF THE CIRCLE

Let R_β denote the counter-clockwise rotation of the unit circle, S, by angle β. Let $Y = \bigcup\{y_n\}_{n \in Z}$, where Y is a full orbit of the point y_0 under R_β, and $y_n = R_\beta(y_0)$.

Let S' and S'' be two copies of S. On S', let $I = \bigcup\{I_n\}_{n \in Z}$ be the union of the closures of the collection of complementary intervals of some Cantor set.

Let $f : S' \longrightarrow S''$ be an onto map whose nondegenerate inverses are precisely the closed intervals $I_n \subset I$, and let $f(I) = X = \bigcup\{x_n\}_{n \in Z}$.

Let $g : S'' \longrightarrow S$ be a homeomorphism taking X onto Y. Re-label the points of X with the same subscripts as their images under g. Then re-label the intervals of I with the same subscripts as their inverses under f.

Let $H_\beta : S' \longrightarrow S'$ be the homeomorphism defined by $H_\beta = (gf)^{-1}R_\beta(gf)$ on $C(I)$, and extend by limits on the endpoints of I, and linearly on the interiors of the intervals of I. H_β is called a **Denjoy homeomorphism** of the circle.

Recall the following definition: Let $h : S^1 \longrightarrow S^1$ be an orientation preserving homeomorphism of the unit circle onto itself. Let $x \in S^1$, and let $H : R \longrightarrow R$ be a lift of h. The **rotation number** of h, denoted $\rho(h)$, is defined by

DENJOY HOMEOMORPHISM

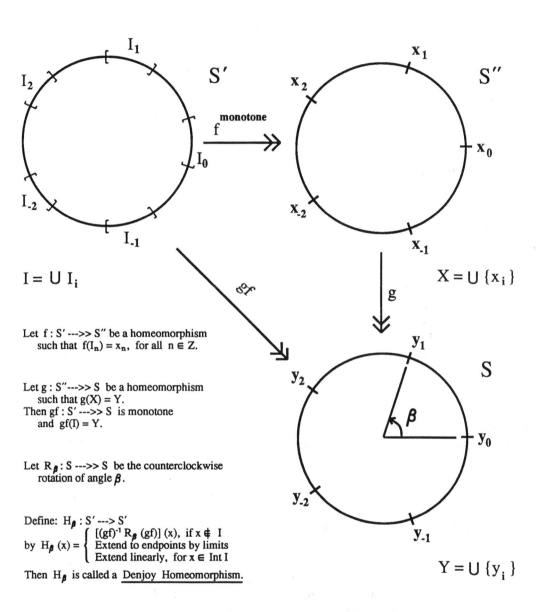

$I = \bigcup I_i$

Let $f : S' \longrightarrow\!\!\!\!> S''$ be a homeomorphism
such that $f(I_n) = x_n$, for all $n \in Z$.

Let $g : S'' \longrightarrow\!\!\!\!> S$ be a homeomorphism
such that $g(X) = Y$.
Then $gf : S' \longrightarrow\!\!\!\!> S$ is monotone
and $gf(I) = Y$.

Let $R_\beta : S \longrightarrow\!\!\!\!> S$ be the counterclockwise
rotation of angle β.

Define: $H_\beta : S' \longrightarrow S'$
by $H_\beta (x) = \begin{cases} [(gf)^{-1} R_\beta (gf)] (x), & \text{if } x \notin I \\ \text{Extend to endpoints by limits} \\ \text{Extend linearly, for } x \in \text{Int } I \end{cases}$

Then H_β is called a <u>Denjoy Homeomorphism.</u>

$X = \bigcup \{x_i\}$

$Y = \bigcup \{y_i\}$

Figure 1

$$\rho(H) = lim_{n \to \infty} \frac{|H^n(\tilde{x})|}{n}$$

where $\tilde{x}$ is any lift of x, and

$$\rho(h) = \rho(H)(mod\, 1)$$

It can be shown that $\rho(h)$ is independent of both x and of the particular lift of h that is chosen. (See [vK] for some properties of the rotation number.)

Then our homeomorphism H_β has the same rotation number as the rotation R_β. Figure 1 illustrates the process described above.

3. THEOREMS

3.1. Theorem. *For each irrational number β, there exists a homeomorphism h_β of the closed unit disk D onto itself such that*

(1) *$h_\beta \upharpoonright Int(D)$ is topologically conjugate to the counterclockwise irrational rotation of angle β on $Int(D)$, and*

(2) *$h_\beta \upharpoonright Bd(D)$ is a Denjoy homeomorphism.*

Proof. The proof becomes transparent, if one refers to Figure 2 while reading the following.

Notation.

Let D be the unit disk with boundary S.

Let R_β be the counterclockwise, irrational rotation of angle β on S, keeping the set Y invariant as in Figure 1.

Let $\hat{R}_\beta$ be the homeomorphism that cones R_β over the disk D, that is, the counterclockwise, irrational rotation of angle β on D.

Let D' be a copy of the unit disk with boundary S'.

Let H_β be a Denjoy homeomorphism on S' "rotating" the arcs $\{I_i\}_{i \in Z}$, as in Figure 1.

Let $\hat{H}_\beta$ be the homeomorphism that cones H_β over the disk D'.

Procedure.

(1) Construct an arbitrary sequence of disks, $\{B'_i\}_{i=1}^{\infty}$, in the "thumb disk" over I_0, as indicated in Figure 2. (We call this a *thumb disk* because it is shaped like the tip of one's thumb.) However, the largest disk of the sequence should be in the closed sector of the unit disk, D', determined by the radii from the origin to the endpoints of I_0.

(2) Copy $\{B'_i\}_{i=n}^{\infty}$ on thumb disks over I_n and I_{-n}, using the Denjoy disk homeomorphism $\hat{H}_\beta$. For example, the thumb disk over I_{-2} would consist of $\hat{H}_\beta^{-2}(B'_i)$ for each $i \geq 2$. Note that the thumb disks are small enough so that the thumb disks over I_i are disjoint from the thumb disks over I_j, for $i \neq j$.

(3) Construct an arbitrary "Hawaiian earring" sequence of disks at y_0; that is, construct a sequence of disks, $\{B_i\}_{i=1}^{\infty}$, as indicated in Figure 2. (We call B_n an *Hawaiian disk* because $\bigcup_{i=n}^{\infty} Bd(B_i)$ looks like the well-known Hawaiian earring example.) However, the disk B_n must be small enough so that the $2n+1$ disks of the collection $\{\hat{R}_\beta^i(B_n)\}_{i=-n}^{n}$ are disjoint.

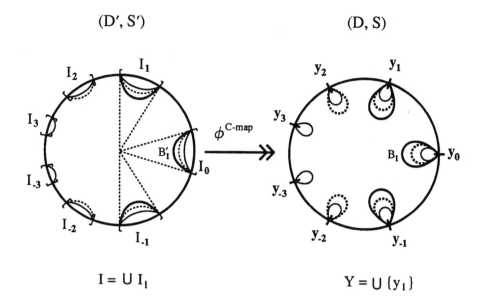

Figure 2

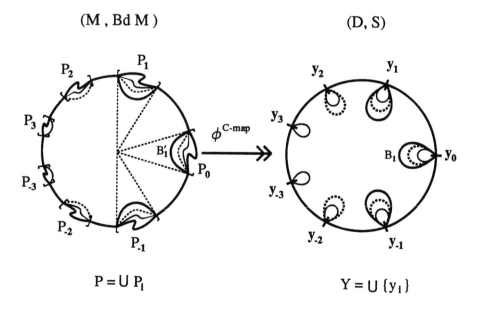

Figure 3

(4) Copy $\{B_i\}_{i=n}^{\infty}$ at the points y_n and y_{-n} using $\hat{R}_\beta$ in the same way that $\hat{H}_\beta$ is used in number (2) above.

(5) Map the thumb disk B'_1 homeomorphically onto the Hawaiian disk B_1, preserving the diagram; that is, the boundary of B'_i maps to the boundary of B_i, for all $i \geq 1$. Call this map ϕ_1.

(6) Copy ϕ_1 to get $\phi_2 : \bigcup_{i=-\infty}^{\infty} \{$thumb disks over $I_i\} \longrightarrow \bigcup_{i=-\infty}^{\infty} \{$Hawaiian disks over $I_i\}$. For example, on the Hawaiian disk over y_{-2}, one would put $\hat{H}_\beta^{-2}(B'_i)$ for each $i \geq 2$.

(7) Extend ϕ_2 to get $\phi^* : (S' \cup \{$thumb disks$\}) \longrightarrow (S \cup \{$Hawaiian disks$\})$.

(8) Observe that $D' - (S' \cup \{$thumb disks$\})$ is an open disk bounded by a simple closed curve, say C. Thus the map ϕ^* can be extended to all of D' by extending $\phi^* \restriction C$ to the disk which it bounds.

Call this map ϕ.

Result.

Then $\phi : D' \longrightarrow D$ is a map which is one-to-one on all of D', except for the intervals of I, which map to the points of Y. We define $h_\beta : D' \longrightarrow D'$ by

$$h_\beta(x) = \begin{cases} \phi^{-1}R_\beta\phi(x) & \text{for} \quad x \in C(I) \\ H_\beta(x) & \text{for} \quad x \in I. \end{cases}$$

Then $h_\beta : D' \longrightarrow D'$ is the desired (onto) homeomorphism. $\square$

3.2 Theorem. *For each β, there exist uncountably many pairs $\{(U_\alpha, h_{\beta_\alpha})\}_\alpha$ such that*

(1) *U_α is a bounded, simply connected domain, with $Cl(U_\alpha) = M_\alpha$*

(2) *$Bd(U_\alpha)$ is not a simple closed curve,*

(3) *$h_{\beta_\alpha} \restriction U_\alpha$ is topologically conjugate to the counterclockwise rotation of angle β on $Int(D)$.*

In fact, the elements of the collection $\{M\alpha\}$ may be taken to be topologically distinct.

Proof. In D' of Figure 2, replace I_0 by a chainable continuum different from an arc, say P_0, and having two endpoints, and "copy" P_0, using the Denjoy homeomorphism $\hat{H}_\beta$ extended to the entire plane, to replace I_i by P_i. This produces a continuum, M, different from the disk, since its boundary is not a simple closed curve.

In order to get uncountably many examples, we may proceed in either of two ways:

(1) In [L], Wayne Lewis describes uncountably many inequivalent embeddings $\{P_\alpha\}$ of the pseudo arc in the plane, each embedding with two endpoints. For each α, we get an example by replacing I_0 with P_α, and then copying the continuum in the other I_i's, using the Denjoy homeomorphism $\hat{H}_\beta$ extended to the entire plane. This produces uncountably many examples, M_α.

(2) In [A], J.J. Andrews describes uncountably many inequivalent chainable continua, each with two endpoints. One could use these continua, as in (1), to obtain uncountably many inequivalent examples.

See Figure 3 for a graphical representation of the result. $\square$

3.3 Theorem. *In each of the examples of Theorem 3.2, the homeomorphism induced by h_{β_α} on the prime end circle associated with $Bd(U_\alpha)$ $(= Bd(M_\alpha))$ is a Denjoy homeomorphism conjugate to H_β.*

Proof. In Theorem 3.1, a C-map (or conformal map) from D' to the unit disk D is approximated by a homeomorphism from the subdisk of D' bounded by C. Closer approximations are obtained by enlarging this subdisk by sequentially moving to the next thumb disk over each I_i at each stage. In the case of Theorem 3.1, the identity homeomorphism is a C-map, so the Denjoy homeomorphism is induced.

In Theorem 3.2, there is a subdisk of M_α corresponding to the subdisk of D' bounded by C. Since M_α is not a disk, we can no longer use the identity homeomorphism into the unit disk. But, clearly, there is a homeomorphism, ψ'' into the unit disk which approximates the diagram. We then extend this homeomorphism to $Int(M)$, by performing a C-map extension to the interior of B_1', and copying the resulting map, say ψ', using $\hat{R}_\beta^n \psi' \hat{H}_3^{-n}$, to the interiors of the copies of B_n' over P_n and P_{-n}. From this, it is clear that the induced homemorphism, ψ, on $Bd(D)$ is a Denjoy homeomorphism. $\square$

REFERENCES

[A] James J. Andrews, *A chainable continuum no two of whose nondegenerate subcontinua are homeomorphic*, Bull. Amer. Math. Soc. **12** (1961), 333-334.

[BG] Marcy Barge and Richard M. Gillette, *Rotation and periodicity in plane separating continua*, Ergodic Theory and Dynamical Systems **11** (1991), 619-631.

[B] Beverly Brechner, *On stable homeomorphisms and imbeddings of the pseudo arc*, Illinois J. Math. **22** (1978), 630-661.

[BGM] B. L. Brechner, M. D. Guay, and J. C. Mayer, *The rotational dynamics of cofrontiers*, These proceedings.

[C] M. Charpentier, *Sur quelques propriétés des courbes de M. Birkhoff*, Bull. Soc. Math. France **62** (1934), 193-224.

[Ha] Michael Handel, *A pathological area preserving C^∞ diffeomorphism of the plane*, Proc. Amer. Soc. **86** (1982), 163-168.

[He] Michael Herman, *Construction of some curious diffeomorphisms of the Riemann sphere*, J. London Math. Soc. **34** (1986), 375-384.

[L] Wayne Lewis, *Embeddings of the pseudo arc in E^2*, Pac. J. Math **93** (1981), 115-120.

[MO] John C. Mayer and Lex G. Oversteegen, *Denjoy meets rotation on an indecomposable cofrontier*, These proceedings (1992).

[Moe] Richard Moeckel, *Rotations of the closures of some simply connected domains*, Complex Variables **4** (1985), 285-294.

[PR] Ch. Pommerenke and B. Rodin, *Intrinsic rotations of simply connected regions, II*, Complex Variables **4** (1985), 223-232.

[R1] James T. Rogers, Jr., *Is the boundary of a Siegel disk a Jordan curve?*, Bull. Amer. Math. Soc. **27** (1992), 284-287.

[R2] James T. Rogers, Jr., *Singularities in the boundaries of local Siegel disks*, preprint (1992).

[vK] E.R. van Kampen, *The topological transformations of a simple closed curve into itself*, Amer. J. Math. **57** (1935), 142-152.

The Rotational Dynamics of Cofrontiers

BEVERLY L. BRECHNER   University of Florida, Gainesville, Florida

MERLE D. GUAY   University of Southern Maine, Portland, Maine

JOHN C. MAYER   University of Alabama at Birmingham, Birmingham, Alabama

ABSTRACT. A *pseudorotation* is defined to be an extendable homeomorphism of a cofrontier $\Lambda \subset \mathbf{R}^2$ whose induced prime end homeomorphisms from both sides are conjugate to a single rotation R_α on S^1. A class $\mathcal{H}$ of cofrontiers in the plane is constructed following the recipe suggested by M. Handel [Proc. AMS 86(1982), 163-168]. The authors study periodicity, almost periodicity, and recurrence properties of pseudorotations of cofrontiers $\Lambda \in \mathcal{H}$, as well as cofrontiers in general. They also study the relationship between each of these properties on Λ and the corresponding property on the induced prime end circles and disks.

1. INTRODUCTION

We construct a class $\mathcal{H}$ of cofrontiers in the plane, together with a class of automorphisms which we call pseudorotations. These examples are modeled on Handel's construction in [Ha]. We are interested in studying both the topological and the embedding-dependent topological properties of these cofrontiers as described by their prime end structure, as well as the dynamical properties of the pseudorotations.

Initially, some (corollaries of) theorems of Hemmingsen [He] and Cartwright [Ca] are observed, characterizing (1) invariant cofrontiers under irrational rotations of the plane and (2) a minimal cofrontier under an almost periodic homeomorphism of the cofrontier, respectively (Theorems 2.7 and 2.9), and relating almost periodicity on an arbitrary cofrontier-plus-interior with almost periodicity on its prime end circle (Theorem 2.8).

Theorem 2.11, and Theorem 2.12 with Corollary 2.13, are prime end theorems, about periodic pseudorotations, and about the prime end structure of indecomposable cofrontiers admitting irrational pseudorotations, respectively. Part of our motivation in studying pseudorotations comes from the well-known Siegel disk problem in complex analytic dynamics. For a discussion of this classical problem see Rogers' paper [Ro2] (these Proceedings). In this independent, and earlier, work Rogers proves theorems (originally announced in [Ro1]) about "extendable intrinsic rotations" of a bounded domain similar to our Theorems 2.12 and 2.13.

Supported in part by NSF-Alabama EPSCoR grant number RII-8610669. A summary version of this paper without proofs appeared in *Contemporary Mathematics* **117** (1991).

Next, we construct the family $\mathcal{H}$ carefully, and obtain some topological, dynamical, and number-theoretic results about cofrontiers $\Lambda \in \mathcal{H}$ and pseudorotations f_α of Λ. (f_α is a pseudorotation on Λ if and only if the induced map on the circle of prime ends on each side of Λ is conjugate to the rotation of angle $2\pi\alpha$ for some $\alpha \in [0, 1)$.) In Theorem 5.1, we prove that *every* $\Lambda \in \mathcal{H}$ admits an irrational pseudorotation f_α, for some α. The main theorems of this paper are Theorems 2.12 and 5.8. Theorem 2.12 describes the prime end structure of indecomposable cofrontiers admitting irrational pseudorotations, including members of our family $\mathcal{H}$, while Theorem 5.8 describes the dynamical properties and a number-theoretic property of certain indecomposable members of $\mathcal{H}$. In Theorem 5.8, we put restrictions on several of the technical conditions in the construction of $\Lambda \in \mathcal{H}$ to show that such restricted Λ are indecomposable, and if f_α is a "standard" pseudorotation on such a restricted $\Lambda \in \mathcal{H}$, then α is irrational and not of constant type, and f_α is recurrent, minimal, and not almost periodic on Λ.

In [BG], Barge and Gillette show that if Λ is a cofrontier left invariant by a homeomorphism f of the plane onto itself, and if the set of rotation numbers of f associated with Λ is nondegenerate, the Λ is indecomposable. This improves a result of Charpentier [C], which says that if the interior and exterior rotation numbers are distinct, then Λ is indecomposable.

2. COFRONTIERS, PRIME ENDS, AND PSEUDOROTATIONS

2.1. Cofrontier. A *continuum* is a compact, connected metric space. A continuum is *indecomposable* iff it cannot be written as the union of two of its proper subcontinua. A *cofrontier* Λ is a continuum in the plane $\mathbf{R}^2$ which irreducibly separates $\mathbf{R}^2$ into exactly two components. That Λ separates $\mathbf{R}^2$ into exactly two components is a topological property, as is the property that a cofrontier is the common boundary of its complementary domains. Two cofrontiers Λ_1 and Λ_2 are said to be *equivalent* iff there is a homeomorphism $h : (\mathbf{R}^2, \Lambda_1) \to (\mathbf{R}^2, \Lambda_2)$. A cofrontier may well have inequivalent embeddings in $\mathbf{R}^2$ (though S^1 does not, of course). We usually denote a cofrontier together with its embedding into $\mathbf{R}^2$, by Λ. We denote the unbounded component of $\mathbf{R}^2 - \Lambda$ by $\text{Ext}(\Lambda)$ and the bounded component by $\text{Int}(\Lambda)$.

2.2. Prime end compactification and prime end rotation number. Suppose U is a bounded, simply connected domain (connected planar open set) with nondegenerate boundary and $f : \overline{U} \to \overline{U}$ is a homeomorphism. The *prime end compactification* $\phi : U \to \mathbf{D}$ (where $\mathbf{D}$ is the open unit disk), and the homeomorphism $f|U$ induce a homeomorphism $F : \overline{\mathbf{D}} \to \overline{\mathbf{D}}$ as the extension of $\phi f \phi^{-1}$. Note that ϕ may be taken to be a conformal homeomorphism, and exists by the Riemann Mapping Theorem. We call $F|\partial\mathbf{D}$ the *induced homeomorphism* on the circle $\partial\mathbf{D}$ of prime ends. (For more on prime ends, see Section 2.10.) Always, $F|\partial\mathbf{D}$ has a well-defined *rotation number* given by

$$\rho(F) = \lim_{n \to \infty} \left(\frac{\widetilde{F}^n(\widetilde{x})}{n} \right) (\text{mod } 1)$$

where $\widetilde{F}$ is any lift of $F|\partial\mathbf{D}$ to the universal cover $\mathbf{R}$, and $\widetilde{x}$ is any point in $\mathbf{R}$. We call this number the *prime end rotation number* of f on U.

Now suppose Λ is a cofrontier and $f : (\mathbf{R}^2, \Lambda) \to (\mathbf{R}^2, \Lambda)$ is a homeomorphism. The *prime end compactification* $\phi_i : \text{Int}(\Lambda) \to \mathbf{D}$, and the homeomorphism $f|\text{Int}(\Lambda)$ induce a homeomorphism $F_i : \overline{\mathbf{D}} \to \overline{\mathbf{D}}$ as the extension of $\phi_i f \phi_i^{-1}$. We call $F_i|\partial\mathbf{D}$ the *induced homeomorphism* on the circle $\partial\mathbf{D}$ of prime ends. See Figure 1. Similarly, the prime end compactification $\phi_e : \text{Ext}(\Lambda) \cup \{\infty\} \to \mathbf{D}$, and the homeomorphism $f|\text{Ext}(\Lambda)$ induce a homeomorphism $F_e : \overline{\mathbf{D}} \to \overline{\mathbf{D}}$ as the extension of $\phi_e f \phi_e^{-1}$. In general, neither $F_i|\partial\mathbf{D}$ nor $F_e|\partial\mathbf{D}$ need be conjugate to any rotation of $\partial\mathbf{D} = S^1$, let alone the same rotation.

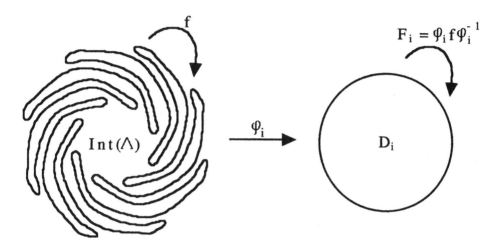

FIGURE 1. PRIME END COMPACTIFICATION OF THE DOMAIN $\text{Int}(\Lambda)$.

2.3. Pseudorotation. A *pseudorotation* of a cofrontier Λ is a homeomorphism $f : \Lambda \to \Lambda$, extendable to a homeomorphism of $\mathbf{R}^2$, such that each of the induced homeomorphisms on the circles of prime ends, corresponding respectively to $\text{Int}(\Lambda)$ and $\text{Ext}(\Lambda)$, is conjugate to the same rotation R_α of S^1 through angle $2\pi\alpha$ for some $\alpha \in [0, 1)$. Since this α is uniquely determined, we write f_α when f is a pseudorotation and R_α is the induced rotation on the circle of prime ends. Since f_α is extendable, we often assume a particular extension $f_\alpha : (\mathbf{R}^2, \Lambda) \to (\mathbf{R}^2, \Lambda)$. A given pseudorotation will have many extensions to homeomorphisms.

2.4. Generalized rotation number. Though a pseudorotation is associated with a single prime end rotation number α via the prime end compactification, it is not necessarily the case that every point of Λ has α as its rotation number. We define the *generalized rotation number* for a cofrontier Λ under a homeomorphism f, when it exists, as follows:

Suppose $0 \in \text{Int}(\Lambda)$ and $f : (\mathbf{R}^2, \Lambda, 0) \to (\mathbf{R}^2, \Lambda, 0)$ is a homeomorphism. Let $\gamma : \mathbf{R}^2 - \{0\} \to S^1$ be radial projection, and let $\widetilde{\gamma} : \mathbf{R}^1 \times (0, 1) \to \mathbf{R}^1$ be a lift of γ to the universal cover $\mathbf{R} \times (0, \infty)$ (of $\mathbf{R}^2 - \{0\} \cong S^1 \times (0, \infty)$) with covering projection $\pi \times \text{id}$. Let $\widetilde{f} : (\mathbf{R}^1 \times (0, \infty), \widetilde{\Lambda}) \to (\mathbf{R}^1 \times (0, \infty), \widetilde{\Lambda})$ be a lift of f. Let $x \in \Lambda$ and let $\widetilde{x} \in \widetilde{\Lambda}$ be such that $(\pi \times \mathrm{id})(\widetilde{x}) = x$. Define

$$\rho(x) := \left\{ \text{accumulation points of} \left\{ \frac{\widetilde{\gamma}\widetilde{f}^k(\widetilde{x})}{k} \right\}_{k=1}^{\infty} \right\} \pmod 1$$

Equivalently,

$$\rho(x) = \left\{ \text{accumulation points of} \left\{ \frac{\widetilde{\gamma}\widetilde{f}^k(\widetilde{x}) - \widetilde{\gamma}(\widetilde{x})}{k} \right\}_{k=1}^{\infty} \right\} \pmod 1$$

In general, though $\rho(x)$ is independent of the choice of lifts $\widetilde{f}$ of f and $\widetilde{\gamma}$ of γ, and of the point $\widetilde{x} \in \pi^{-1}(x)$, it may not be a single real number, but rather a set, and $\rho(x)$ may vary with $x \in \Lambda$. We define the *rotation set* of Λ by $\rho(\Lambda) := \bigcup_{x \in \Lambda} \rho(x)$. We say the *generalized rotation number* of $x \in \Lambda$ is α iff there is a real number α such that $\rho(x) = \alpha$. Roughly, α is the average angular displacement of x under successive iterates of f. We say the *generalized rotation number* of Λ is α iff for all $x \in \Lambda$, $\rho(x) = \alpha$. We write $\rho(\Lambda) = \alpha$. In case Λ is homeomorphic to S^1 (by a homeomorphism $h : (\mathbf{R}^2, \Lambda) \to (\mathbf{R}^2, S^1)$), the generalized rotation number of f on Λ is just the rotation number of the conjugate hfh^{-1}.

One does have the following well-known relationship between the prime end rotation number and the generalized rotation number:

2.5. Theorem. *Suppose $f : (\mathbf{R}^2, \Lambda) \to (\mathbf{R}^2, \Lambda)$ is a homeomorphism of a cofrontier Λ. Then the prime end rotation number of f on $\mathrm{Int}(\Lambda)$ (respectively, $\mathrm{Ext}(\Lambda)$) is the same as the generalized rotation number of any point of Λ accessible from $\mathrm{Int}(\Lambda)$ (respectively, $\mathrm{Ext}(\Lambda)$).*

2.6. Recurrent, almost periodic, and minimal homeomorphisms.

A homeomorphism $h : X \to X$ is said to be *recurrent* iff for every $\epsilon > 0$, there is an $n \in \mathbf{Z}$ such that $\mathrm{d}(h^n, \mathrm{id}_X) < \epsilon$. Equivalently, h is recurrent iff $\{h^n | n \in \mathbf{Z}\}$ has a subsequence converging uniformly to the identity.

A homeomorphism $h : X \to X$ is said to be *almost periodic* iff for every $\epsilon > 0$, there is a subsequence $\{h^{n_i} | i \in \mathbf{Z}\}$ of $\{h^n | n \in \mathbf{Z}\}$ with bounded gaps such that for all i, $\mathrm{d}(h^{n_i}, \mathrm{id}_X) < \epsilon$. On a compact metric space, h is almost periodic iff $\{h^n | n \in \mathbf{Z}\}$ is an equicontinuous family of iterates. Clearly, h periodic implies h almost periodic implies h recurrent.

Recall that the recurrent homeomorphisms of S^1 are almost periodic, and the almost periodic homeomorphisms of S^1 are conjugate to rotations, with the rational rotations being periodic and the irrational rotations being nonperiodic, almost periodic. In particular, recurrent is equivalent to almost periodic on S^1.

A homeomorphism $h : X \to X$ is *minimal* iff X is the only nonempty closed invariant subset of X under h. It follows that $h : X \to X$ is minimal iff every orbit is dense. Recall that an irrational rotation of S^1 is minimal.

In [He], Hemmingsen shows that topological disks and annuli are the only plane continua with interior points which (continua) admit nonperiodic, almost periodic homeomorphisms (i.e. homeomorphisms whose families of iterates are equicontinuous), and that any such homeomorphism must be conjugate to an irrational rotation on the disk or annulus. The first two theorems below may be seen as corollaries of Hemmingsen's theorem, although the first one is very easy to prove directly. The third is an observation of Cartwright [Ca] based on Hemmingsen's results.

2.7. Theorem. *Let* $h : \mathbf{R}^2 \to \mathbf{R}^2$ *be conjugate to an irrational rotation, and let* Λ *be a cofrontier. If* $h(\Lambda) = \Lambda$, *then* Λ *is homeomorphic to* S^1.

2.8. Theorem. *Let* h *be an almost periodic, nonperiodic homeomorphism of* $\Lambda \cup \mathrm{Int}(\Lambda)$ *onto itself, where* Λ *is a cofrontier in* $\mathbf{R}^2$. *Then* Λ *is a simple closed curve and* $h|\Lambda$ *is conjugate to an irrational rotation. Thus, the induced homeomorphism on the prime end disk will also be conjugate to an irrational rotation, and therefore almost periodic on the circle of prime ends.*

2.9. Theorem. *Let* $h : (\mathbf{R}^2, \Lambda) \to (\mathbf{R}^2, \Lambda)$ *be a homeomorphism, where* Λ *is a nondegenerate continuum. If* h *is both almost periodic and minimal on* Λ, *then* Λ *is a simple closed curve and* $h|\Lambda$ *is conjugate to an irrational rotation.*

2.10. Prime ends. Basic definitions and theory of prime ends, and further references, may be found in [Br2,P]. Suppose U is a bounded domain with nondegenerate boundary. For convenience, we identify $\mathbf{R}/\mathbf{Z} = \partial \mathbf{D}$, and by η we denote both the point $\eta \in \partial \mathbf{D}$ and the prime end η of U corresponding to that point. Suppose $\{Q_i\}_{i=1}^\infty$ is a chain of crosscuts of U defining η, and for each i, U_i is the complementary subdomain of U cut off by Q_i that contains Q_{i+1}. We denote the *impression* of a prime end η by $I(\eta)$ and the *principal continuum* of η by $P(\eta)$. By definition

$$P(\eta) = \{w \in \mathbf{R}^2 \mid \exists \text{ chain } \{Q_i\}_{i=1}^\infty \text{ of crosscuts of } U \text{ defining } \eta \text{ with } Q_i \underset{i}{\to} w\}$$

$$I(\eta) = \bigcap_{i=1}^\infty \overline{U_i} = \bigcap_{i=1}^\infty \partial U_i,$$

Note that $P(\eta) \subset I(\eta)$.

Consider the conditions

(1) $P(\eta)$ is degenerate.
(2) $I(\eta) = P(\eta)$.

A prime end η is of the *first, second, third, or fourth kind* accordingly, as η satisfies both (1) and (2), (1) only, (2) only, or neither (1) nor (2). A prime end of the third kind whose impression is all of Λ is said to be a *simple dense canal*.

A point $b \in \partial U$ is called *accessible with respect to the domain* U iff there is an endcut $[a, b)$ such that $[a, b) \subset U$ and $\overline{[a, b)} = [a, b]$. We call both the half-open arc $[a, b)$ and the arc $[a, b]$, an *arc of accessibility to* b. The point $b \in \partial U$ is said to be *accessible with respect to the prime end* η *of* U iff there exist both a chain of crosscuts $\{Q_i\}_{i=1}^\infty$ defining η and an arc of accessibility $[a, b]$ to b such that $Q_i \cap (a, b)$ is a (necessarily different) singleton for each i. Since endcuts go to endcuts under a C-map, it follows that the map ϕ may be extended (though not generally continuously) to include the accessible points of U. Thus, our definition of arc of accessibility is not ambiguous.

We note that if η is a prime end of the first or second kind, there is exactly one principal point. In each case, this point is both accessible from U and accessible with respect to η.

Simple examples show that a point p may be accessible from U, yet be the principal point of two (or more) distinct prime ends. For example, let U be the interior of the unit circle in the plane minus the arc $[0,1] \times \{0\}$. Then the point $(\frac{1}{2}, 0)$ is accessible from U, and is also accessible with respect to two distinct prime ends, but evidently not by the same endcut in U.

We obtain the following additional theorems about pseudorotations in general.

2.11. Theorem (Periodic pseudorotations). *Suppose Λ is a cofrontier, $h : (\mathbf{R}^2, \Lambda) \to (\mathbf{R}^2, \Lambda)$ is a homeomorphism, and $H : \overline{\mathbf{D}} \to \overline{\mathbf{D}}$ denotes the induced homeomorphism on either the interior or exterior prime end compactification. Then $h|\Lambda$ is periodic of period n iff $H|\partial \mathbf{D}$ is periodic of period n.*

Proof. Fix U, one of the complementary domains of Λ. Assume $h|\Lambda$ is periodic of period $n > 1$. Let b be a point of Λ such that b is accessible from U and is of period $n > 1$. Let $K = [a, b)$ be an endcut to b in U. Then $h^n(K)$ is also an endcut to b, since $h^n(b) = b$. Thus, K and $h^n(K)$ are endcuts to the same acessible point b in ∂U. By the proof of Lemma 3.11 of [Br2], since Λ is an irreducible separator of $\mathbf{R}^2$, each accessible point of ∂U is associated with exactly one prime end of U, so that K and $h^n(K)$ define the same prime end η of U. It follows that $\phi(K)$ and $\phi(h^n(K))$ are endcuts to the same point $b' \in \partial \mathbf{D}$. But $\phi(h^n(K)) = \phi(h^n(\phi^{-1}(\phi(K)))) = H^n(\phi(K))$. Therefore, both $\phi(K)$ and $H^n(\phi(K))$ are endcuts to the same point b' in $\partial \mathbf{D}$. Thus, $H^n(b') = b'$. Since distinct accessible points of ∂U map to distinct points of $\partial \mathbf{D}$, b' is a periodic point of period n of $\partial \mathbf{D}$. By Theorem 3.7 and Corollary 3.8 of [Br2], the set A of points of Λ accessible from U is dense in Λ, while the set A' of points of $\partial \mathbf{D}$ corresponding to the points of A is dense in $\partial \mathbf{D}$. Thus, $H|\partial \mathbf{D}$ is periodic of period n.

Conversely, assume that $H|\partial \mathbf{D}$ is of period n, and let $b' \in \partial \mathbf{D}$ correspond to an accessible (from U) point $b \in \partial U$. Let K be an endcut to b. Then $\phi(K)$ is an endcut to $b' \in \partial \mathbf{D}$, and therefore, $H^n(\phi(K))$ is an endcut to $H^n(b') \in \partial \mathbf{D}$. But $H^n(b') = b'$ by hypothesis. So $H^n(\phi(K))$ is also an endcut to b'. Now $h^n(K) \subset \overline{U}$ is an endcut to $h^n(b)$ in ∂U. But $\phi(h^n(K)) = \phi h^n \phi^{-1}(\phi(K)) = H^n(\phi(K))$ is an endcut to b'. So $h^n(K)$ is an endcut to b. That is, $h^n(b) = b$. Since the accessible points are dense in $\Lambda = \partial U$, $h|\Lambda$ is periodic of period n. ∎

The following theorem and corollary, which follow from theorems of Rutt [Ru] on prime ends, show that the fine structure of an indecomposable cofrontier admitting an irrational pseudorotation must be quite complicated. Rogers [Ro1-2] has obtained theorems similar to Theorem 2.12 and Corollary 2.13 for extendable intrinsic rotations of a bounded domain. Related theorems are proved for quadratic Julia sets in [MR].

2.12. Theorem (Prime end structure). *Suppose Λ is a cofrontier which admits an irrational pseudorotation. Then the following are equivalent:*

(1) *Λ is indecomposable.*
(2) *Some prime end of $\mathrm{Int}(\Lambda)$ (respectively, $\mathrm{Ext}(\Lambda)$) has as its impression all of Λ.*
(3) *There exists a dense subset $\mathcal{D}$ of the circle of prime ends of $\mathrm{Int}(\Lambda)$ (respectively, $\mathrm{Ext}(\Lambda)$) such that for every $\eta \in \mathcal{D}$ the impression of η is all of Λ.*
(4) *Every prime end of $\mathrm{Int}(\Lambda)$ (respectively, $\mathrm{Ext}(\Lambda)$) has as its impression all of Λ.*

(5) *The set of prime ends of* $\text{Int}(\Lambda)$ *(respectively,* $\text{Ext}(\Lambda)$*) corresponding to simple dense canals is second category in the circle of prime ends.*

(6) *Some prime end of* $\text{Int}(\Lambda)$ *(respectively,* $\text{Ext}(\Lambda)$*) is a simple dense canal.*

Proof. We refer to $\text{Int}(\Lambda)$ in the proofs below, but the statements apply to $\text{Ext}(\Lambda)$ as well. Throughout we assume $\phi : \text{Int}(\Lambda) \to \mathbf{D}$ is the prime end compactification of $\text{Int}(\Lambda)$, $f_\alpha : (\mathbf{R}^2, \Lambda) \to (\mathbf{R}^2, \Lambda)$ is the irrational pseudorotation, and F the induced homeomorphism on the circle S^1 of prime ends which is conjugate to R_α.

(1) $\Longrightarrow$ (2). Rutt's proofs, and hence theorems [Ru], hold for cofrontiers, as well as for the nonseparating plane continua for which he originally stated the theorems. By Rutt's Theorem 1, if Λ is indecomposable, then some prime end of $\text{Int}(\Lambda)$ has as its impression all of Λ.

(2) $\Longrightarrow$ (3). Suppose prime end η has $I(\eta) = \Lambda$. For all $n \in \mathbf{Z}$, $I(F^n(\eta)) = \Lambda$, since f^n is a homeomorphism of $\mathbf{R}^2$, and therefore of $\overline{\text{Int}(\Lambda)}$. Let $\mathcal{D} = \{F^n(\eta) \mid n \in \mathbf{Z}\}$. Since F is conjugate to an irrational rotation, $\mathcal{D}$ is dense in S^1.

(3) $\Longrightarrow$ (4). Suppose $\mathcal{D}$ is a set of prime ends dense in S^1 such that $I(\eta) = \Lambda$ for all $\eta \in \mathcal{D}$. Let μ be a prime end of $\text{Int}(\Lambda)$ and $\{Q_i\}_{i=1}^\infty$ a chain of crosscuts of $\text{Int}(\Lambda)$ defining μ. Fix i, and consider the crosscut $\phi(Q_i)$ of $\mathbf{D}$. Since $\mathcal{D}$ is dense in $\partial \mathbf{D} = S^1$, there is an element $\eta_i \in \mathcal{D}$ such that η_i is in the subdomain $\phi(U_i)$ of $\mathbf{D}$ cut off by $\phi(Q_i)$ that contains $\phi(Q_{i+1})$. Consequently, the corresponding subdomain U_i of $\text{Int}(\Lambda)$ cut off by Q_i has $I(\eta_i) = \Lambda$ in its boundary. Since this holds for all i, $I(\mu) = \Lambda$.

(4) $\Longrightarrow$ (5). Suppose every prime end of $\text{Int}(\Lambda)$ has Λ as its impression. Then $\text{Int}(\Lambda)$ contains no prime ends of the first kind. By a theorem of Cartwright [P], the prime ends of the first and third kinds together form a set of second category in S^1. Consequently, the set of prime ends corresponding to simple dense canals is second category in S^1.

(5) $\Longrightarrow$ (6). Trivial.

(6) $\Longrightarrow$ (1). By [Ru, Theorem 3], the existence of a simple dense canal implies Λ is indecomposable. ∎

2.13. Corollary. *Suppose* Λ *is an indecomposable cofrontier admitting an irrational pseudorotation. Then each composant of* Λ *contains the principal continuum of at most one prime end of* $\text{Int}(\Lambda)$ *(respectively,* $\text{Ext}(\Lambda)$*). Further, no composant can contain the principal continuum of both a prime end,* η*, of* $\text{Int}(\Lambda)$ *and another prime end,* γ*, of* $\text{Ext}(\Lambda)$*.*

Proof. Let U be one of the complementary domains of Λ, and suppose K is a composant of Λ such that $K \supset X \cup Y$, where X and Y are the principal continua of the distinct prime ends η and γ, respectively, of U. Let Z be the irreducible continuum containing $X \cup Y$; note that $Z \subset K$. Let $\pi : (\mathbf{R}^2, \Lambda, Z) \to (\mathbf{R}^2, \Lambda, p)$ be the quotient map of $\mathbf{R}^2$ taking Z to the point p.

We first note that by Theorem 2.12 above and by Iliadis' theorem (see [Br1]), no composant contains more than one accessible point; for otherwise, there would exist prime ends whose impresions were proper in Λ. Thus, it follows that any crosscut in a chain of crosscuts defining η (or γ) must have its endpoints outside K. It follows from this, that chains of crosscuts defining different prime ends must go to chains of crosscuts defining different prime ends under π.

Let $U = \text{Int}(\Lambda)$. Since π is monotone, $\pi(\Lambda)$ is an irreducible separator of $\mathbf{R}^2$, so $\pi(U) = \text{Int}(\Lambda/Z)$. Let A and B be rays defining the prime ends η and γ, respectively, such that A converges to X and B converges to Y. Then $\pi(A)$ and $\pi(B)$ are each endcuts to p, with $\pi(A) \cap \pi(B) = \emptyset$. Further, $\pi(A)$ and $\pi(B)$ represent different prime ends of $\pi(U)$, since A and B represent different prime ends of U. This follows from the remarks in the above paragraph. However, $\pi(A)$ and $\pi(B)$ are endcuts to the same point, p, of $\pi(\Lambda)$. As before, it follows from the proof of Lemma 3.11 of [Br2] that each accessible point of $\pi(\Lambda)$ corresponds to exactly one prime end of $\pi(U)$. This is a contradiction, and the first part of the theorem is proved.

For the second part, let U and V be the complementary domains of Λ, and suppose K is a composant of Λ such that $K \supset X \cup Y$, where X and Y are the principal continua of the distinct prime ends η of U and γ of V, respectively. Let Z be the irreducible continuum containing $X \cup Y$; note that $Z \subset K$. As before, let $\pi : (\mathbf{R}^2, \Lambda, Z) \to (\mathbf{R}^2, \Lambda, p)$ be the quotient map of $\mathbf{R}^2$ taking Z to the point p. Let A and B be rays defining the prime ends η and γ, respectively, such that A converges to X and B converges to Y. Then $\pi(A)$ and $\pi(B)$ are each endcuts to p from the disjoint domains $\pi(U)$ and $\pi(V)$ in $\mathbf{R}^2$. Thus, p is a local cut point of $\pi(\Lambda)$.

Now, Λ admits an irrational pseudorotation, h. Thus, under h, no prime end of either complementary domain is fixed. Therefore, $H(\eta)$ and $H(\gamma)$ are distinct from η and γ, respectively. It follows from the first part of the corollary that X and $h(X)$ (respectively, Y and $h(Y)$) lie in different composants of Λ. Thus, $h(Z)$ is a proper subcontinuum of Λ, lying in a different composant from the one containing Z.

We now let θ be a monotone map from $(\mathbf{R}^2, \Lambda, p, h(Z)) \to (\mathbf{R}^2, \Lambda, p, q)$ taking $h(Z)$ to the point q. Then $\theta \circ \pi$ is one-to-one everywhere, except that it takes Z to p and $h(Z)$ to q. Since p is a local cut point of $\pi(\Lambda)$, both p and q are local cut points of $W = \theta(\pi(\Lambda))$. But W must be an indecomposable continuum, since it is the monotone image of an indecomposable continuum. However, W contains two local cut points, each leading from $\theta(\pi(U))$ to $\theta(\pi(V))$. Thus, W is decomposable. This contradiction proves the second part of the corollary. ∎

3. Construction of the family $\mathcal{H}$ of cofrontiers

We are interested in looking at a class $\mathcal{H}$ of cofrontiers, including indecomposable cofrontiers, which are constructed in a regular fashion, and which are a potential source of examples and counterexamples.

3.1. Outline. A cofrontier $\Lambda \in \mathcal{H}$ is constructed as the intersection $\Lambda = \bigcap_{k=1}^{\infty} A_k$ of annuli A_k, where A_{k+1} is embedded essentially in A_k and is invariant under certain topological rotations of $A_1, \ldots, A_k$. Refer to Figure 2 for an example of the construction outlined below.

(1) A_k is divided into q_k *links* (closed topological disks) of diameter $\leq \epsilon_k$.

(2) The embedding of A_{k+1} in A_k is induced by lifting an embedding of an annulus $\bar{A}_{k+1}$ in a "single link" annulus A (where A is the image of A_k under a q_k–to–one covering map).

(3) Each segment of A_{k+1} that *spans* (that is, extends across) a link of A_k is divided into $r_{k+1} > 1$ links. This can be done via the lift. (In the example, $r_{k+1} = 2$.)

(4) A *basic repetitive unit* of A_{k+1} spans some number, say s_{k+1}, of links of A_k. (In the example, $s_{k+1} = 2$, and a basic repetitive unit is high-lighted.)

(5) The "fractional part" of A_k that a basic repetitive unit of A_{k+1} spans is $c_{k+1} = \frac{s_{k+1}}{q_k}$. (In the example, $c_{k+1} = \frac{2}{8} = \frac{1}{4}$.)

See the first paragraph of Section 4 for the significance of the s_k's and the c_k's.

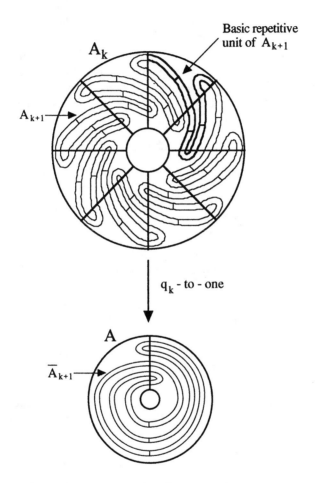

FIGURE 2. CONSTRUCTION OF $\Lambda \in \mathcal{H}$ AS AN INTERSECTION OF ANNULI A_k.

3.2. Details. We will want to draw some conclusions about how rapidly q_k grows relative to choices for r_j, s_j, and c_j for $j < k$. We also want to make sure that A_{k+1} is invariant under certain topological rotations of $A_1, \ldots, A_k$. To this end, we need to describe the above construction more precisely. We will impose certain restrictions on r_k, s_k, and c_k, for all k.

We suppose each annulus A_k is divided into a number of *links* so that it is a circular chain of closed topological disks, which meet adjacent links in a separating arc of their union (as in Figure 2, where $q_k = 8$). We use A_k somewhat ambiguously to denote both the annulus and the collection of links whose union is the annulus.

Let q_1 denote the number of links of A_1. We suppose that A_1 is $S^1 \times [0,1]$ embedded as the annulus between the circle of diameter $1/2$ and the circle of diameter 1 centered at the origin, and that $q_1 > 1$. Let g_1 be a periodic homeomorphism of A_1 of period q_1 that carries links to links, moving forward one link. (At this stage we may take g_1 to be a rigid rotation of A_1.) Let A denote the annulus $S^1 \times [0,1]$ with T the arc $\{0\} \times [0,1]$ in A. We suppose A_1 is mapped onto A by a q_1-to-one covering map h_1 that takes each component of the intrinsic boundary of each link of A_1 onto the arc T, and maps the outer boundary component of A_1 to the outer boundary component of A. We require that $h_1 = h_1 g_1$. (At this stage we may take h_1 to be $\exp(q_1 z) \times \mathrm{id}$.)

The embedding of A_2 in A_1 is induced by lifting an embedding of an annulus $\bar{A}_2$ in A whose algebraic intersection number with T is 1. Choose consistent orientations on $\bar{A}_2$ and A. For simplicity, we suppose $\bar{A}_2$ wraps forward some number of times around the hole in A spiraling in toward the inner boundary of A, and wraps backward one less time spiraling toward the outer boundary of A, with no extra bending. (In other constructions, we may want to allow more bending in order to permit, for example, the pseudocircle to be one of our cofrontiers.) Since the algebraic intersection number of $\bar{A}_2$ with T is 1, $\bar{A}_2$ is a retract of A by a map which is homotopic to the identity.

Let c_2 denote the ratio of the number s_2 of forward wraps of $\bar{A}_2$ in A to the number q_1 of links of A_1. If it were the case that $s_2 = c_2 q_1 = 1$, then $\bar{A}_2$ would go once around A with no bending. In our construction, we require that $s_2 > 1$ and that $c_2 \le 1$. We lift $\bar{A}_2$ via covering map h_1 to define A_2 contained in A_1. Thus, A_2 is symmetric with respect to the previously designated topological rotations of A_1 generated by g_1.

Observe that a "stretched out copy" of $\bar{A}_2$ will have been repeated q_1 times, each link of A_1 containing the first link of such a copy, which we call a *basic repetitive unit* of A_2, and each basic repetitive unit will itself *span* (that is, extend through) $s_2 = c_2 q_1$ links of A_1, turning around in the next link, and returning to the other boundary of the link of A_1 in which it started, turning around again to begin the next basic repetitive unit. A basic repetitive unit is outlined more heavily in Figure 2. Thus, c_2 is the *fractional part* of A_1 that a basic repetitive unit of A_2 *spans*.

We subdivide A_2 into links by first subdividing $\bar{A}_2$ and lifting this subdivision. Thus, the links of A_2 are invariant under g_1. Let r_2 denote the number of links into which each full wrap (forward or backward) of $\bar{A}_2$ is subdivided. We suppose the two bends where the forward and backward wraps of $\bar{A}_2$ join consist of just one link each. In our construction, we require that $r_2 > 1$. We show below how to compute q_2 from c_2, r_2, and q_1. For now, suppose $q_2 = n_2 q_1$.

The distinction that arises in embedding A_3 in A_2, and which we will have to carry forward inductively, is that we want A_3 to be invariant, not only under certain topological rotations of A_2 that carry links to links, but also under the topological rotations of A_1 generated by g_1. To this end, let g_2 be a periodic homeomorphism of A_2 carrying links to links, moving forward one link, and which generates a group (of cardinality q_2) of topological rotations of A_2 that commutes with g_1 on A_2. This can be done by the Lemma below. As a result, $g_1|A_2$ will be an element of the group of topological rotations of A_2, namely $g_2^{n_2}$. Choose a q_2-to-one covering map $h_2 : A_2 \to A$ so that it commutes with the previously designated topological rotation g_2 of A_2, and, therefore, with the topological rotation g_1 of A_1. Then $A_3 \subset A_2$ may be defined by lifting, via h_2, an embedding of an annulus $\bar{A}_3 \subset A$ whose algebraic intersection number with T is 1; simultaneously, s_3, c_3, and r_3 are defined in a similar manner to the above. In particular, A_3 will be invariant under the designated topological rotations of A_2 and A_1.

In general, given A_k invariant under the topological rotations $g_1, \ldots, g_{k-1}$, of $A_1, \ldots, A_{k-1}$, with $q_1, \ldots, q_k, s_2, \ldots, s_k, c_2, \ldots, c_k, r_2, \ldots, r_k$, and covering maps $h_1, \ldots, h_{k-1}$ defined as above, we choose a periodic generator g_k of a group (of cardinality q_k) of topological rotations of A_k carrying links to links, moving forward one link, so as to commute with the topological rotations $g_1, \ldots, g_{k-1}$ of $A_1, \ldots, A_{k-1}$, respectively. We can do this by the Lemma below. Then we choose a q_k-to-one covering map $h_k : A_k \to A$, so as to commute with the topological rotations $g_1, \ldots, g_k$ of $A_1, \ldots, A_{k-1}, A_k$. The embedding of A_{k+1} in A_k is induced by lifting, via h_k, an embedding of an annulus $\bar{A}_{k+1}$ in A whose algebraic intersection number with T is 1. Let c_{k+1} denote the ratio of the number s_{k+1} of forward wraps of $\bar{A}_{k+1}$ in A to the number q_k of links of A_k. We lift $\bar{A}_{k+1}$ via the covering map h_k to define A_{k+1} contained in A_k. Thus, A_{k+1} is symmetric with respect to the topological rotations $g_1, \ldots, g_k$ of $A_1, \ldots, A_k$, respectively. A basic repetitive unit of A_{k+1} will have been repeated q_k times, each link of A_k containing the first link of a basic repetitive unit which will span $s_{k+1} = c_{k+1}q_k$ links of A_k, turning around in the $s_{k+1} + 1$ link, and returning to the other boundary of the link of A_k in which it started, turning around again to begin the next basic repetitive unit. We subdivide A_{k+1} into links by first subdividing $\bar{A}_{k+1}$ and lifting this subdivision. Let r_{k+1} denote the number of links into which each full wrap (forward or backward) of $\bar{A}_{k+1}$ is subdivided. In our construction, we require that $r_{k+1} > 1$, $s_{k+1} > 1$, and $c_{k+1} \leq 1$, for all k.

Since $s_{k+1} = c_{k+1}q_k > 1$, and since there are a total of $2c_{k+1}q_k - 1$ full wraps in $\bar{A}_{k+1}$, each wrap having been subdivided into r_{k+1} links, plus two bend links, the number of links of A_{k+1} is given by

$$q_{k+1} = (2r_{k+1}c_{k+1}q_k - r_{k+1} + 2)q_k$$
$$= (r_{k+1} + 2)c_{k+1}q_k^2 + (r_{k+1} - 2)q_k(c_{k+1}q_k - 1)$$

Since $c_{k+1}q_k$ is a positive integer, and $r_{k+1} > 1$, the second term in the line above is nonnegative. Therefore, we have for all k,

(1) $$q_{k+1} \geq (r_{k+1} + 2)c_{k+1}q_k^2$$

Note that q_k divides q_{k+1}, and that it takes r_{k+1} links of A_{k+1} to *span* a link (that is, go from one boundary component to the opposite boundary component of the link) of A_k.

Let ϵ_k denote the maximum diameter of a link of A_k. We may assume that $\epsilon_1 \leq \frac{2\pi}{q_1}$, and that ϵ_k is on the close order of

$$\epsilon_1 \left(\prod_{i=2}^{k} r_i \right)^{-1}$$

Since $r_k > 1$, we may assume that the ϵ_k's are summable, with sum at most 1.
The intersection

$$\Lambda = \bigcap_{k=1}^{\infty} A_k$$

of the annuli described above is a cofrontier. Depending upon the choices we make for
c_{k+1} and r_{k+1}, Λ may be homeomorphic to the circle S^1 (as we show in Theorem 4.2),
indecomposable (as we show in Theorem 4.1), or possibly something in between (as we ask
in Question 6.4). If we allow more bending, Λ can be made hereditarily indecomposable,
i.e. the pseudocircle.
   The necessary lemma that we use above is the following.

3.2.1. Lemma.

   (1) *Suppose A_1, A_2, and g_1 are given as above, with A_1 and A_2 invariant under the
       period q_1 homeomorphism g_1 of A_1 onto itself, taking outer boundary of A_1 to
       outer boundary of A_1, intrinsic boundary segments to such segments, and links of
       A_i to links of A_i for $i = 1, 2$. Assume that each basic repetitive unit of A_2 has n_2
       links, where $q_2 = n_2 q_1$. Then there exists a period q_2 homeomorphism g_2 of A_2
       carrying links of A_2 to links of A_2, moving forward one link, such that g_2 generates
       a cyclic group of topological rotations of A_2 which commute with the topological
       rotations of A_1 generated by g_1.*
   (2) *Suppose $A_1, \ldots, A_k$, and $g_1, \ldots, g_{k-1}$ are given as above, with A_k invariant under
       the topological rotations $g_1, \ldots, g_{k-1}$ of periods $q_1, \ldots, q_{k-1}$, respectively, respect-
       ing the boundaries appropriately. Then there exists a period $q_k = n_k q_{k-1}$ homeo-
       morphism g_k of A_k carrying links of A_k to links of A_k, moving forward one link,
       such that g_k generates a cyclic group of topological rotations of A_k which commute
       with the topological rotations of A_i generated by g_i, for $i = 1, \ldots, k - 1$.*

   Proof of (1).   Let R_1 be the period q_1 clockwise rotation of A onto itself through angle
$\frac{2\pi}{q_1}$. Let $\phi_2 : A_2 \to A$ be a homeomorphism of A_2 onto A, taking outer boundary onto
outer boundary, inner boundary onto inner boundary, and intrinsic boundary segments of
links onto the iterates of $0 \times [0,1]$ under the period q_2 clockwise rotation R_2 of A onto itself
through angle $\frac{2\pi}{q_2}$. We may choose ϕ_2 so that $R_1 \phi_2 = \phi_2 g_1$ on A_2. Name the links of A_2, in
clockwise order and in chunks consisting of the basic repetitive units, as $a_{2,1,1}, \ldots, a_{2,q_1,n_2}$,
where $a_{i,j,k}$ is the kth link of the jth basic repetitive unit of A_i.
   Define g_2 on the jth basic repetitive unit of A_2, $1 \le j \le q_1$, as follows: on link $a_{2,j,k}$,
for $1 \le k \le n_2 - 1$, let $g_2 = \phi_2^{-1} R_2 \phi_2$; on link a_{2,j,n_2}, let $g_2 = \phi_2^{-1} R_1 R_2^{-(n_2-1)} \phi_2 = g_1 \phi_2^{-1} R_2^{-(n_2-1)} \phi_2$. The last equality follows because ϕ_2 conjugates $g_1 | A_2$ to R_1 on A. In
this way, g_2 is not only periodic of period q_2 on A_2, but $g_2^{n_2} = g_1 | A_2$ and g_2 commutes
with g_1 on A_2.
   Proof of (2).   At the next stage, we need only insure that $g_3^{n_3} = g_2 | A_3$ and that g_3
commutes with g_2 on A_3 in our construction. It will automatically follow that g_3 also
commutes with g_1. We leave it to the reader to complete the induction. ∎
   We use the following version of a well-known theorem of Cook and Ingram [CI] to detect
indecomposability in our examples:

3.3. **Theorem.** *Let* $\Lambda = \bigcap_{k=1}^{\infty} A_k$ *be a cofrontier constructed as above. Then* Λ *is indecomposable iff there is a subsequence* $\{A_{k_j}\}_{j=1}^{\infty}$ *such that*

(*) *for each pair of links* $a, b \in A_{k_j+1}$, *at least one of the maximally coherent subcollections of* $A_{k_j+1} - \{a, b\}$ *meets every link of* A_{k_j}.

If A_j in A_i satisfies condition (*) of Theorem 3.3, we say that A_j is *folded* in A_i.

3.4. **Inducing pseudorotations on** $\Lambda \in \mathcal{H}$. We induce a pseudorotation f_α on $\Lambda \in \mathcal{H}$ as the uniform limit of topological rotations f_k of the annuli A_k. The existence of the required topological rotations has been established in our construction of Λ in Section 3.2. Refer to Figure 3 for an example.

(1) Homeomorphism $f_1 = g_1^{m_1}$ rotates A_1 onto itself by the fractional amount $\frac{m_1}{q_1}$, taking links to links, and leaving $A_2, A_3, \ldots$ invariant, so $f_1|\Lambda$ is a homeomorphism.

(2) Homeomorphism f_2 first does f_1, then rotates A_2 onto itself by the amount $\frac{m_2}{q_2}$ within its own annular structure, taking links to links, and leaving $A_3, A_4, \ldots$ invariant. Thus, $f_2 = g_2^{m_2} f_1 = g_2^{m_2} g_1^{m_1}$ rotates A_2 the total amount $\frac{m_1}{q_1} + \frac{m_2}{q_2}$, and is a homeomorphism on Λ.

(3) In general, homeomorphism f_k first does f_{k-1}, then rotates A_k an additional amount $\frac{m_k}{q_k}$ within its own annular structure leaving $A_{k+1}, A_{k+2}, \ldots$ invariant. Thus, $f_k = g_k^{m_k} f_{k-1} = g_k^{m_k} g_{k-1}^{m_{k-1}} \ldots g_1^{m_1}$ rotates A_k the total amount

$$\frac{p_k}{q_k} = \sum_{j=1}^{k} \frac{m_j}{q_j}$$

and is a homeomorphism on Λ.

(4) Assume that the m_k's are not too large, i.e. $m_k < r_k$, and recall that the ϵ_k's are summable. In the limit,

$$\frac{p_k}{q_k} \to \alpha = \sum_{j=1}^{\infty} \frac{m_j}{q_j}$$

and the homeomorphisms f_k converge uniformly to the pseudorotation f_α on Λ.

We call any homeomorphism on Λ, induced via uniform limits of the designated (in Section 3.2) topological rotations of the A_k's as above, a *standard* pseudorotation.

It is easy to extend f_α to a homeomorphism of $\mathbf{R}^2$. We provide only a sketch (for more details see [Ha]). Since f_1 is a rigid rotation of A_1, we may extend it to the corresponding rotation of $\mathbf{R}^2 - \text{Int}(A_1)$. On $A_1 - \text{Int}(A_2)$, we may continuously deform $f_1|\partial A_1$ to $f_2|\partial A_2$ by using an embedding of an isotopy between appropriate rigid rotations of S^1. Proceeding inward, we deform f_2 to f_3 as we move in toward the boundary of A_3. This procedure can be continued inductively so as to define a homeomorphism on $\mathbf{R}^2$ whose restriction to Λ is f_α. Note that this extension of f_α is not topologically conjugate to a rotation because of the differing amounts of rotation on the boundaries of the A_k's. However, this extension does induce a topological rotation on the inner and outer circles of prime ends. (To see this, one may construct the prime end uniformizations stage-by-stage in tandem with the extension of f_α to $\mathbf{R}^2$ so that the topological rotations on the circles of prime ends are each a limit of topological rotations on circles converging to the boundary of the unit disk. We recall for the reader that a prime end uniformization need not be conformal, but only a C-map [UY].)

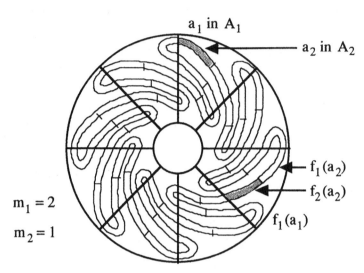

FIGURE 3. INDUCING A PSEUDOROTATION BY A SEQUENCE OF RATIO-
NAL ROTATIONS OF THE DEFINING ANNULI A_k.

4. PROPERTIES OF COFRONTIERS $\Lambda \in \mathcal{H}$

The "length" of basic repetitive units determines whether Λ is indecomposable or home-omorphic to S^1. There are two measures of this length: s_{k+1}, the absolute number of links of A_k that a basic repetitive unit of A_{k+1} spans, and c_{k+1}, the fractional part of A_k that a basic repetitive unit of A_{k+1} spans. We made certain assumptions about various numbers in the construction, of which we remind the reader: $q_1 > 1$, $c_{k+1} \leq 1$, $c_{k+1}q_k = s_{k+1} > 1$, and $r_{k+1} > 1$.

4.1. Theorem (Λ indecomposable). *If $\sum_{k=2}^{\infty} c_k = \infty$, then Λ is indecomposable.*

Proof. It suffices to show that for all i, there is a j such that A_j is folded in A_i. Without loss of generality, suppose $i = 1$. For all $k \geq 1$, $c_{k+1}q_k > 1$. Let D_k be a chain of dq_k links of A_k, where $\frac{1}{q_k} \leq d \leq 1$, where the fraction d is an integral multiple of $\frac{1}{q_k}$. We call D_k the d *(fractional) part of* A_k.

Claim: We need at least one basic repetitive unit, and at most the first $(d - c_{k+1})q_k$, basic repetitive units of A_{k+1} to meet the d part of A_k.

Proof of claim: Refer to Figure 4, where we illustrate the following argument with $d = \frac{3}{4}$ and $c_{k+1} = \frac{1}{4}$. Since $c_{k+1}q_k > 1$, each basic repetitive unit of A_{k+1} meets $c_{k+1}q_k + 1$ links of A_k (one more than it spans). Hence, the basic repetitive unit of A_{k+1} which begins in link number $(d - c_{k+1})q_k$ of D_k meets the last $c_{k+1}q_k + 1$ links of D_k. The basic repetitive

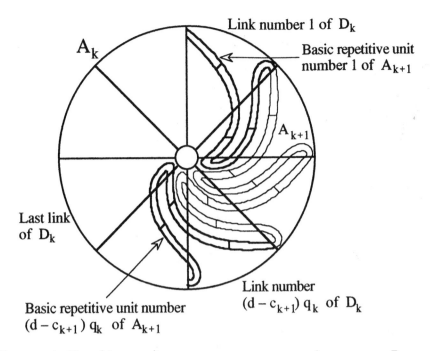

FIGURE 4. THE $(d - c_{k+1})q_k$ FRACTIONAL PART OF A_{k+1} MEETS D_k, THE d FRACTIONAL PART OF A_k.

unit which begins in link m of D_k (for all $m < (d - c_{k+1})q_k$) certainly meets the link in which it starts. Therefore, the first $(d - c_{k+1})q_k$, basic repetitive units of A_{k+1} meet the entire d part of A_k.

Suppose $d = 1$ and $k = 1$; it follows that we need at least one, and at most $(1 - c_2)q_1$ consecutive basic repetitive units of A_2 in order to meet all the links of A_1. It follows by induction with $d = 1 - c_2 - c_3 - \cdots - c_k$ that for each $k > 1$, we need at least one, and at most $(1 - c_2 - c_3 - \cdots - c_{k+1})q_k$, consecutive basic repetitive units of A_{k+1} in order to meet $(1 - c_2 - c_3 - \cdots - c_k)q_k$ consecutive links of A_k. Consequently, we need at most $(1 - c_2 - c_3 - \cdots - c_{k+1})q_k$ consecutive basic repetitive units of A_{k+1} in order to meet all of A_1. Since the c_k's are not summable, there is a j such that $\sum_{i=2}^{j} c_i \geq \frac{1}{2}$. Therefore, we need at most half of the basic repetitive units of A_j to meet every link of A_1. It follows by the symmetry of A_j in A_1, that A_j is folded in A_1. ∎

4.2. Theorem ($\Lambda \cong S^1$). *If the s_k's are uniformly bounded above, then Λ is homeomorphic to S^1.*

Proof. Refer to Figure 5, which illustrates the proof below. In that diagram, two basic repetitive units of A_2 in A_1 are shown, as are three basic repetitive units of A_3 in A_2. Note that a basic repetitive unit of A_3 spans two links of A_2, but meets three links of A_2. Note also that a basic repetitive unit of A_2 contains 8 links.

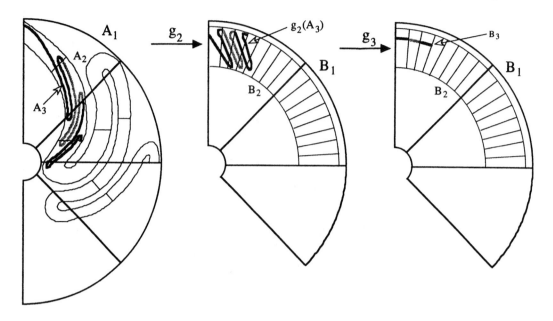

FIGURE 5. "STRAIGHTENING-OUT" A_2 IN A_1, AND A_3 IN A_2, TO SHOW
THAT Λ IS A SIMPLE CLOSED CURVE WHEN THE s_k'S ARE BOUNDED
ABOVE.

Let B_i be a sequence of concentric, circular chains, with B_{i+1} contained in B_i as B_2 is
contained in B_1 in Figure 5. In particular, if there are t_{i+1} links in a basic repetitive unit
of A_{i+1}, then each link of B_i contains t_{i+1} links of B_{i+1}.

Let $g_1 : A_1 \to B_1$ be a one-to-one, order-preserving correspondence between the links.
Let $g_{i+1} : A_{i+1} \to B_{i+1}$ be a one-to-one, order-preserving function from the links of A_{i+1}
to the links of B_{i+1}, taking the links of a basic repetitive unit of A_{i+1} to the links of B_{i+1}
which refine a link of B_i. Since basic repetitive units begin at one side of a link of the
previous stage and end at the other side (see Figure 5 again), we may preserve containment
with respect to the first and last links of each basic repetitive unit, and straigten out the
intermediate links so that they correspond to a straight chain refinement of B_{i+1} in B_i.

In the standard manner, every $x \in \Lambda$ can be written as the intersection of a tower of
links (or consolidations of links) of the A_i's with diameters tending to zero. Then let $g(x)$
be defined as the intersection of the tower of links of the B_i's corresponding under the g_i's.
Clearly, g is one-to-one.

To see that g is also continuous (and thus a homeomorphism), let m be an upper bound
for the s_k's. Thus for all k, a basic repetitive unit of A_{k+1} spans no more than m links
of A_k. Recall that each $r_k \geq 2$, so that each segment of A_{k+1} that spans a link of A_k is
divided into at least two links. There is an integer k such that $2^k > (m+1)$. Thus, a basic
repetitive unit of A_{2^k+1} refines a single link of A_1 (and perhaps meets one neighbor), so

that all subsequent "straightening-out" of basic repetitive units is done to sets of diameter less than twice the diameter of links of A_1.

Let $A_{t_1} = A_1$, and let $A_{t_2} = A_{2^k+1}$. Similarly, we can find A_{t_3} such that a basic repetitive unit of A_{t_3} refines a single link of A_{t_2}. As before, a basic repetitive unit of A_{t_3} refines a single link of A_{t_2} (and perhaps meets one neighbor), so that all subsequent "straightening-out" of basic repetitive units is done to sets of diameter less than twice the diameter of links of A_{t_2}. This process may be continued inductively to see that, for any $\epsilon > 0$, there is a stage A_{t_j}, so that from that point on, "straightening-out" is done only to sets of diameter less than ϵ. Thus, g is continuous, and the theorem follows. ∎

4.3. Remark. We note that in Theorem 5.1 below, we prove that every $\Lambda \in \mathcal{H}$ admits an irrational pseudorotation. Thus it follows that *for every indecomposable member of $\mathcal{H}$, the conclusions of Theorem 2.12 hold.*

5. PROPERTIES OF STANDARD PSEUDOROTATIONS

Assume f_α is a standard pseudorotation induced on $\Lambda \in \mathcal{H}$. We made assumptions about various numbers in the construction, of which we remind the reader: $q_1 > 1$, $c_k \leq 1$, $c_{k+1}q_k = s_{k+1} > 1$, $r_k > 1$, the ϵ_k's are summable, and $0 \leq m_k < r_k$. We obtain the following theorems detailing some of the dynamical properties of these standard pseudorotations.

5.1. Theorem (α irrational). *Assume f_α is a standard pseudorotation on $\Lambda \in \mathcal{H}$. If Λ is not homeomorphic to S^1 (e.g., Λ is indecomposable) and for infinitely many values of k, $m_k > 0$, then α is irrational. Thus every $\Lambda \in \mathcal{H}$ admits an irrational pseudorotation.*

Proof. Recall that by construction, we have inequality 3.2(1) $q_{k+1} \geq (r_{k+1} + 2)c_{k+1}q_k^2$; hence, q_k increases to ∞. Suppose by way of contradiction that

$$\alpha = \lim \frac{p_k}{q_k} = \sum_{k=1}^{\infty} \frac{m_k}{q_k} = \frac{p}{q}$$

is rational. Then, we have for each k,

$$\frac{1}{qq_k} \leq \frac{p}{q} - \frac{p_k}{q_k} = \sum_{j=k+1}^{\infty} \frac{m_j}{q_j} \leq \sum_{j=k+1}^{\infty} \frac{m_j}{(r_j + 2)c_jq_{j-1}^2}$$

Since $0 \leq m_j < r_j$ for all j, this becomes

$$\frac{1}{qq_k} \leq \sum_{j=k+1}^{\infty} \frac{1}{c_jq_{j-1}^2}$$

Since for infinitely many values of k, $m_k > 0$, the above inequality holds for all k. We proceed to expand the right-hand-side of the inequality as follows:

$$\frac{1}{qq_k} \leq \frac{1}{c_{k+1}q_k^2} + \frac{1}{c_{k+2}q_{k+1}^2} + \frac{1}{c_{k+3}q_{k+2}^2} + \dots$$

$$\leq \frac{1}{(c_{k+1}q_k)q_k} + \frac{1}{(c_{k+2}q_{k+1})q_{k+1}} + \frac{1}{(c_{k+3}q_{k+2})q_{k+2}} + \dots$$

$$\leq \frac{1}{(c_{k+1}q_k)q_k} + \frac{1}{(c_{k+2}q_{k+1})(r_{k+1}+2)(c_{k+1}q_k)q_k}$$

$$+ \frac{1}{(c_{k+3}q_{k+2})(r_{k+2}+2)(c_{k+2}q_{k+1})(r_{k+1}+2)(c_{k+1}q_k)q_k} + \dots$$

Therefore, multiplying by q_k, factoring out $\frac{1}{c_{k+1}q_k}$, and simplifying (using $c_jq_{j-1} > 1$ and $r_j > 1$, for all j), we have

$$\frac{1}{q} \leq \frac{1}{c_{k+1}q_k}\left(1 + \frac{1}{2} + \frac{1}{4} + \dots\right) \leq \frac{2}{c_{k+1}q_k}$$

Since $c_{k+1}q_k = s_{k+1}$, and since Λ not homeomorphic to S^1 implies $\limsup s_{k+1} = \infty$ by Theorem 4.2, it follows that this cannot hold for all k. Therefore, α is irrational. ∎

If the number of subdivisions of a spanning segment is large, and the incremental rotations ($\frac{m_k}{q_k}$) are small, then f_α is recurrent on Λ, and Λ has a well-defined generalized rotation number, for any Λ in $\mathcal{H}$ (whether Λ is indecomposable or not). More precisely,

5.2. Theorem (f_α recurrent). *Assume f_α is a standard pseudorotation on $\Lambda \in \mathcal{H}$. If for some $M > 0$ and for all k, $r_{k+1} \geq q_k$ and $m_k \leq M$, then $f_\alpha|\Lambda$ is recurrent. Moreover, $\rho(\Lambda) = \alpha$.*

Proof. We first prove a useful lemma.

5.2.1. Lemma. *The hypotheses of Theorem 5.2 imply that*

(1) *there is a constant $C > 0$ such that $d(f_{k+1}^i, f_k^i) < C\epsilon_k$ for all i with $0 \leq i \leq q_k$.*

Consequently, we have

(2) *$d(f_\alpha^i, f_k^i) \leq C\sum_{j=k}^\infty \epsilon_j$ for all i with $0 \leq i \leq q_k$.*

Proof. First we show condition (1) implies (2). By the uniform convergence of f_k^i to f_α^i, and the triangle inequality, for each $\beta > 0$, there is an $m > 0$ such that

$$d(f_\alpha^i, f_k^i) < \beta + \sum_{j=k}^{k+m} C\epsilon_j < \beta + \sum_{j=k}^\infty C\epsilon_j$$

Thus, in the limit as β goes to 0,

$$d(f_\alpha^i, f_k^i) \leq \sum_{j=k}^\infty C\epsilon_j \leq C\sum_{j=k}^\infty \epsilon_j$$

That the hypotheses of Theorem 5.2 imply (1) follows from the construction: Choose j so that $M < q_j$. Choose $C = \max\{M, \epsilon_j^{-1}\}$. If $k < j$ the conclusion follows trivially. So suppose $k \geq j$. Each iteration of f_{k+1} moves a link of A_{k+1}, $m_{k+1} \leq M$ links of A_{k+1} away from what f_k would. Since $r_{k+1} \geq q_k$ and $M < q_k$, f_{k+1} moves a link of A_{k+1} at most 1 link of A_k away in the annular structure of A_k from what f_k would. In general, f_{k+1}^i moves a link of A_{k+1} at most $\lceil \frac{Mi}{q_k} \rceil$ links of A_k away in the annular structure of A_k. Since the diameter of a link of A_k is at most ϵ_k, it follows that for all $i \leq q_k$, $d(f_{k+1}^i, f_k^i) < M\epsilon_k$. ∎

Now we will show that for all $\epsilon > 0$, there is a positive integer i such that for every $x \in \Lambda$, $d(f_\alpha^i(x), x) < \epsilon$. Let C be the constant of Lemma 5.2.1. Since the ϵ_k's are summable, we may choose k such that $C \sum_{j=k}^{\infty} \epsilon_j < \epsilon$. Let $x \in \Lambda$. Then there is a link $a \in A_k$ such that $x \in a$. By construction, $f_k^{q_k}$ is the identity on A_k, so $f_k^{q_k}(x) = x$. By Lemma 5.2.1,

$$d(f_\alpha^{q_k}(x), f_k^{q_k}(x)) \leq C \sum_{j=k}^{\infty} \epsilon_j < \epsilon$$

Consequently,

$$d(f_\alpha^{q_k}(x), x) < \epsilon$$

Hence, $f_\alpha|\Lambda$ is recurrent.

We want to show that the generalized rotation number for Λ is well-defined. For each k, let γ_k denote the projection of A_k to its core circle S^1. Let $\widetilde{\gamma}_k : \widetilde{A}_k \to \mathbf{R}^1$ denote the lift of γ_k to the respective universal covers (with π_k and θ the respective covering projections). Then $\widetilde{\Lambda} = \bigcap_{k=1}^{\infty} \widetilde{A}_k$ covers Λ under each projection π_k. Let $\widetilde{f}_k$ and $\widetilde{f}_\alpha$ denote the lifts of f_k and f_α, respectively. Since two lifts of the same map may differ by an integer, we choose the "simplest" lift in each case. We want to show that the generalized rotation number at each point of Λ is α. It suffices to show that for all $\widetilde{x} \in \widetilde{\Lambda}$

$$\frac{\widetilde{\gamma}_1 \widetilde{f}_\alpha^i(\widetilde{x}) - \widetilde{\gamma}_1(\widetilde{x})}{i} \xrightarrow[i]{} \alpha$$

Since $c_k \leq 1$ for all k, it follows that $|\widetilde{\gamma}_{k+1} - \widetilde{\gamma}_k| \leq 2$. Consequently,

(1) $$|\widetilde{\gamma}_k - \widetilde{\gamma}_1| \leq 2k$$

Since f_k is, by definition, rotation of A_k by the fractional amount $\alpha_k = \frac{p_k}{q_k}$, it follows that

(2) $$\widetilde{\gamma}_k \widetilde{f}_k^i(\widetilde{x}) - \widetilde{\gamma}_k(\widetilde{x}) = i\alpha_k$$

Since $q_k \geq 2^k$, it follows by Lemma 5.2.1 that for all sufficiently large k, for all i with $0 \leq i \leq 2^k$,

(3) $$|\widetilde{\gamma}_1 \widetilde{f}_\alpha^i - \widetilde{\gamma}_1 \widetilde{f}_k^i| < 1$$

Claim: For all sufficiently large i, there exists $k = k(i)$ with $2^{k-1} \leq i < 2^k$ such that

(4)
$$\left| \frac{\tilde{\gamma}_1 \tilde{f}_\alpha^i(\tilde{x}) - \tilde{\gamma}_1(\tilde{x})}{i} - \alpha \right| \leq |\alpha - \alpha_k| + \frac{4k+1}{i}$$

Proof of Claim: It follows from (2) that

$$|\tilde{\gamma}_k \, \tilde{f}_k^i(\tilde{x}) - \tilde{\gamma}_k(\tilde{x}) - i\alpha| = |i\alpha_k - i\alpha|$$

Therefore, by (1) applied twice, and the triangle inequality

$$|\tilde{\gamma}_1 \tilde{f}_k^i(\tilde{x}) - \tilde{\gamma}_k(\tilde{x}) - i\alpha| \leq |i\alpha_k - i\alpha| + 2k$$

$$|\tilde{\gamma}_1 \tilde{f}_k^i(\tilde{x}) - \tilde{\gamma}_1(\tilde{x}) - i\alpha| \leq |i\alpha_k - i\alpha| + 4k$$

So by (3) and the triangle inequality,

$$|\tilde{\gamma}_1 \, \tilde{f}_a^i(\tilde{x}) - \tilde{\gamma}_1(\tilde{x}) - i\alpha| \leq |i\alpha_k - i\alpha| + 4k + 1$$

Then (4) follows by dividing by i.

Since $|\alpha - \alpha_{k(i)}| \underset{i}{\to} 0$ and

$$\frac{4k+1}{2^k} < \frac{4k+1}{i} \leq \frac{4k+1}{2^{k-1}}$$

it follows that the right-hand-side of (4) goes to 0 as $i \to \infty$. ∎

5.3. Irrationals of constant type. The irrational α is of *constant type* iff there is an $\epsilon > 0$ such that for all rationals $\frac{p}{q}$ in lowest terms,

$$\left| \alpha - \frac{p}{q} \right| \geq \frac{\epsilon}{q^2}$$

Thus, in an informal sense, α is poorly approximated by rationals.

5.4. Theorem (α not of constant type). *Assume f_α is a standard pseudorotation on $\Lambda \in \mathcal{H}$. If*

(1) *Λ is not homeomorphic to S^1,*
(2) *for infinitely many values of k, $m_k > 0$,*
(3) *for some $M > 0$ and for all k, $m_k \leq M$, and*
(4) *for all k, $r_{k+1} \geq q_k$,*

then the irrational α is not of constant type; that is, in an informal sense, α is well-approximated by rationals.

Proof. We have shown in Theorem 5.1 that α is irrational. Now suppose by way of contradiction that α is of constant type. Then there exists an $\epsilon > 0$ such that for every rational $\frac{p}{q}$, $|\alpha - \frac{p}{q}| \geq \frac{\epsilon}{q^2}$. By hypotheses (2) and (3) and inequality 3.2(1), we have for every k,

$$\frac{\epsilon}{q_k^2} \leq \alpha - \frac{p_k}{q_k} = \sum_{j=k+1}^{\infty} \frac{m_j}{q_j} \leq M \sum_{j=k+1}^{\infty} \frac{1}{q_j} \leq M \sum_{j=k+1}^{\infty} \frac{1}{(r_j + 2)c_j q_{j-1}^2}$$

Expanding the sum, applying 3.2(1) and hypothesis (4), and simplifying, we have

$$\frac{\epsilon}{q_k^2} \leq M \left(\frac{1}{(r_{k+1}+2)(c_{k+1}q_k^2)} + \frac{1}{(r_{k+2}+2)(c_{k+2}q_{k+1}^2)} + \frac{1}{(r_{k+3}+2)(c_{k+3}q_{k+2}^2)} + \cdots \right)$$

$$\leq M \left(\frac{1}{(r_{k+1}+2)(c_{k+1}q_k^2)} + \frac{1}{(r_{k+2}+2)(c_{k+2}q_{k+1})(r_{k+1}+2)(c_{k+1}q_k^2)} \right.$$

$$\left. + \frac{1}{(r_{k+3}+2)(c_{k+3}q_{k+2})(r_{k+2}+2)(c_{k+2}q_{k+1})(r_{k+1}+2)(c_{k+1}q_k^2)} + \cdots \right)$$

$$\leq M \left(\frac{1}{c_{k+1}q_k^3} + \frac{1}{2(c_{k+1}q_k^3)} + \frac{1}{2^2(c_{k+1}q_k^3)} + \cdots \right)$$

$$\leq \frac{M}{c_{k+1}q_k^3} \left(1 + \frac{1}{2} + \frac{1}{4} + \cdots \right)$$

$$\frac{\epsilon}{q_k^2} \leq \frac{2M}{c_{k+1}q_k^3}$$

Consequently, multiplying by q_k^2 and dividing by $2M$, we have

$$\frac{\epsilon}{2M} \leq \frac{1}{c_{k+1}q_k}$$

Since $c_{k+1}q_k = s_{k+1}$, and since $\limsup s_{k+1} = \infty$ by hypothesis (1) and Theorem 4.2, it follows that this cannot hold for all k. Therefore, α is not of constant type. ∎

The following example serves two purposes: it shows that the equivalence of almost periodic and recurrent does not hold for all cofrontiers, and it shows that Theorem 2.11 cannot be extended to almost periodic.

5.5. Example (f_α not almost periodic). *follow the construction recipe:*

(1) $q_1 = 4$,
(2) $c_{k+1} = \frac{1}{2}$ *for all* $k \geq 1$, *and*
(3) $r_{k+1} = q_k$ *for all* $k \geq 1$,

with minimal bending.

The cofrontier $\Lambda \in \mathcal{H}$ obtained will be indecomposable by (2) and Theorem 4.1. Induce a pseudorotation f_α on Λ by choosing $m_k = 1$ for all k. Then by Theorems 5.1-5.3, f_α is recurrent on Λ and α is an irrational not of constant type. By Theorem 5.6 below, f_α is minimal on Λ. It then follows by Theorem 2.9 that f_α is **not** almost periodic on Λ.

5.6. Theorem (Λ minimal under f_α). *Suppose f_α is a standard pseudorotation of $\Lambda \in \mathcal{H}$. If $r_{k+1} \geq q_k$ and $0 < m_k \leq M$ for all k, then f_α is minimal on Λ.*

Proof. The first condition in Lemma 5.2.1 is the analog of one of the conditions used in [Ha] to deduce the minimality of f_α on Λ. The hypotheses of Theorem 5.6 seem to be the most direct way of realizing that condition in a specific construction. (The other hypothesis in [Ha] used to obtain minimality, that f_k is transitive on the links of A_k, is superfluous, as we show below in Lemma 5.6.1.)

We will show that Λ is the closure of the orbit of each of its points. First we need the following lemma:

5.6.1. Lemma. *The hypotheses of Theorem 5.6 imply that for sufficiently large k, for any link $a \in A_{k+1}$, the sequence of iterates $\{(f_{k+1}^{q_k})^s(a)\}_{s=1}^t$, where $t = \frac{q_{k+1}}{q_k}$, meets every link of A_{k-1}.*

Proof. Since $q_k \to \infty$, we may choose k such that $M < q_{k-1}$. Let $a \in A_{k+1}$. Then $f_{k+1}^{q_k}(a)$ is at least q_k and at most Mq_k links of A_{k+1} away from its starting link in the annular structure of A_k. It takes at least q_{k-1} links of A_k to span a link of A_{k-1}, and at least q_k links of A_{k+1} to span a link of A_k. Since $Mq_k < q_{k-1}q_k$, $f_{k+1}^{q_k}(a)$ is within one link of its starting link in A_{k-1}. Recall that $f_{k+1}^{q_{k+1}}$ is the identity on A_{k+1}. Therefore, $f_{k+1}^{sq_k}(a)$ steps through the links of A_{k-1} at most one link at a time, with increasing s, meeting every link of A_{k-1} before $sq_k = q_{k+1}$. ∎

Let $x \in \Lambda$ and let U be an open set in the plane such that $U \cap \Lambda \neq \emptyset$. Let C be the constant of Lemma 5.2.1. Since the ϵ_k's are summable, we may choose $k - 1$ such that there is a link $a \in A_{k-1}$ and $\epsilon > 0$ such that $B(a, \epsilon) \subset U$ and $C \sum_{j=k+1}^\infty \epsilon_j < \epsilon$. By Lemma 5.6.1, there is a positive integer $i < q_{k+1}$ and a link $b \in A_{k+1}$ such that $x \in b$ and $f_{k+1}^i(b) \subset a$. By Lemma 5.2.1,

$$d(f_\alpha^i(x), f_{k+1}^i(x)) \leq C \sum_{j=k+1}^\infty \epsilon_j < \epsilon$$

Since $f_{k+1}^i(x) \in a$, it follows that

$$f_\alpha^i(x) \in B(a, \epsilon) \subset U \quad \blacksquare$$

We have shown that if the number of subdivisions of a spanning segment is large, and the incremental rotations are small, then every orbit of $f_\alpha|\Lambda$ is dense in Λ. We leave the proof of the following theorem, showing that Example 5.5 is far from unique in $\mathcal{H}$, to the reader.

5.7. Theorem (f_α recurrent, minimal, and not almost periodic). *Suppose f_α is a standard pseudorotation of $\Lambda \in \mathcal{H}$. If there are constants c and M such that $c_{k+1} \geq c > 0$, $r_{k+1} \geq q_k$, and $0 < m_k \leq M$ for all k, then f_α is recurrent and minimal, but not almost periodic, on Λ.*

The following theorem summarizes the preceding results concerning cofrontiers in $\mathcal{H}$:

5.8. Theorem. *Suppose f_α is a standard pseudorotation on $\Lambda \in \mathcal{H}$. Suppose further that*

    (a) *there exists $c > 0$ such that for all k, $c_{k+1} \geq c$,*
    (b) *for all k, $r_{k+1} \geq q_k$,*
    (c) *for some $M > 0$ and for all k, $0 < m_k \leq M$.*

Then

    (1) *Λ is indecomposable, by (a) and Theorem 4.1,*
    (2) *Λ is a minimal set under f_α, by (b), (c), and Theorem 5.6,*

(3) f_α is recurrent, by (b), (c), and Theorem 5.2,

(4) f_α is not almost periodic, by (a), (b), (c), and Theorem 5.7, or by conclusions (1) and (2) and Theorem 2.9,

(5) α is irrational, by (a), (c), and Theorem 5.1, and

(6) α is not of constant type, by (a), (b), (c), and Theorem 5.4.

6. QUESTIONS

We conclude with some questions about cofrontiers and pseudorotations, both arbitrary ones and those in our class $\mathcal{H}$.

6.1. Question. Does there exist an indecomposable cofrontier Λ admitting an irrational pseudorotation h_α for which α is of constant type?

6.2. Question. Suppose Λ is a cofrontier and h_α is an irrational pseudorotation of Λ. Must $h_\alpha|\Lambda$ be minimal? Recurrent?

Note that M. Barge has asked the "minimal" part. The example which motivated Barge's question is constructed in [Wa]: the "split hairy circle" is a decomposable cofrontier X invariant under a homeomorphism g of the plane on which the induced map on one circle of prime ends is a rigid rotation, but on the other circle is a Denjoy map. Thus, g is a "one-sided pseudorotation." However, g is neither minimal nor recurrent on X. So rotation-like from both sides is necessary in 6.2. In [MO] (these Proceedings), an indecomposable cofrontier Y is constructed invariant under a homeomorphism h of the plane with the induced prime end maps being a rotation on one side and a Denjoy map on the other. In this case, h is minimal on Y, though h does not appear to be recurrent on Y; still, it is only a one-sided pseudorotation.

6.3. Question. Let $\mathcal{A}$ be the set of all irrationals α such that there exists an indecomposable cofrontier Λ (in $\mathcal{H}$ or arbitrary) and a pseudorotation h_α on Λ. Is $\mathcal{A}$ a proper subset of the irrationals? Does there exist a number-theoretic characterization of $\mathcal{A}$?

6.4. Question. Are there any cofrontiers in $\mathcal{H}$ that are neither indecomposable, nor homeomorphic to S^1?

In an attempt to answer Question 6.4, one might try to construct a $\Lambda \in \mathcal{H}$ for which $\limsup s_k = \infty$ and $\sum_{k=2}^{\infty} c_k < \infty$ (so neither of Theorems 4.1 or 4.2 applies).

6.5 Recurrent and analytic extensions. We have briefly outlined (see Section 3.4, last paragraph) one way in which one may extend the induced pseudorotations we obtain to $\mathbf{R}^2$. There are many possible extensions. Another such procedure is outlined in Handel's construction of an area-preserving diffeomorphism of $\mathbf{R}^2$ leaving a cofrontier Λ ($\cong$ pseudocircle) invariant [Ha]. Similar procedures can be applied in the authors' constructions. However, neither Handel's nor the authors' extensions are recurrent on $\text{Int}(\Lambda)$, nor conjugate on $\text{Int}(\Lambda)$ to the rotation R_α on the open unit disk $\mathbf{D}$.

6.6. Question. Suppose f_α is a standard pseudorotation on $\Lambda \in \mathcal{H}$, and that $f_\alpha|\Lambda$ is recurrent. Can $f_\alpha|\Lambda$ be extended so as to be recurrent on $\text{Int}(\Lambda)$? Can $f_\alpha|\Lambda$ be extended to $\text{Int}(\Lambda)$ so as to be conjugate to the rotation R_α on the open disk $\mathbf{D}$?

M. R. Herman [Hr] obtains pseudorotations of some cofrontiers in S^2 that are complex analytically conjugate on $\text{Int}(\Lambda)$ to R_α on the complex unit disk **D** (and C^∞ conjugate on $\text{Ext}(\Lambda)$).

6.7. Question. In general, when can a pseudorotation of a cofrontier $\Lambda \subset S^2$ be extended so as to be recurrent on $\text{Int}(\Lambda)$ (and $\text{Ext}(\Lambda)$)? (Complex analytically) conjugate on $\text{Int}(\Lambda)$ to R_α on **D**?

REFERENCES

[BG]   M. Barge and R. M. Gillette, *Rotation and periodicity in plane separating continua*, Ergodic Theory and Dynamical Systems **11** (1991), 619–631.

[Br1]   B. L. Brechner, *On stable homeomorphisms and imbeddings of the pseudo arc*, Illinois J. of Math. **22** (1978), 630–661.

[Br2]   ———, *Extendable periodic homeomorphisms on chainable continua*, Houston J. of Math. **7** (1981), 327–344.

[BGM]   B. L. Brechner, J. C. Mayer, and M. D. Guay, *Rotational dynamics on cofrontiers*, Contemporary Mathematics **117** (1991), 39–48.

[Ca]   M. L. Cartwright, *Equicontinuous mappings of plane minimal sets*, Proc. London Math. Soc. (3) **14A** (1965), 51–54.

[C]   M. Charpentier, *Sur quelques proprites des courbes de M. Birkoff*, Bull. Soc. Math. France **62** (1934), 193–224.

[CI]   H. Cook and W. T. Ingram, *A characterization of indecomposable compact continua*, Proceedings of Topology Conference (held at Arizona State University, Tempe, AZ, March, 1967), 168–169.

[Ha]   M. Handel, *A pathological area-preserving diffeomorphism of the plane*, Proceedings AMS **86** (1982), 163–168.

[He]   E. Hemmingsen, *Plane continua admitting non-periodic autohomeomorphisms with equicontinuous iterates*, Math. Scand. **2** (1954), 119–141.

[Hr]   M. R. Herman, *Construction of some curious diffeomorphisms of the Riemann sphere*, J. London Math. Soc. (2) **34** (1986), 375–384.

[MO]   J. C. Mayer and L. G. Oversteegen, *Denjoy meets rotation on an indecomposable cofrontier*, these Proceedings.

[MR]   J. C. Mayer and J. T. Rogers, Jr., *Indecomposable continua and the Julia sets of polynomials* (preprint).

[P]   G. Piranian, *The boundary of a simply connected domain*, Bull. Amer. Math. Soc. **64** (1958), 45–55.

[Ro1]   J. T. Rogers, Jr., *Rotations of simply connected regions and circle-like continua*, Contemporary Math. **117** (1991), 139–148.

[Ro2]   ———, *Indecomposable continua, prime ends, and Julia sets*, these Proceedings.

[Ru]   N. E. Rutt, *Prime ends and indecomposability*, Bull. A. M. S. **41** (1935), 265–273.

[UY]   H. D. Ursell and L. C. Young, *Remarks on the theory of prime ends*, Mem. Amer. Math. Soc. **3** (1951), 1–29.

[Wa]   R. Walker, *Basin boundaries with irrational rotations on prime ends*, Transactions AMS **324** (1991), 303–317.

A Periodic Homeomorphism of the Plane

MORTON BROWN   University of Michigan, Ann Arbor, Michigan

1.   Introduction

The homeomorphism described in this note has proved to be interesting for dynamics, topology, and combinatorics. For this reason, I have been persuaded to elevate it to the status of a formal publication.

2.   The homeomorphism H .

Let H denote the planar homeomorphism

$$H(x,y) = (|x| - y , x).$$

It is useful to observe that H can be represented as a composition $H = FoG$ where G is the $90°$ rotation

$$G(x,y) = (-y,x) ,$$

and F is the horizontal pseudo-shear

$$F(x,y) = (x + |y|, x) .$$

Obviously, H is a piecewise linear homeomorphism with the origin as its unique fixed point. Furthermore, H is area preserving and "reversible" [3] in that the sense that

$$HoR = RoH^{-1}$$

where R is the reflection $R(x,y) = (y,x)$.

The remarkable property of H is that it is periodic of
period 9 . A purely combinatorial proof of this periodicity, due
to Knuth, is given in the last section of this paper. In the next
section I will give a "geometric" proof. (Although the
consequence of "reversibility" does not appear in my proof, it is
interesting to observe how it appears in Knuth's proof).

3.   Proof of the periodicity of H .

Observe that $H(cx,cy) = cH(x,y)$ for all non-negative
constants c , so it suffices to find an invariant simple closed
curve surrounding the origin on which H has period 9 . One can
easily check (with very little arithmetic complication) that H
is periodic with rotation number 2/9 on the polygonal curve of
Figure 1 . The vertices of the polygon are:
$(1,0),(2,1),(1,1),(1,2),(0,1),(-1,1),(-1,0),(0,-1),(-1,-1)$.

4.   Remarks.

There is a resemblance between the homeomorphism H and the
Lozi map $L(x,y) = (1 + By-A|x|,x)$, particularly the case
B = -1, A = -1 , whereupon L is area preserving. This latter
case was studied by R. Devaney [2] . It has some remarkable
properties, and interesting dynamics. Although not periodic, it
is periodic (with varying periods) at each rational point in the
plane. In fact we can view H as the limiting base k=0 of the
"Lozi" map $L(x,y) = (k + By - A|x|,x)$.

Historical note.

In 1983 I submitted this problem [1] to the AMS Monthly:
Prove that a sequence of real numbers satisfying the recursion
relation

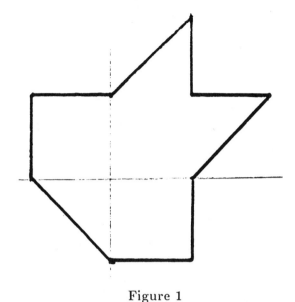

Figure 1

$$a_{n+2} = |a_{n+1}| - a_n$$

has period 9 . A number of solutions were offered. The most
elegant one was due to Donald Knuth [4]:

Notice that $a_{n-1} = |a_n| - a_{n+1}$, so we can run the
recurrence backward or forward. Since $a_n < 0$ implies that
$a_{n+2} \geq 0$, the sequence contains two consecutive non-negative
terms x and y . If $y \geq x$, the sequence must therefore be
$(\ldots,x,y,y-x,-x,2x-y,|2x-y|+x,|2x-y|+y-x,y-2x,x-y,x,y,\ldots)$;
and if $y \leq x$, the same idea applies, but in reverse:
$(\ldots x,y,y-x,x-2y,|2y-x|+x-y,|2y-x|+y,2y-x,-y,x-y,x,y,\ldots)$.

Professor Knuth kindly sent me a copy of a letter [5] that he included with his solution. As it partly motivates my publishing this paper, I quote the letter in full:

"When I saw advanced problem 6439, I couldn't believe that it was 'advanced': a result like that has to be either false or elementary!"

"But I soon found that it wasn't trivial. There is a simple proof, yet I can't figure out how on earth anybody would discover such a remarkable result. Nor have I discovered any similar recurrence relations having the same property."

"So in a sense I have no idea how to solve the problem properly. Is there an 'insightful' proof, or is the result simply true by chance?"

REFERENCES

1. Morton Brown, *American Mathematical Monthly*, Vol. 90 (1983), p. 569.

2. R. L. Devaney, A Piecewise Linear Model for the Zones of Instability of an Area Preserving Map, *Physica* 10D (1984), p. 387-393.

3. R. Devaney, Reversible Diffeomorphisms and Flows, *Trans. Amer. Math. Soc.* Vol. 218 (1976), 89-113.

4. Donald Knuth, *American Mathematical Monthly* Vol. 92 (1985) p. 218.

5. Donald Knuth, Correspondence with the Problems Editor of the *Amer. Math. Monthly*.

6. R. Lozi, Un Attracteur Etrange(?) Du Type Attracteur De Henon, *J. Phys.* 39 (1978), 9-10.

7. A. F. Beardon, S. R. Ballett, P. S. Rippon, Uniformly Periodic Difference Equations, (Pre-Print).

8. Aharanov, D. and Elias, U., Invariant Curves Around a Parabolic Fixed Point at Infinity, *Ergodic Theory and Dynamical Systems* 10 (1990), 209-229.

Dynamical Connections Between a Continuous Map and Its Inverse Limit Space

LIANG CHEN   Memphis State University, Memphis, Tennessee

SHIHAI LI   University of Florida, Gainesville, Florida, and National University of Singapore, Singapore

Abstract.   This is a survey about the dynamical connections between a continuous self-mapping f on a compact metric space X and the shift map on the inverse limit space of the sequence

$$X \xleftarrow{f} X \xleftarrow{f} X \xleftarrow{f} \dots$$

Denote the inverse limit space by $\varprojlim(X, f)$. Let σ_f denote the shift map on $\varprojlim(X, f)$. We collect some known dynamical properties that f has these properties if and only if σ_f has them. Properties covered in this survey are positive topological entropy, chaos, shadowing property and so on. We then introduce a theorem which relates the set of the chain recurrent points $CR(f)$ of f and that of σ_f (i.e. $CR(\sigma_f) = \varprojlim(CR(f), f)$). Similar properties are also covered for nonwandering points, recurrent points, ω-limit points and almost periodic points.

Let X be a compact metric space and let $f : X \to X$ be continuous. The inverse limit space of the sequence

$$X \xleftarrow{f} X \xleftarrow{f} X \xleftarrow{f} \dots$$

is defined to be the set of points $\underline{x} = (x_0, x_1, x_2, \dots)$ satisfying $f(x_{i+1}) = x_i$ with the metric

$$\underline{d}(\underline{x}, \underline{y}) = \sum_{i=0}^{\infty} \frac{d(x_i, y_i)}{2^i}$$

where d is a metric on X. Let $l(X, f)$ denote the inverse limit space. The shift map $\sigma_f : l(X, f) \to l(X, f)$ is defined by $\sigma_f((x_0, x_1, \dots)) = (f(x_0), x_0, x_1, \dots)$. Obviously, σ_f is a homeomorphism and $\sigma_f^{-1}((x_0, x_1, x_2, \dots)) = (x_1, x_2, \dots)$. Without causing confusion, we denote σ_f by σ simply. The projection maps $\pi_i : l(X, f) \to X$ are defined by $\pi_i((x_0, x_1, \dots, x_i, \dots)) = x_i$ for $i = 0$, 1, $\dots$. They are continuous.

One can study the dynamical connections between f and σ_f. For example, Bowen [Bo] has proved that the topological entropy of f equals the topological entropy of σ_f. Then f has positive topological entropy if and only if σ_f has positive toplogical entropy. Block [Bl] has shown that $\overline{Per(\sigma_f)} = l(\overline{Per(f)}, f)$ where $Per(f)$ is the collection of periodic points and $\overline{Per(f)}$ is the closure of $Per(f)$. In this survey, we review more properties like these. Most of them are recent results. As an application, we obtain some theorems for the shift maps induced by continuous maps on the interval.

1. Sets of recurrence.

A point $x \in I$ is called an ω-limit point if there is a $y \in I$ such that x is an accumulation point of $\{f^n(y)\}$. Denote the set of ω-limit points of x by $\omega(x, f)$ or $\omega(x)$. The collection of all ω-limit points is denoted by $\Lambda(f)$. A point x is called a chain recurrent point if for any number $\epsilon > 0$ there exists a finite sequence of points, $x_0 = x, x_1, \ldots, x_{n-1}, x_n = x$, such that $|f(x_i) - x_{i+1}| < \epsilon$ for $i = 0, 1, \ldots, n-1$. The collection of chain recurrent points is denoted by $CR(f)$ or CR. A point x is called a nonwandering point if for any neighborhood V of x there is an integer $n > 0$ such that $f^n(V) \cap V \neq \phi$. The collection of nonwandering points is denoted by $\Omega(f)$ or Ω. Ω is a closed set. A point is called a *recurrent point* if it is an ω-limit point of itself. The collection of recurrent points is denoted by $R(f)$ or R. A point x is called an *almost periodic point* if, for any neighborhood U, $\{n \geq 0; \ f^n(x) \in U\}$ has bounded gaps, i.e. if we write $\{n \geq 0; \ f^n(x) \in U\}$ as $\{0 = n_0 < n_1 < n_2 < \ldots\}$, then $\{n_{k+1} - n_k\}$ is bounded. Denote the collection of almost periodic points by $AP(f)$ or AP.

For better understanding of these sets, see Chapter 1 in [Li3]. Similar to the property $\overline{Per(\sigma_f)} = l(\overline{Per(f)}, f)$, in [Li2], the second author has proved the following property.

THEOREM 1.1. *Let* $f : X \to X$ *be continuous. Properties* $1) - 7)$ *hold.*

1) *Let* $x \in X$ *and let* $\underline{x} \in l(X, f)$ *satisfy* $\underline{x} = (x, x_1, x_2, \ldots)$. *Then* $\omega(\underline{x}, \sigma) = l(\omega(x, f), f)$.

2) $R(\sigma) = l(R(f), f)$.

3) $AP(\sigma) = l(AP(f), f)$.

4) $CR(\sigma) = l(CR(f), f)$.

5) *When f is onto, $\Omega(\sigma) = l(\Omega(f), f)$.*

6) *If f is onto and $\Lambda(\sigma)$ is closed, then $\Lambda(\sigma) = l(\Lambda(f), f)$.*

7) *The condition "f is onto" in 5) and 6) is not necessary if $X = [0, 1]$. And $\Lambda(\sigma) = \Omega(\sigma)$ when $\Lambda(\sigma)$ is closed.*

The following question is open.

QUESTION 1.2. *Given a continuous, onto map $f : X \to X$ with $\Lambda(f)$ closed, is $\Lambda(\sigma)$ closed?*

Note that if $\Lambda(\sigma) = l(\Lambda(f), f)$ and $\Lambda(f)$ is closed, then $\Lambda(\sigma)$ is closed. But we are not able to prove $\Lambda(\sigma) = l(\Lambda(f), f)$ without the assumption that $\Lambda(\sigma)$ is also closed, see [Li2]. If $X = [0, 1]$, then $\Lambda(f)$ is closed [Sh]. It is not known if $\Lambda(\sigma)$ is closed.

Define $\Lambda^n(f) = \Lambda(f|_{\Lambda^{n-1}(f)})$, i.e., $x \in \Lambda^n(f)$ if and only if there is a $y \in \Lambda^{n-1}(f)$ such that $x \in \omega(y, f)$. Then we have the following corollary.

COROLLARY 1.3. *If f is onto, and both $\Lambda^i(f)$ $(0 < i < n)$ and $\Lambda(\sigma_f)$ are closed, then $\Lambda^n(\sigma_f) = l(\Lambda^n(f), f)$.*

PROOF: First we prove that $\Lambda^2(\sigma_f) = l(\Lambda^2(f), f)$. Let $\underline{x} \in \Lambda^2(\sigma_f)$. Then there is a point $\underline{y} \in \Lambda(\sigma_f)$ such that $\underline{x} \in \omega(\underline{y}, \sigma_f)$. It is obvious that $\omega(\pi_0(\underline{y}), f) = \omega(\pi_i(\underline{y}), f)$. By 1) of Theorem 1.1, for each $i \geq 0$, $\pi_i(\omega(\underline{y}, \sigma_f)) = \omega(\pi_0(\underline{y}), f)$. Then $\pi_i(\omega(\underline{y}, \sigma_f)) = \omega(\pi_i(\underline{y}), f)$. Hence $\pi_i(\underline{x}) \in \omega(\pi_i(\underline{y}), f)$. Since $\underline{y} \in \Lambda(\sigma_f)$, by 6) of Theorem 1.1, $\pi_i(\underline{y}) \in \Lambda(f)$. Thus $\pi_i(\underline{x}) \in \Lambda^2(f)$ for all $i \geq 0$. Therefore $\underline{x} \in l(\Lambda^2(f), f)$.

On the other hand, we suppose that $\underline{x} \in l(\Lambda^2(f), f)$. Then $\pi_i(\underline{x}) \in \Lambda^2(f)$ for all $i \geq 0$. There is a point $y \in \Lambda(f)$ such that $\pi_i(\underline{x}) \in \omega(y, f)$. Since f maps $\Lambda(f)$ onto itself, there is a point $\underline{y} \in l(\Lambda(f), f) = \Lambda(\sigma_f)$ with $\pi_0(\underline{y}) = y$. Then $\pi_i(\underline{x}) \in \omega(y, f) = \omega(\pi_o(\underline{y}), f) = \omega(\pi_i(\underline{y}), f) = \pi_i(\omega(\underline{y}, \sigma_f))$ for all $i \geq 0$. Therefore $\underline{x} \in \omega(\underline{y}, \sigma_f) \subset \Lambda^2(\sigma_f)$.

By 6) of Theorem 1.1, $\Lambda(\sigma_f) = l(\Lambda(f), f)$. Assume $\Lambda^{n-1}(\sigma_f) = l(\Lambda^{n-1}(f), f|_{\Lambda^{n-1}(f)})$. By assumption, $\Lambda^{n-1}(f)$ is closed. Then $\Lambda^{n-1}(\sigma_f)$ is closed. Note that $f|_{\Lambda^{n-1}(f)}$ is an onto map. By repeating the above proof, we can see that $\Lambda^n(\sigma_f)$ is closed. $\qquad\square$

In the rest of this section, we assume that f is a continuous self-mapping defined on the interval $[0,1]$. It is known [Xiong1] that

$$P(f) \subset R(f) \subset \Lambda^n(f) = \cdots = \Lambda^2(f) \subset \overline{P(f)} \subset \Lambda(f) \subset \Omega(f) \subset CR(f).$$

Then by Theorem 1.1, we have

THEOREM 1.4. *The following inclusions hold.*

$$P(\sigma_f) \subset R(\sigma_f) \subset \Lambda^n(\sigma_f) = \cdots = \Lambda^2(\sigma_f) \subset \overline{P(\sigma_f)} \subset \Lambda(\sigma_f) \subset \Omega(\sigma_f) \subset CR(\sigma_f).$$

REMARK 1.5. *It is known that* $\Omega(f) \setminus \Lambda(f)$ *and* $\Lambda(f) \setminus \overline{P(f)}$ *are countable [Xiong2]. Thus* $\Omega(\sigma_f) \setminus \overline{P(\sigma_f)}$ *is also countable.*

Similar to the property $\Lambda(\sigma_f) = \Omega(\sigma_f)$, one can ask if $\overline{P(\sigma_f)} = \Lambda(\sigma_f)$ too. The answer is no. The example constructed by Block and Coven in Example 1 of [BC] serves as a counterexample. Since the original example is lengthy we don't copy it. Example 1.6: The map $f : [-1, 2] \to [-1, 2]$ is the map constructed by Block and Coven [BC]. Then $-\frac{2}{3} \in \Lambda \setminus \overline{P}$. Let $x_0 = -\frac{2}{3}$. Since $f(\Lambda) = \Lambda$, there is a point $x_1 \in \Lambda$ such that $f(x_1) = x_0$. $x_1 \notin \overline{P}$ since $x_0 \notin \overline{P}$. Similarly we can find a sequence of points $x_i \in \Lambda \setminus \overline{P}$ for $i = 2, 3, \ldots$ such that $f(x_i) = x_{i-1}$. Let $\underline{x} = x_0 x_1 \ldots x_i \ldots$. Then $\underline{x} \in l(\Lambda(f), f)$, and $\underline{x} \notin l(\overline{P}, f) = \overline{P(\sigma)}$. We claim that $\Lambda(\sigma) \neq \overline{P(\sigma)}$. Otherwise $\Lambda(\sigma) = \overline{P(\sigma)}$, and thus $\Lambda(\sigma)$ is closed. By 7) of Theorem 1.1, $\Lambda(\sigma) = l(\Lambda(f), f)$. Then $\underline{x} \in \Lambda(\sigma)$. But $\underline{x} \notin \overline{P(\sigma)}$. This is a contradiction. Therefore $\Lambda(\sigma) \neq \overline{P(\sigma)}$.

In a survey given by Xiong [Xiong2], the following properties are equivalent. (1) $CR(f) = P(f)$; (2) $\Omega(f) = P(f)$; (3) $\Lambda(f) = P(f)$; (4) $\overline{P(f)} = P(f)$; (5) $R(f) = P(f)$; (6) $AP(f) = P(f)$; (7) for any $x \in I$, $\omega(x, f)$ is a periodic orbit. By Theorem 1.1, it is easy to check that the following theorem holds.

THEOREM 1.7. *The following are equivalent.*

(1) $CR(\sigma_f) = P(\sigma_f)$; *(2)* $\Omega(\sigma_f) = P(\sigma_f)$; *(3)* $\Lambda(\sigma_f) = P(\sigma_f)$; *(4)* $\overline{P(\sigma_f)} = P(\sigma_f)$; *(5)* $R(\sigma_f) = P(\sigma_f)$; *(6)* $AP(\sigma_f) = P(\sigma_f)$; *(7) for any* $\underline{x} \in l(I, f)$, $\omega(\underline{x}, \sigma_f)$ *is a periodic orbit.*

2. Shadowing properties.

Given $\delta > 0$, say a sequence $x_0, \ldots, x_n, \ldots$ of points in X is a *δ-pseudo orbit* of f if $d(x_{n+1}, f(x_n)) \leq \delta$ for every $n = 0, 1, \ldots$. For a given positive number $\epsilon > 0$, say a sequence $\{x_0, \ldots, x_n, \ldots\}$ of points in X is ϵ-*shadowed* under f provided that there exists $y \in X$ satisfying $d(x_n, f^n(y)) \leq \epsilon$ for all $n \geq 0$.

DEFINITION 2.1[Ano]. *We say f has the shadowing property if for any $\epsilon > 0$, there exists $\delta > 0$ such that every δ-pseudo orbit is ϵ-shadowed.*

THEOREM 2.2 [CL]. *f has the shadowing property if and only if σ_f has the shadowing property.*

Let X be a continuum. We say that X is *chainable* if , for any $\epsilon > 0$, there are finitely many open sets with diameter smaller than ϵ, say O_1, O_2, ..., O_n, which cover X and $O_i \cap O_j \neq \phi$ iff $|i - j| \leq 1$. X is also called *snakelike*. We say that X is *decomposable* if X can be written as the union of two proper subcontinua. If X is not decomposable, then X is *indecomposable*. X is *hereditarily indecomposable* if each of its subcontinua is indecomposable. The *pseudo-arc* is a continuum which is homeomorphic to a nondegenerate hereditarily indecomposable chainable continuum.

By using Theorem 2.2 and a known example of Henderson [Hen], we obtain a homeomorphism on the pseudo-arc possesing the shadowing property (for details, see [CL]). This homeomorphism has zero topological entropy since the map given by Henderson has zero entropy. It is an open question if there is a homeomorphism of positive topological entropy on the pseudo-arc with shadowing property. We conjecture that the answer is yes. Minc and Transue [MT] have proved that the inverse limit spaces of certain topologically transitive maps on the interval are pseudo-arcs. If one can construct such transitive maps possessing shadowing property, then the respective shift maps are what we want by Theorem 2.2 and the fact that transitive maps on the interval have positive entropy. For the shadowing properties of continuous maps on the interval with positive entropy, see [Ch].

3. Chaos.

A continuous map $f : X \to X$ is called *topologically transitive* if for any open sets U and V, there exists an integer $n > 0$ such that $f^n(U) \cap V \neq \emptyset$. We say f has *sensitive dependence on initial conditions* if there exists $\delta > 0$ such that, for any $x \in X$ and any neighborhood N of x, there exist a point $y \in N$ and an integer $n \geq 0$ such that $d(f^n(x), f^n(y)) > \delta$.

DEFINITION 3.1. *Let X be a metric space. Let $f : X \to X$ be a continuous map. f is said to be chaotic in X in the sense of Devaney if there is a closed invariant set $D \subset X$ such that the following conditions are satisfied.*

 1. *$f|_D$ is topologically transitive.*

 2. *$f|_D$ has sensitive dependence on initial conditions.*

 3. *The periodic points of f in D are dense in D.*

The set D is called a chaotic set.

REMARK 3.2. *It has been recently known, see [BBCDS] and [Li, 1990], that if a continuous map on a compact metric space has transitivity and a dense set of periodic points, it is sensitively dependent on initial conditions. Therefore it is sufficient to have the conditions 1 and 3 in the above definition.*

THEOREM 3.3 [Li2]. *f is chaotic if and only if σ_f is chaotic.*

Note that if a continuous map on the interval is transitive, then the periodic points of f are dense in $[0,1]$. Minc and Transue [MT] have shown that there are transitive maps on the interval whose inverse limit spaces are the pseudo-arc. Then by the above theorem and remark, we obtain a chaotic homeomorphism on the pseudo-arc (see [Li2] for details). Kennedy [Ken] has also constructed a chaotic homeomorphism on the pseudo-arc.

DEFINITION 3.4 [Li1]. *We say that S is an ω-scrambled set if, for any $x, y \in S$ with $x \neq y$,*

 1) *$\omega(x, f) \setminus \omega(y, f)$ is uncountable;*

 2) *$\omega(x, f) \cap \omega(y, f)$ is nonempty;*

 3) *$\omega(x, f)$ is not contained in the set of periodic points.*

We say that f is ω-chaotic if there exists an ω-scrambled set.

THEOREM 3.5 [Li2]. *f is ω-chaotic if and only if σ_f is ω-chaotic.*

REMARK 3.6. *Let $f : [0,1] \to [0,1]$ be a continuous map. the second author [Li1] has shown that the following statements are equivalent*

1) f has positive topological entropy.

2) f is chaotic in the sense of Devaney.

3) f is ω-chaotic.

It is also known [Bl, Mis] that 1) is equivalent to the following statement

4) f has a periodic point of period $m2^n$ with $m > 1$ odd and $n \geq 0$.

By Theorem 3.5, it is not hard to see that if we replace f by σ_f, the above statements are also equivalent.

4. Ergodicity and hyperbolicity.

Let $(X, \mathcal{B}, \mu)$ be a probability space. The set $\{\pi^{-1}(B) : B \in \mathcal{B}\}$ generates a σ-algebra on $l(X, f)$. Define $\nu = \mu\pi$. Then ν is a probability measure on $l(X, f)$. Brown [Br1, Br2] has proved the following theorem.

THEOREM 4.1. *f is ergodic (strongly mixing, weakly mixing) if and only if σ_f is ergodic (strongly mixing, weakly mixing).*

Next let us consider hyperbolic canonical coordinates for self-covering mappings. Let M be a compact connected Riemannian manifold. For $f \in C^0(M, M)$ and a compact invariant set $\Lambda \subseteq M$ under f, define

$$\Lambda^f = \{\bar{x} : \ \bar{x} = \{x_i\}_{-\infty}^{+\infty}, x_i \in \Lambda, f(x_i) = x_{i+1}, i \in Z\}$$

and

$$\bar{d}(\bar{x}, \bar{y}) = \sum_{-\infty}^{+\infty} \frac{1}{2^{|i|}} d(x_i, y_i).$$

Then $(\Lambda^f, \bar{d})$ becomes a metric space.

Let $q_f : \Lambda^f \to \Lambda^f$ be the left shift, and $Cov^r(M)(r \geq 0)$ be the space of C^r self-covering maps. Then q_f is a homeomorphism.

Let $f \in Cov^0(M)$. For $\epsilon > 0$ and $\overline{x} \in \Lambda^f$, denote

$$W_\epsilon^s(\overline{x}, f) = \{y_0 \in M | d(f^n(x_0), f^n(y_0)) < \epsilon, \text{ for all } n \in Z^+\}$$

and

$$W_\epsilon^u(\overline{x}, f) = \{z_0 \in M | \exists \{z_{-n}\}_{n=0}^{+\infty} \subset M \text{ such that } f(z_{-n-1}) = z_{-n} \text{ and}$$

$$d(z_{-n}, x_{-n}) < \epsilon \text{ for all } n \in Z^+\}.$$

If for any $\epsilon > 0$ and $f \in Cov^0(M)$, there exists a $\delta(\epsilon) > 0$ such that, for any $\overline{x}, \overline{y} \in \Lambda^f$, if $\overline{d}(\overline{x}, \overline{y}) < \delta(\epsilon)$, we have $W_\epsilon^s(\overline{x}, f) \cap W_\epsilon^u(\overline{y}, f) \neq \phi$, then we say that f is of *canonical coordinates on* Λ. We say f is of *hyperbolic canonical coordinates* if furthermore $\exists \epsilon^* > 0, 0 < \lambda < 1$ and $c \geq 1$, for any $\overline{x} \in \Lambda^f$ with $p(\overline{x}) = x_0$, and for $y_0 \in W_{\epsilon^*}^s(\overline{x}, f)$, we have

$$d(f^n(x_0), f^n(y_0)) \leq c\lambda^n d(x_0, y_0)$$

for all $n \in Z^+$; Also if $y_0 \in W_{\epsilon^*}^u(\overline{x}, f)$, then $\exists \{y_{-n}\}_{n=0}^{+\infty}$ such that $f(y_{-n-1}) = y_n$ and $d(x_{-n}, y_{-n}) \leq c\lambda^n d(x_0, y_0)$ for all $n \in Z^+$.

Niu [Niu] has proved the following property.

THEOREM 4.2. *If f has hyperbolic canonical coordinates, then so is q_f.*

Note that q_f is conjugate to σ_f. Therefore if f has hyperbolic canonical coordinates, so is σ_f.

REFERENCES

[Ano] Anosov, D.V., *Geodesic flows on closed Riemannian manifolds with negative curvature,* Tr. Institute of math. AN SSSR **90** (1967).

[BBCDS] Banks, J., Brooks, J., Cairns, G., Davis, G., and Stacey, P., *On Devaney's Definition of Chaos,* preprint (1990).

[BC] L. Block and E. M. Coven, *ω-limit sets for maps of the interval,* Ergod. Th. Dynam. Sys. **6** (1986), 335-344.

[Bl] Block, L., *Homoclinic points of mappings of the intervals,* Proc. Amer. Math. Soc. **72** (1978), 576-580.

[Bo] Bowen, R., *Topological entroy and Axiom A,* "Global Analysis," Amer. Math. Soc., Providence, 1970, pp. 23-41.

[Br1] Brown, J.R., *Inverse limits, entropy and weak isomorphism for discrete dynamical systems,* Trans Amer. Math. Soc. **164** (1972), 55-66.

[Br2] Brown, J.R., "Ergodic Theory and Topological Dynamics," Academic Press, Inc., New York, 1976.

[Ch] Chen, L., *Linking and the shadowing property for piecewise monotone maps*, Proc. Amer. Math. Soc. **113** (1991), 251-263.

[CL] Chen, L. and Li, S.H., *Shadowing Properties on the Inverse Limit Spaces*, Proc. Amer. Math. Soc. **115** (1992) (to appear).

[Ken] Kennedy, J., *The Construction of Chaotic Homeomorphisms on Chainable Continua*, Trans. Amer. Math. Soc. (to appear).

[Li1] Li, S.-H., *ω-chaos and topological entropy*, Trans. Amer. Math. Soc. (1992) (to appear).

[Li2] Li, S.-H., *Dynamical Properties of the Shift Maps on the Inverse Limit Spaces*, Ergod. Th. Dynam. Sys. **12** (1992) (to appear).

[Li3] Li, S.-H., *Recurrence, Chaos and Inverse Limit Space*, Dissertation (1991).

[MT] P. Minc and W. R. R. Transue, *A Transitive Map on* [0, 1] *Whose Inverse Limit is the Pseudo-arc,*, Preprint.

[Mis] Misiurewicz, M., *Horseshoes for continuous mappings of an interval*, Bull.Acad.Pol.Sci., **27** (1979), 167-169.

[Niu] Niu, D-X, *Self-covering mappings with hyperbolic canonical coordinates. (Chinese).*, J. Math. Res. Exposition **9** (1989), 423-426. MR: 91a: 58098.

[Sh] Sharkovskii, A. N., *On some properties of discrete dynamical systems*, "Proc. Interat. Colloq. on Iterat. Theory and its Appl.," Univ. Paul Sabatier, Toulouse, 1982, pp. 153-158.

[Xiong1] J.-C. Xiong, *The attracting center of a continuous self-map of the interval*, Ergod. Theor. and Dynam. Syst. **8** (1988), 205-213.

[Xiong2] J.-C. Xiong, *Dynamics for maps on the interval: nonwandering set, topological entropy and chaos (survey)*, preprint (1986).

Horseshoelike Mappings and Chainability

JAMES F. DAVIS   West Virginia University, Morgantown, West Virginia

Abstract. In this paper we give a geometric description of a horseshoe-type mapping of a certain plane disk into itself whose attracting set is not chainable. In contrast, the attracting set of the Smale Horseshoe Map constructed by the standard geometric description is a chainable continuum.

1. Introduction. The Smale horseshoe map, first described by Stephen Smale in [6], is a homeomorphism, F, which maps a square disk in the plane, D, over itself in the manner shown in left side of Figure 1. In order to study the iterates of this mapping it must be extended to a domain D' which includes $f(D')$. A geometric approach to doing this is to place "caps", A and B, on the ends of D and extend the map F so that it maps A and B into the interior of A as shown in the right side of Figure 1. If this extension is done in a certain manner, the attracting set of F, the intersection of all forward images, under F, of $A \cup D \cup B$, is Knaster's indecomposable continuum with one end point (see the paper of M. Barge [1]).

If the extension of F to A and B is done in a slightly different manner, the attracting set is not homeomorphic with the Knaster continuum. We now outline what we mean by a "different manner." Choose a vertical section, S_B, of constant width running from the top to the bottom of B, and running *between* the two arms of the image of $A \cup D \cup B$ under the

---

Presented at the Chico Topology Conference, Chico CA, June 1989, and in poster session at the AMS-IMS-SIAM Joint Summer Research Conference on Continuum Theory and Dynamical Systems, June 1989.

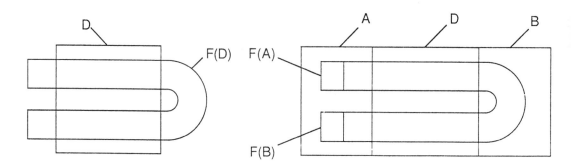

Figure 1

second iterate of F, $F^2(A \cup D \cup B)$. Choose a vertical section, S_A, of the same constant width as S_B, running from top to bottom in $A - F(D)$ (see Figures 2 and 4). Extend F to A so that $F(A) \subset \text{int}(A)$, S_A runs through $F(\text{int}(A))$, and so that F maps S_A onto $F(A) \cap S_A$, preserving horizontal distances in S_A. Extend F to B so that $F(B) \subset \text{int}(A)$, S_A runs through $F(\text{int}(B))$, and so that F maps S_B onto $F(B) \cap S_A$, preserving horizontal distances in S_B (see Figure 2). The attracting set of the mapping so defined is shown in Figure 3. This continuum is not homeomorphic with the Knaster continuum, but is homeomorphic with a continuum constructed by W. T. Ingram and this author in [5].

The purpose of this paper is to give a careful definition of this modified horseshoe map, and to construct a homeomorphism between the attracting set of the map and the example in [5]. In doing so we will produce a homeomorphism between the attracting set of the map F and an inverse limit of simple triods and a homeomorphism between that inverse limit and the one defined in [5]. The first of these homeomorphisms is a conjugacy between the restriction of F to its attracting set and the shift map on the inverse limit. The techniques we use to construct this homeomorphism are those used by Barge in [1].

2. Basic definitions and constructions. All spaces considered in this paper are metric. A *continuum* is a compact connected space and

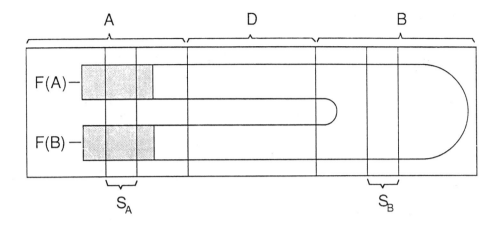

Figure 2

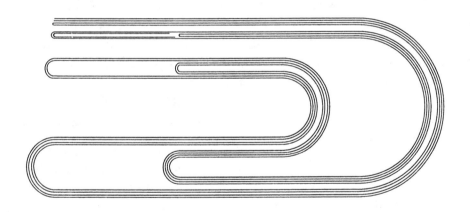

Figure 3

The Attracting Set of F

a *mapping* is a continuous function. Suppose that X and Y are spaces, and $f:X \to X$ and $g:Y \to Y$ are mappings. The mappings f and g are *conjugate* (respectively *semiconjugate*) provided there is a homeomorphism (resp. mapping) $h:X \to Y$ such that $gh = hf$. The homeomorphism (resp.

mapping) h is said to be a *conjugacy* (resp. *semiconjugacy*) between f and g.

If X and Y are spaces and $f: X \to X$ and $p: X \to Y$ are mappings, f *respects the fibers* of p provided $f(p^{-1}(p(x))) \subset p^{-1}(p(f(x))$ for all $x \in X$. If f and p are such mappings, then there is an *induced mapping* $f^*: Y \to Y$, defined by $f^*(y) = p(f(p^{-1}(y)))$, such that $pf = f^*p$ (see [3, Theorem VI.4.3, p. 126]).

An *inverse sequence* is a sequence of spaces $X_1, X_2, X_3, \ldots$, called the *factor spaces*, together with mappings $f_i: X_{i+1} \to X_i$ called the *bonding maps*. The *inverse limit*, $\lim\{X_i, f_i\}$, of this inverse sequence is the subspace of all points $(x_1, x_2, x_3, \ldots) \in \prod_{i>0} X_i$ such that $f_i(x_{i+1}) = x_i$ for all i. Of special interest in this paper are inverse limits $\lim\{X, f\}$ with a single factor space, X, and a single bonding map $f: X \to X$. Letting $Y = \lim\{X, f\}$ be such an inverse limit, we define the *right shift map*, $\hat{f}: Y \to Y$ by $\hat{f}(x_1, x_2, \ldots) = (f(x_1), x_1, x_2, \ldots)$. The mapping $\hat{f}$ is a homeomorphism.

In [1] Barge defines a horseshoe map on a disk as a homeomorphism which respects the fibers of a certain projection to the unit interval and which has several other properties (see Definition 1 below). He obtains a conjugacy between the horseshoe map on its attracting set and the inverse limit on the unit interval with the map induced by the horseshoe map via the projection as a bonding map. The theorem in which he produces this conjugacy ([1, Theorem 1]) is true in a more general setting, and the proof given by Barge holds in that more general setting.

Theorem 1. *(Barge [1]) Suppose that X and Y are continua, $g: X \to Y$ is a mapping of X onto Y, and $F: X \to X$ is a homeomorphism of X into itself which respects the fibers of g. Let $f: Y \to Y$ be the mapping of Y induced by F (i.e. $f = F^*$) and let $Z = \bigcap_{n>0} F^n(X)$. The mapping $\tilde{g}: Z \to \lim\{Y, f\}$ given by*

$$\tilde{g}(z) = (g(z), g(F^{-1}(z)), g(F^{-2}(z)), \ldots)$$

is a semi-conjugacy between $F|Z$ and $\hat{f}$, the right shift map on $\lim\{Y, f\}$. Moreover, if $\operatorname{diam}(F^n(g^{-1}(y))) \to 0$ uniformly for $y \in Y$ as $n \to \infty$, then $\tilde{g}$ is a conjugacy.

3. Horseshoe maps. We adopt notation similar to that used in [1]. Let $I = [-1, 1]$, $\alpha = [-3, -1]$, and $\beta = [1, 3]$. Define subsets of $R \times R$ by

$A = \alpha \times I$, $B = \beta \times I$, $C = I \times I$, and $D_0 = A \cup C \cup B$. Define $P: D_0 \to I$ by

$$P(x, y) = \begin{cases} -1, & \text{if } (x, y) \in A, \\ x, & \text{if } (x, y) \in C, \\ 1, & \text{if } (x, y) \in B. \end{cases}$$

Definition 1. A homeomorphism $F: D_0 \to D_0$ is an *H-map* provided
(1) F respects the fibers of P,
(2) $F(A) \subset \text{int}(A)$; $F(B) \subset \text{int}(A)$, and
(3) $P^{-1}(x) \cap F(D_0)) \cap (C \cup A)$ has exactly 2 components for all $x \in I$.
If, in addition, F satisfies
(4) $\text{diam}(F^n(P^{-1}(P(z)))) \to 0$ uniformly in z as $n \to \infty$,
then F is called a *(2-fold) horseshoe map* (see [1, p. 29]).

Suppose that $F: D_0 \to D_0$ is an H-map. If n is a positive integer, let $D_n = F^n(D_0)$. We define the *attracting set* of F as $\Lambda = \bigcap_{n>0} D_n$. For $z \in D_1$ let $V(z)$ be the component of $P^{-1}(P(z)) \cap D_1$ which contains z.

We will describe the modifications to be made to the horseshoe map by starting with an H-map. Propositions 1 and 2 below give facts which we need to know about H-maps in order to make these modifications. Both of these propositions follow easily from the geometry of H-maps and their proofs are left to the reader. Before we state the propositions we establish some notation: If $F: D_0 \to D_0$ is an H-map and $z \in D_1$ define $V(z)$ to be the component of $P^{-1}(P(z)) \cap D_1$ which contains z.

Proposition 1. *Suppose that $F: D_0 \to D_0$ is an H-map.*
(a) If $z_1, z_2 \in D_0$, $P(z_1) < P(z_2)$ and $V(F(z_1)) = V(F(z_2))$ then

$$F(P^{-1}([P(z_1), P(z_2)])) \subset V(F(z_1)).$$

(b) If $z \in D_0$ and $F(z) \in D_1 - (A \cup B)$ then $F(P^{-1}(P(z))) = V(F(z))$.
(c) There is a subinterval $\gamma \subset \text{int}(I)$ such that $F^{-1}(B) = P^{-1}(\gamma) = \gamma \times I$.
(d) There are just two components of $D_1 \cap (A \cup C)$, one containing $F(A)$, the other containing $F(B)$.

Denote by L_A and L_B be the components of $(A \cup C) \cap D_1$, guaranteed by Proposition 1(d), which contain, respectively, $F(A)$ and $F(B)$. Denote the restrictions of P to L_A and L_B by P_A and P_B, respectively.

We note here that the "fiber component function" V suffers from the following defect when restricted to $L_A \cup L_B$: If $P(z) = 1$ and $z \in L_A$ then $(V|L_A \cup L_B)(z) \not\subset L_A$ and has two components. A similar problem arises if $z \in L_B$. To remedy this, for $z \in L_A \cup L_B$ $(= (A \cup C) \cap D_1)$, define

$$V'(z) = \begin{cases} V(z) \cap L_A, & \text{if } z \in L_A, \\ V(z) \cap L_B, & \text{if } z \in L_B. \end{cases}$$

It is easily seen that $V'(z) = V(z)$ if $P(z) < 1$ and that, for all $z \in L_A \cup L_B$, $V'(z)$ is the component of $P^{-1}(P(z)) \cap D_1 \cap (A \cup C)$ which contains z. The following analogue of Proposition 1(b) holds for V'.

Proposition 2. *If* $F \colon D_0 \to D_0$ *is an H-map then* $P_A^{-1}(P(z)) = V'(z)$ *for* $z \in L_A$, *and* $P_B^{-1}(P(z)) = V'(z)$ *for* $z \in L_B$.

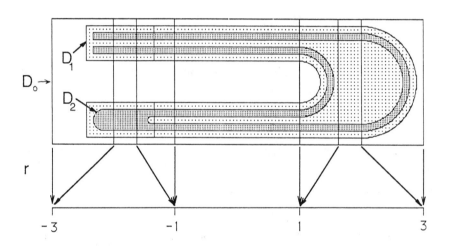

Figure 4.

4. Modifications of horseshoe maps. Let $\beta_0 = [\mu, \nu]$ be a subinterval of $\text{int}(\beta)$. Let $\beta_1 = [1, \mu]$, $\beta_3 = [\nu, 3]$, $\alpha_0 = [-\nu, -\mu]$, $\alpha_{-1} = [-\mu, -1]$ and $\alpha_{-3} = [-3, -\nu]$. For $i = 0, 1, 3$ let $B_i = \beta_i \times I$ and $A_{-i} = \alpha_{-i} \times I$. Define $r \colon D_0 \to \alpha \cup I \cup \beta$ to be the mapping such that

$$r(x, y) = \begin{cases} P(x, y) = x, & \text{if } (x, y) \in C, \\ i, & \text{if } (x, y) \in A_i \text{ for } i \in \{-1, -3\}, \\ i, & \text{if } (x, y) \in B_i \text{ for } i \in \{1, 3\}, \end{cases}$$

and such that r maps B_0 linearly onto β and A_0 linearly onto α (see Figure 4).

Definition 2. Suppose $F: D_0 \to D_0$ is an H-map. Let γ, L_A, and L_B be those sets guaranteed by Proposition 1(a) and (b). Let r_A and r_B be the restrictions of r to L_A and L_B respectively, and let $C_0 = \gamma \times I$. Define F to be an *elongated horseshoe map* provided that, in addition to satisfying conditions (1)–(3) of Definition 1, F also satisfies the following:

(5) $F(r^{-1}(r(z))) \subset r^{-1}(r(F(z)))$ for $z \notin C_0$;

(6) $F(r_A^{-1}(r(z))) \subset r^{-1}(r(F(z)))$ for $z \in C_0 \cap L_A$;

(7) $F(C_0 \cap L_B) \subset B_1$;

(8) $F(A_0) \subset A_0$; $F(A_{-1}) \subset A_{-1}$; $F(A_{-3}) \subset A_{-3}$; $F(B_0) \subset A_0$; $F(B_1) \subset A_{-1}$; $F(B_3) \subset A_{-3}$;

(9) For $x \in \alpha \cup I$, $(r^{-1}(x) \cap D_1) \cap (A \cup C)$ has just two components;

(10) For $x \in \beta$, $r^{-1}(x) \cap D_2 \cap B_0$ has just two components;

(11) $\mathrm{diam}(F^n(r^{-1}(r(z)))) \to 0$ uniformly in $z \in \mathrm{cl}(D_0 - C_0)$ as $n \to \infty$.

An illustration of the embedding of D_2 and D_1 in D_0 for such a map is given in Figure 4. For $z \in D_1 \cap (A \cup C)$ let $V_r(z)$ denote the component of $r^{-1}(r(z)) \cap D_1 \cap (A \cup C)$ which contains z. The analog to Proposition 2 holds for the projection r.

Proposition 3. *If $z \in L_A$ then $r_A^{-1}(r(z)) = V_r(z)$ and if $z \in L_B$ then $r_B^{-1}(r(z)) = V_r(z)$.*

Proof. First observe that each of L_A and L_B is a continuum which intersects $P^{-1}(1)$. Since $L_A \subset A \cup C$, we have $P^{-1}(1) \cap L_A \subset (A \cup C) \cap B \subset r^{-1}(1)$. Thus $L_A \cap r^{-1}(1) \neq \emptyset$. Similarly, $L_B \cap r^{-1}(1) \neq \emptyset$. From (8) and the definitions of L_A and L_B, it follows that L_A and L_B each intersect $r^{-1}(-3)$.

Suppose $z \in L_A$. To prove that $V_r(z) \subset r_A^{-1}(r(z))$ note that $V_r(z) \subset L_A$ and if $w \in V_r(z)$, then $r(z) = r(w)$. Thus, if $w \in V_r(z)$ then $w \in r_A^{-1}(r(z))$.

Before proving that $r_A^{-1}(r(z)) \subset V_r(z)$, we observe that there is a point $y \in L_B$ such that $r(y) = r(z)$. If $r(z) = -3$ or $r(z) = 1$, we have seen above that the point y exists. If $-3 < r(z) < 1$ the point y must exist, since, otherwise, $r^{-1}(r(z))$ would separate L_B. Now $V_r(y) \subset L_B$, so $V_r(y) \cap L_A = \emptyset$, and, thus $V_r(y) \cap V_r(z) = \emptyset$. Therefore, the two components of $r^{-1}(r(z))$ are $V_r(z)$ and $V_r(y)$.

Now suppose $w \in r_A^{-1}(r(z))$. Then $r(w) = r(z)$ and $w \in L_A$. Thus $V_r(w) \subseteq L_A$. Since $V_r(w)$ is a component of $r^{-1}(r(w)) = r^{-1}(r(z))$, either $V_r(w) = V_r(z)$ or $V_r(w) = V_r(y)$. The latter cannot hold since $V_r(y) \cap L_A = \emptyset$. Therefore, $w \in V_r(z)$.

By a similar argument we have that $r_B^{-1}(r(z)) = V_r(z)$ for $z \in L_B$. ∎

Let T be the simple triod in the plane defined by

$$T = \{ (0, y) \mid 0 \le y \le 2 \} \cup \{ (x, 0) \mid -4 \le x \le 0 \} \cup \{ (x, 0) \mid 0 \le x \le 4 \}.$$

Define $g : D_1 \to T$ by

$$g(z) = \begin{cases} (0, r(z) - 1), & \text{if } z \in B, \\ (r(z) - 1, 0), & \text{if } z \in L_B, \\ (1 - r(z), 0), & \text{if } z \in L_A, \end{cases}$$

It is easily seen that g is continuous.

Proposition 4. *The map F respects the fibers of g.*

Proof. If z and z' are both in one of $B \cap D_1$, L_A, or L_B, it is easily seen that $g(z) = g(z')$ if and only if $r(z) = r(z')$. Consequently, if $z \in D_1$ then

$$g^{-1}(g(z)) = \begin{cases} r^{-1}(r(z)), & \text{if } z \in B, \\ r_B^{-1}(r(z)), & \text{if } z \in L_B, \\ r_A^{-1}(r(z)), & \text{if } z \in L_A, \end{cases}$$

If $z \in B$, $g^{-1}(g(z)) = r^{-1}(r(z))$ is clearly connected. If $z \in L_A \cup L_B$ then $g^{-1}(g(z))$ is connected by Proposition 3.

Suppose $z \in D_1$ and $z \notin C_0$. Then, by (5),

$$F(g^{-1}(g(z))) \subset F(r^{-1}(r(z))) \subset r^{-1}(r(F(z))).$$

Since $z \notin C_0, z \in L_A \cup L_B$. Suppose $z \in L_A$. Since $F(g^{-1}(g(z)))$ is connected and $F(z) \in F(g^{-1}(g(z)))$, we have $F(g^{-1}(g(z))) \subset L_A$. Since $r(F(g^{-1}(g(z)))) \subset r(F(z))$, we obtain

$$F(g^{-1}(g(z))) \subset r_A^{-1}(r(F(z))) = g^{-1}(g(F(z))).$$

If $z \in L_B$, a similar argument establishes $F(g^{-1}(g(z))) \subset g^{-1}(g(F(z)))$.

Now suppose $z \in D_1 \cap C_0$. Then $F(z) \in B$. If $z \in L_B$ then $g^{-1}(g(z)) = r_B^{-1}(r(z)) \subset L_B$ and, using (7), $F(g^{-1}(g(z))) \subset B_1 = r^{-1}(r(F(z)))$ (the last equality holds by the definition of r, since $F(z) \in B_1$). Thus $F(g^{-1}(g(z))) \subset g^{-1}(g(F(z)))$. If $z \in L_A$ then, by virtue of (6),

$$F(g^{-1}(g(z))) = F(r_A^{-1}(r(z))) \subset r^{-1}(r(F(z)) = g^{-1}(g(F(z))).$$

Hence, F respects the fibers of g. ∎

Proposition 5. *If F is an elongated H-map then* $\mathrm{diam}(F^n(g^{-1}(t)) \to 0$ *uniformly in $t \in T$ as $n \to \infty$.*

Proof. Suppose that $\epsilon > 0$. By (11) there is an integer N such that, if $N \leq n$ and $z \in \mathrm{cl}(D_0 - C_0)$, then $\mathrm{diam}(F^{n+1}(r^{-1}(r(z))) < \epsilon$. Suppose that $t \in T$ and $N + 1 \leq n$.

If $t = (0, s)$ with $0 \leq s \leq 2$ then $g^{-1}(t) = r^{-1}(1 + s)$. Choose $z \in r^{-1}(1 + s)$. Then $g^{-1}(t) = r^{-1}(r(z))$, and

$$\mathrm{diam}(F^n(g^{-1}(t))) = \mathrm{diam}(F^{n-1+1}(r^{-1}(r(z)))) < \epsilon.$$

If $t = (s, 0)$ with $-4 \leq s \leq 0$ then $g^{-1}(t) = r_A^{-1}(1 - s)$. Choose $z \in r_A^{-1}(1 - s)$. If $z \in \mathrm{cl}(D_0 - C_0)$, then $r_A^{-1}(1 - s) = r_A^{-1}(r(z)) \subset r^{-1}(r(z))$ and

$$\mathrm{diam}(F^n(g^{-1}(t))) = \mathrm{diam}(F^{n-1+1}(r_A^{-1}(r(z))))$$
$$\leq \mathrm{diam}(F^{n-1+1}(r(^{-1}(r(z))) < \epsilon.$$

Now suppose that $z \notin \mathrm{cl}(D_0 - C_0)$. By the choice of z, $z \in L_A$, hence $z \in D_1$. Let $y = F^{-1}(z)$. Since $F(C_0) \subset B$, $F(B) \subset A$, and $F(A) \subset A$, it follows that $y \in C - C_0$. Hence $r^{-1}(r(y)) = P^{-1}(P(y))$ and $F(r^{-1}(r(y))) = V(F(y)) = V(z) = P_A^{-1}(P(z)) = r_A^{-1}(r(y))$, the first of these equalities following from Proposition 1(d), and the third from Proposition 2. Thus

$$\mathrm{diam}(F^n(g^{-1}(t))) = \mathrm{diam}(F^n(r_A^{-1}(r(z)))$$
$$= \mathrm{diam}(F^n(V(z)))$$
$$= \mathrm{diam}(F^n(F(V(y))))$$
$$\leq \mathrm{diam}(F^{n+1}(r^{-1}(r(y))) < \epsilon.$$

The proof for the case $t = (0, s)$ with $0 \leq s \leq 4$ is identical to the foregoing one after replacing r_A by r_B and P_A by P_B. Thus, in any case $\text{diam}(F^n(g^{-1}(t))) < \epsilon$. ∎

Let $f: T \to T$ be the mapping induced by $F: D_1 \to D_1$ such that $fg = gF$. Let $X = \lim\{T, f\}$ and let $\hat{f}: X \to X$ be the right shift map on X. The next theorem follows directly from Theorem 1 and Propositions 3 and 4.

Theorem 2. *The maps $\hat{f}$ and F are conjugate on Λ. In particular the mapping $\tilde{g}: \Lambda \to \Lambda$ defined by*

$$\tilde{g}(z) = (g(z), g(F^{-1}(z)), g(F^{-2}(z)), \ldots)$$

is a homeomorphism from Λ onto X and $F\tilde{g} = \tilde{g}\hat{f}$.

Let $s: I \to I$ be the mapping such that $s([-1, -\frac{2}{3}]) = -1$, $s([-\frac{1}{3}, \frac{1}{3}]) = 1$, $s([\frac{2}{3}, 1]) = -1$, and which is linear on $[-\frac{2}{3}, -\frac{1}{3}]$ and $[\frac{1}{3}, \frac{2}{3}]$. The mapping s_2 of [1, p. 32] is conjugate to s, and $\lim\{I, s\}$ is homeomorphic with the Knaster continuum (see [4]).

Theorem 3. *There is a mapping $h: \Lambda \to \lim\{I, s\}$ which is a semiconjugacy between F and $\hat{s}$ (the right shift map on $\lim\{I, s\}$) on Λ.*

Proof. The map $F: D_1 \to D_1$ respects the fibers of the projection P of Definition 1 since it is an H-map. Let $t: I \to I$ be the mapping induced on I by F via the mapping P. Then, by Theorem 1, there is a semiconjugacy h_1 between $F|\Lambda$ and the shift map $\hat{t}$ on $\lim\{I, t\}$. Condition (2) of Definition 1 and Propositions 1(a) and (d) imply that there are numbers $-1 = a_0 < a_1 < a_2 < \ldots < a_5 = 1$ such that $t([a_0, a_1]) = t([a_4, a_5]) = -1$, $t([a_2, a_3]) = 1$, and such that t is a homeomorphism on $[a_1, a_2]$ and $[a_3, a_4]$. Thus, by [1,Theorem 2], there is a semiconjugacy $h_2: \lim\{I, t\} \to \lim\{I, s\}$ between $\hat{t}$ and $\hat{s}$. The mapping $h = h_2 h_1$ is therefore a semiconjugacy between F and $\hat{s}$ on Λ.

5. The non-chainability of Λ. Let Y denote the continuum constructed in [5] as an inverse limit. This continuum has positive span and thus is not chainable. We now prove that Λ is homeomorphic with Y. It should be noted, however, that this homeomorphism is not a conjugacy between $F|\Lambda$ and the shift map on Y.

Theorem 4. *The attracting set Λ is homeomorphic with Y and, hence, Λ is not chainable.*

Proof. If $v_1, v_2 \in T$, denote the unique arc in T with endpoints v_1 and v_2 by $\langle v_1, v_2 \rangle$. We now define simplicial subdivisions of T which are associated with the mapping f. Condition (10) implies that there exists a point $z_B \in B_3 \cap D_2 = r^{-1}(3) \cap D_2$. Let $w_A = F^{-1}(z_B)$. Note that $w_A \in L_A \cap C_o$ since $f(C_o \cap L_B) \subset B_1$. Recall that there are numbers $-1 < t_1 < t_2 < 1$ such that $C_o = [t_1, t_2] \times I$ where $C_o = F^{-1}(B)$. Because L_A is a continuum intersecting both A and B and each of $r^{-1}(t_1)$ and $r^{-1}(t_2)$ separate A and B in D_o, we have the existence of points w'_A, w''_A in $L_A \cap C_o$ such that $r(w'_A) = t_1$ and $r(w''_A) = t_2$. Similarly L_B is also a continuum intersecting both A and B and $r^{-1}(r(w_A))$ separates A and B in D_o, so there is a point $w_B \in L_B \cap C_o$ such that $r(w_B) = r(w_A)$. Let $z'_B = F(w_B)$, $z_A = F(z_B)$ and $z'_A = F(z'_B)$. Note the following:

$$z_A \in A_{-3}, F(z_A) \in A_{-3} \cap L_A; z'_A \in A_{-1}, F(z'_A) \in A_{-1} \cap L_A;$$
$$z_B \in B_3, F(z_B) \in A_{-3} \cap L_B; z'_B \in B_1, F(z'_B) \in A_{-1} \cap L_B.$$

These points are shown in Figure 5.

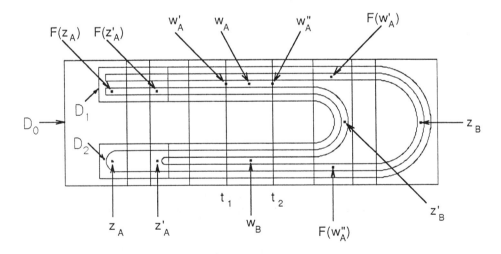

Figure 5.

Define points in T as follows: $a_4 = g(F(z_A)) = (4, 0)$, $a_2 = g(F(z'_A)) = (2, 0)$, $b_0 = g(z'_B) = (0, 0)$, $b_4 = g(F(z_B)) = g(z_A) = (-4, 0)$, $b_2 =$

$g(F(z'_B)) = g(z'_A) = (-2,0)$, and $c = g(z_B) = (0,2)$. Further let $b_1 = g(w_B)$, $a_1 = g(w_A)$, $a'_1 = g(w'_A)$, and $a''_1 = g(w''_A)$. Let

$$W = \{a_2, a_4, b_0, b_2, b_4, c\}, \text{ and}$$
$$V = W \cup \{a_1, a'_1, a''_1, b_1\}.$$

These points are shown in Figure 6.

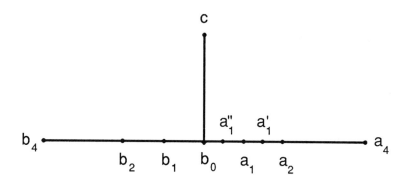

Figure 6.

Recall that $f: T \to T$ is the mapping induced, *via* g, by $F: D_1 \to D_1$. We make the following observations concerning the action of f on the points in V: Since $F(z_A) \in A_{-3}$ and $F(z'_A) \in A_{-1}$ we have $F^2(z_A) \in A_{-3} \cap L_A$ and $F^2(z'_A) \in A_{-1} \cap L_A$. Thus $f(a_4) = a_4$ and $f(a_2) = a_2$. Since $F^2(z_B) = F(z_A)$ and $F^2(z'_B) = F(z'_A)$, we have $f(b_4) = a_4$ and $f(b_2) = a_2$. Since $F(z_B) = z_A$ and $F(z'_B) = z'_A$, $f(c) = b_4$ and $f(b_0) = b_2$. We note that $F(w_A) = z_B$ implies $f(a_1) = c$ and that $F(w'_A)$, $F(w''_A) \in B_1$ implies $f(a'_1) = f(a''_1) = b_0$. Finally we have $f(b_1) = b_0$, because $F(w_B) = z'_B$.

The sets $\mathcal{V}$ and $\mathcal{W}$ are vertex sets of simplicial subdivisions of T. The 1-simplexes of these subdivisions are

$$\mathcal{W}^1 = \{ \langle w_1, w_2 \rangle \mid w_1, w_2 \in \mathcal{W} \text{ and } \langle w_1, w_2 \rangle \cap \mathcal{W} = \{w_1, w_2\}\}, \text{ and}$$
$$\mathcal{V}^1 = \{ \langle v_1, v_2 \rangle \mid v_1, v_2 \in \mathcal{V} \text{ and } \langle v_1, v_2 \rangle \cap \mathcal{V} = \{v_1, v_2\}\}.$$

The observations above show that $\theta = f|\mathcal{V}$ is a vertex map from $\mathcal{V}$ to $\mathcal{W}$. We also note that θ is an *admissible* vertex map in the sense that if $\langle v_1, v_2 \rangle \in \mathcal{V}^1$ then $\langle \theta(v_1), \theta(v_2) \rangle \in \mathcal{W}^1$. Moreover f is *compatible* with θ in the sense that if $\langle v_1, v_2 \rangle \in \mathcal{V}^1$ then f maps $\langle v_1, v_2 \rangle$ monotonically onto $\langle \theta(v_1), \theta(v_2) \rangle$ with $f(v_1) = \theta(v_1)$ and $f(v_2) = \theta(v_2)$.

Let $\varphi: T \to T$ be the piecewise linear mapping determined by θ. Let T' be the triod in [5] and let $\psi: T' \to T'$ be the bonding map of the inverse limit considered in that paper. Then (using the notation of [5] for points of T') φ and ψ are conjugate via the piecewise linear homeomorphism $h: T' \to T$ determined by the following vertex map: $h(A) = c$, $h(B) = b_4$, $h(C) = a_4$, $h(0) = b_0$, $h(\frac{1}{2}B) = b_2$, $h(\frac{1}{4}B) = b_1$, $h(\frac{1}{2}C) = a_2$, $h(\frac{3}{8}C) = a'_1$, $h(\frac{1}{4}C) = a_1$, and $h(\frac{1}{8}C) = a''_1$. Thus $\lim\{T, \varphi\}$ is homeomorphic with $Y = \lim\{T', \psi\}$.

Since $f|\langle v_1, v_2 \rangle$ is monotone for each $\langle v_1, v_2 \rangle \in \mathcal{V}^1$, there is a sequence $\varphi_1, \varphi_2, \varphi_3, \ldots$ of mappings which converges to f uniformly such that, for each $\langle v_1, v_2 \rangle \in \mathcal{V}^1$, $\varphi_n|\langle v_1, v_2 \rangle$ is a homeomorphism from $\langle v_1, v_2 \rangle$ onto $\langle \theta(v_1), \theta(v_2) \rangle$ with $\varphi_n(v_i) = f(v_i) = \theta(v_i)$ for $i = 1, 2$. It is easily seen that $\lim\{T, \varphi_n\}$ is homeomorphic with $\lim\{T, \varphi\}$. By [2, Theorem 3, p. 481], $\lim\{T, f\}$ is homeomorphic with $\lim\{T, \varphi_n\}$, and, hence, with $\lim\{T, \varphi\}$. Consequently, by Theorem 2, Λ is homeomorphic with Y, the continuum of [5]. Hence, by [5, Theorem 3, p. 23], Λ is not chainable. ∎

REFERENCES

1. Marcy Barge, *Horseshoe maps and inverse limits*, Pacific J. Math., Vol. 121(1986), pp. 29–39.
2. Morton Brown, *Some applications of an approximation theorem for inverse limits*, Proc. Amer. Math. Soc., Vol. 11(1960), pp. 478–483.
3. James Dugundji, *Topology*, Allyn and Bacon (1966), Boston.
4. James F. Davis, *Confluent mappings on [0,1] and inverse limits*, Topology Proceedings, Vol. 15(1990), pp. 1–9.
5. James F. Davis and W. T. Ingram, *An atriodic tree-like continuum with positive span which admits a monotone mapping to a chainable continuum*, Fund. Math., Vol. 131 (1988), pp. 13–24.
6. Stephen Smale, *Diffeomorphisms with many periodic points*, in *Differential and Combinatorial Topology* (S. S. Cairus, ed.) Princeton Univ. Press, 1965, pp. 63–80.

Iterated Function Systems, Compact Semigroups, and Topological Contractions

P. F. DUVALL, JR.   University of North Carolina at Greensboro, Greensboro, North Carolina

JOHN WESLEY EMERT   Ball State University, Muncie, Indiana

LAURENCE S. HUSCH   University of Tennessee, Knoxville, Tennessee

Abstract

In this paper, semigroups generated by maps topologically conjugate to contractions are considered. Results of Hutchinson and Barnsley are generalized to this situation. Generally, many invariant sets for these semigroups are shown to exist, but among these exists a unique minimal invariant set. A group naturally arises which carries information about the symmetry of the minimal invariant set. We study the structure of this group. Two large classes of examples are studied: invariant sets for stochastic matrices and invariant sets with symmetry. Conditions which imply that a semigroup generated by a finite number of linear maps has compact closure are studied.

1   Introduction

Recall that a map f of a metric space (X, d) to itself is a *contraction* if there exists $s \in [0, 1)$ such that $d(f(x), f(y)) \leq s \cdot d(x, y)$ for all $x, y \in X$. If (X, d) is complete, then, by the contraction mapping theorem, there exists unique $a \in X$ such that $\lim_{n \to \infty} f^n(x) = a$ for all $x \in X$. If Y is a compact subset of X then it can also be shown that $\lim_{n \to \infty} f^n(Y) = a$.

We define a map f of a complete metric space (X, d) to itself to be a *weak topological contraction* if there exists $a \in X$ such that $\lim_{n \to \infty} f^n(x) = a$ for all $x \in X$. f is a *strong topological contraction* if, for each compact subset $Y \subset X$, $\lim_{n \to \infty} f^n(Y) = a$.

Recall that two maps f, g of X to itself are *topologically equivalent* if there exists a homeomorphism h of X to itself such that $fh = hg$. If f and g are two topologically equivalent

This work was supported in part by a Ball State University Research Grant.

Part of the research was done while L. Husch was a Distinguished Visiting Professor at the University of North Carolina, Greensboro; support is gratefully acknowledged.

maps of X to itself and if f is a contraction, then g need not be a contraction; however, it is easily seen that g is a strong topological contraction. If we let X be m–dimensional Euclidean space $\mathbf{R}^m$ with its usual metric, then, as will be shown in the next section, results of [3], [4] and [6] imply the following characterization of strong topological contractions.

Theorem 1.1 *A homeomorphism f of $\mathbf{R}^m$ onto itself is a strong topological contraction if and only if f is topologically equivalent to the standard contraction $x \to x/2$.*

Suppose that we have a finite number of contractions $f_i, i = 1, 2, ..., n$, of the complete metric space (X, d) to itself. Barnsley [1] introduced the terminology that $\{X; f_1, ..., f_n\}$ is a *hyperbolic IFS (iterated function system)*. Hutchinson [5] associated to each hyperbolic *IFS* a unique invariant set called the *attractor* of the *IFS*. The properties of hyperbolic *IFS*'s have been used extensively to study the "fractal" properties [7] of these attractors. It can be shown that many of the results about attractors follow from the fact that the semigroup of mappings of X to itself generated by the f_i's has compact closure in the space of all continuous maps of X to itself with, for example, the compact-open topology (in the rest of this section, it is convenient to say in this situation that the f_i's *generate a compact semigroup*). In this paper, we replace the condition that the f_i's are contractions by the condition that they are (strong) topological contractions. We are motivated by the construction that if $\{X; f_1, ..., f_n\}$ is a hyperbolic *IFS* and if h is a homeomorphism of X onto itself then the collection $\{hf_ih^{-1}, i = 1, 2, ..., n\}$ of strong topological contractions need not be a hyperbolic *IFS* but will also have a unique invariant set associated to it which is the image of the attractor of the f_i's under h. The problem initially considered was whether there was a characterization of hyperbolic *IFS*'s in the spirit of Theorem 1.1. As will be seen later, this theory is much more than a straightforward generalization; for example, as will be shown in section 3 (Example 3.3), the closure of the semigroup generated by a finite number of strong topological contractions need not be compact. As a consequence, we will restrict our study to *IFS*'s of maps which generate compact semigroups. We are then able to show that there exist invariant sets for such *IFS*'s. But the invariant sets need not be unique; however, in the case when at least one of the maps is a strong topological contraction there does exist a unique minimal invariant set.

In section 2, we begin with a review of equicontinuity and a proof of Theorem 1.1. In section 3, we show that there is no distinction between strong and weak topological contractions for linear and affine maps. We look at the role of various norms for matrices in discussing compactness of semigroups generated by linear and affine maps. We introduce the concept of Z-semigroup, which can be regarded as the topological analogue of the semigroup generated by a hyperbolic *IFS*. In section 4, we look at Hutchinson's construction of a map F induced by an *IFS* $\{X; f_1, ..., f_n\}$ on the space of compacta of X and investigate its properties under the assumption that the f_i's generate a compact semigroup. The limit points of the semigroup generated by F form a compact group which we shall call the *kernel* of the *IFS*. Invariant sets are characterized (Proposition 4.5) by the kernel. In section 5, we apply the theory developed so far to the case of linear maps represented by column stochastic matrices. The kernel is shown to be the trivial group and we completely classify the invariant sets. In sections 6, 7 and 8 we address the problem of determining the structure of the kernel. In section 6, we discuss the consequences of assuming that the identity element of the kernel is the identity map on the space of compacta. In section 7, we show that the kernel is the homomorphic image of the compact semigroup generated by the f_i's in the special case when this semigroup turns out to be a compact Abelian group. In section 8, we investigate the richness of structure of the kernel in the special case when the f_i's generate a finite group. In section 9, we apply the results of the previous sections when it is assumed that at least one of the f_i's is a topological contraction. We show the existence of the minimal invariant set and discuss how the kernel expresses the symmetry of this set.

We express our gratitude to P. Bowers, B. Brechner, J. Dijkstra, O. Karakashian, J. Mogilski and H. Row for discussions and suggestions regarding this paper.

2 Equicontinuity.

Let $\mathcal{F}$ be a collection of functions of a metric space (X, d) to itself. $\mathcal{F}$ is said to be *equicontinuous* if, given $x \in X$ and $\epsilon > 0$, there exists $\delta > 0$ such that whenever $d(x, y) < \delta$ then

$d(f(x), f(y)) < \epsilon$ for each $f \in \mathcal{F}$. The proof of the following is straightforward.

Proposition 2.1 *If* $\{X; f_1, ..., f_n\}$ *is a hyperbolic IFS then the semigroup of maps of X to itself generated by* $\{f_1, ..., f_n\}$ *is an equicontinuous family of maps.*

Proposition 2.2 *Let f be a strong topological contraction of a locally compact metric space X; then the collection,* $\{f^i\}_{i \geq 0}$, *of non-negative iterates of f is equicontinuous at each point of X.*

Proof. Let $x \in X$ and let $\epsilon > 0$ be given. Let U be a compact neighborhood of x; choose $\delta_0 > 0$ so that the δ_0 neighborhood of x lies in U. Since f is a strong topological contraction, $\lim_{n \to \infty} f^n(U) = a$ for some $a \in X$; *i.e*, there exists N such that for $n > N$, $f^n(U)$ is contained in the $\epsilon/2$ neighborhood of a. For $i \leq N$ choose $\delta_i > 0$ so that if $d(x, y) < \delta_i$ then $d(f^i(x), f^i(y)) < \epsilon$. The minimum of $\delta_0, \delta_1, ..., \delta_N$ is the desired δ. $\square$

Proposition 2.3 *Let f be a weak topological contraction of a locally compact metric space X such that the collection,* $\{f^i\}_{i \geq 0}$, *of non-negative iterates of f is equicontinuous at each point of X. The f is a strong topological contraction.*

Proof. Let a be the limit point given by the definition of weak topological contraction and let Y be a compact subset of X. For each $x \in Y$ choose $\delta_x > 0$ such that $d(x, y) < \delta_x$ implies that $d(f^n(x), f^n(y)) < \epsilon/3$ for all i. Choose a finite subset $\{x_1, x_2, ..., x_p\} \subset Y$ such that the δ_{x_i} neighborhoods of x_i cover Y. For $i = 1, 2, ..., p$, choose N_i so that if $n > N_i$ then $d(f^n(x_i), a) < \epsilon/3$. Let N be the maximum of the N_i's, let $n > N$ and $y \in Y$; it is easily seen that $d(f^n(y), a) < \epsilon$ from which it follows that $\lim_{n \to \infty} f^n(Y) = a$. $\square$

As a consequence of this proposition, whenever we are working with functions which are elements of an equicontinuous semigroup we can drop the adjectives "strong" and "weak". In general, there is a distinction between strong and weak topological contractions. V. I. Opoitsev [10] has given an example of a homeomorphism of the plane which is a weak contraction but is not a strong contraction.

Lemma 2.1 *Let f be a homeomorphism of the locally compact metric space (X, d) onto itself which is a strong topological contraction and let a be the limit of images of compact sets under iterates of f. Then, for each $x \in X \setminus \{a\}$, the set $\{f^{-i}(x) : i \geq 0\}$ has no limit points in X.*

Proof. Suppose that there exists $x \in X \setminus \{a\}$ such that $\{f^{-i}(x) : i = 0, 1, ...\}$ has a limit point z. For some subsequence $\{i_k\}$ of the positive integers, $\lim_{k \to +\infty} f^{-i_k}(x) = z$. By Proposition 2.2, the collection of positive iterates of f is equicontinuous at z; hence, given $\epsilon > 0$ there exists $\delta > 0$ such that for all positive integers i, $d(f^i(z), f^i(w)) < \epsilon$ whenever $d(z, w) < \delta$. If we choose K such that $d(z, f^{-i_k}(x)) < \delta$ for $k > K$, then $d(f^{i_k}(z), x) < \epsilon$ which implies $\lim_{i \to +\infty} f^{i_k}(\{z\}) = x$ and we contradict the hypothesis that $\lim_{i \to +\infty} f^i(\{z\}) = a$. $\square$

Proof of Theorem 1.1. Suppose that f is a strong topological contraction on $\mathbf{R}^m$; let $a \in \mathbf{R}^m$ be the point given by the definition of strong topological contraction. Therefore, for each $x \in \mathbf{R}^m$, $\lim_{i \to +\infty} f^i(x) = a$ and, by the previous lemma, for each $x \in \mathbf{R}^m \setminus \{a\}$, $\lim_{i \to +\infty} f^{-i}(x) = \infty$. The conclusion now follows from the topological characterization of contractions given in [3] [4] [6]. $\square$

One of the important uses of equicontinuity is the following characterization of compactness of a collection of maps (see [9, p.290]). Let X be a locally compact metric space and let $\mathcal{C}(X)$ be the space of continuous maps from X to itself with the compact–open topology. Let $\mathcal{F} \subset \mathcal{C}(X)$ and let $x \in X$; we define $\mathcal{F}(x) = \{f(x) : f \in \mathcal{F}\}$.

Theorem 2.1 (Ascoli-Arzela) *The closure of $\mathcal{F}$ in $\mathcal{C}(X)$ is compact if and only if $\mathcal{F}$ is equicontinuous and, for each $x \in X$, the closure of $\mathcal{F}(x)$ in X is compact.*

Although the Ascoli-Arzela theorem requires that the closure of $\mathcal{F}(x)$ in X be compact for all $x \in X$, this restriction can be relaxed with additional hypotheses. In the following corollary, the compactness of $\overline{\mathcal{F}}(x_0)$ implies the compactness of $\overline{\mathcal{F}}(x)$ for all $x \in X$. The proof is left to the reader.

Corollary 2.1 *Suppose that X is a connected, locally compact metric space such that the compact subsets of X are characterized as those subsets which are closed and bounded. Suppose* that $\mathcal{F} \subset \mathcal{C}(X)$ such that $\mathcal{F}$ is equicontinuous at each $x \in X$ and suppose that there exists $x_0 \in X$ such that the closure of $\mathcal{F}(x_0)$ in X is compact. Then closure of $\mathcal{F}$ in $\mathcal{C}(X)$ is compact.

As a consequence of Theorem 2.1 and Proposition 2.1, we have the following.

Corollary 2.2 *If $\{X; f_1, ..., f_n\}$ is a hyperbolic IFS of a locally compact complete metric space X then the semigroup generated by the f_i's has compact closure in the space of continuous maps of X to itself with the compact-open topology.*

Recall that equicontinuity depends upon the metric for X when X is a non–compact space; *e.g.*, the semigroup generated by the function $\vec{x} \rightarrow 2\vec{x}$ fails to be equicontinuous at all $\vec{x} \in \mathbf{R}^m \setminus \{\vec{0}\}$ with the restriction of the usual metric of $\mathbf{R}^m$. However, if we use the restriction of a metric induced from the one–point compactification of $\mathbf{R}^m$, the above mapping is equicontinuous on $\mathbf{R}^m \setminus \{\vec{0}\}$.

We also recall the fact that if two metrics are equivalent then equicontinuity with respect to one of these metrics implies equicontinuity with respect to the other metric.

3 Linear and Affine Mappings

Fixing a basis in $\mathbf{R}^m$, we can represent a linear mapping f of $\mathbf{R}^m$ to itself by matrix multiplication $f(\vec{x}) = A\vec{x}$ where A is an $m \times m$ matrix and $\vec{x}$ is regarded as an $m \times 1$ matrix. The *spectral radius* of A, $\rho(A)$, is the maximum of the magnitudes of the eigenvalues of A; the *spectral radius* of f, $\rho(f)$ is defined to be $\rho(A)$. From matrix theory [12, p.284], we know that $\lim_{i \to \infty} A^i$ is the zero matrix if and only if $\rho(A) < 1$. Using the fact that the columns of A^i are the images of the basis elements under f^i, one can show the following.

Proposition 3.1 *A linear map f of $\mathbf{R}^m$ to itself is a weak topological contraction if and only if $\rho(f) < 1$.*

Corollary 3.1 *A linear transformation f of $\mathbf{R}^m$ to itself is a weak topological contraction if and only if it is a strong topological contraction.*

Proof. Suppose that f is a weak topological contraction with $f(\vec{x}) = A\vec{x}$. Since $\rho(A) < 1$, as mentioned above, $\lim_{i \to \infty} A^i$ is the zero matrix. Therefore, the semigroup generated by f has compact closure in the space of continuous mappings of $\mathbf{R}^m$ to itself with the compact-open topology. If $Y \subset \mathbf{R}^m$ is compact and if $U \subset \mathbf{R}^m$ is an open set which contains the zero vector, then the collection of maps of $\mathbf{R}^m$ whose image of Y lies in U is a neighborhood of the zero linear map and as such must contain all but a finite number of the f^i's. It follows that f is a strong topological contraction. $\square$

As a consequence of this corollary, when referring to a linear mapping f which is a weak or strong topological contraction, we will drop off the adjectives "strong" and "weak" and write that f is *linear topological contraction*.

Recall that the *singular values* of a matrix A are the non–negative square roots of the eigenvalues of the product $A^t A$ where A^t is the transpose of A. Let $\mathbf{S}^{m-1} = \{\vec{x} : \|\vec{x}\| = 1\}$ denote the unit sphere in $\mathbf{R}^m$ where the norm $\| \cdot \|$ of a vector is the usual norm. Later in this section, we shall consider other norms on $\mathbf{R}^m$. Recall that the image of $\mathbf{S}^{m-1}$ under f is an ellipsoid and the singular values represent one-half the lengths of the axes of this ellipsoid. Recall also that the norm of f, $\|f\|$, which is defined by $\|f\| = \max\{\|f(\vec{x})\| : \vec{x} \in \mathbf{S}^{m-1}\}$ is equal to the maximum of the singular values of A; as a consequence of these comments, we have the following.

Proposition 3.2 *A linear transformation f of $\mathbf{R}^m$ to itself is a contraction if and only if $\|f\| < 1$.*

Example 3.1 Consider the following linear maps of $\mathbf{R}^2$:

$$f(x, y) = (2y, x/4) \quad \text{and} \quad g(x, y) = (5y, x/6)$$

and their composition

$$fg(x, y) = (x/3, 5y/4).$$

Note that $\rho(f) = \sqrt{2}/2$, $\rho(g) = \sqrt{5/6}$ and $\rho(fg) = 5/4$. By Proposition 3.1, f and g are both topological contractions and fg is not a topological contraction. Since the collection of positive iterates of fg fails to be equicontinuous at $\vec{0}$, the fixed point of fg, the semigroup generated by f and g fails to be equicontinuous and, hence, its closure fails to be compact.□

The following is a consequence of Corollary 2.1 since $\mathcal{S}(\vec{0})$, the orbit of $\vec{0}$, consists of a single point and, hence, is compact.

Proposition 3.3 *Suppose that a semigroup $\mathcal{S}$, generated by a collection of linear maps of $\mathbf{R}^m$ to itself, is equicontinuous. Then the closure of $\mathcal{S}$ in $\mathcal{C}(\mathbf{R}^m)$ with the compact–open topology is compact.*

Suppose that $\mathcal{S}$ is the semigroup generated by maps, $f_1, f_2, ..., f_n$, of a space X to itself and suppose that p is a positive integer. We let $\mathcal{S}^{(p)}$ designate the subset of $\mathcal{S}$ which consists of all possible compositions $g = f_{i_1} f_{i_2} ... f_{i_p}$ of precisely p elements of $\{f_1, f_2, ..., f_n\}$; we say that the *length* of g is p. Of course, the length of g need not be unique; *e.g.*, if some f_i is the identity map then the length of each g can be any arbitrarily large positive integer.

Proposition 3.4 *Suppose that $\mathcal{S}$ is the semigroup generated by a finite number of linear maps, $f_1, f_2, ..., f_n$, of $\mathbf{R}^m$ to itself such that, for some positive integer p and for each $g \in \mathcal{S}^{(p)}$, $\|g\| \leq 1$. Then $\mathcal{S}$ is equicontinuous and its closure in $\mathcal{C}(\mathbf{R}^m)$ with the compact–open topology is compact.*

Proof. Let

$$M = \max\{\|h\| : h \in \bigcup_{i=1}^{p} \mathcal{S}^{(i)}\}.$$

Given $\epsilon > 0$, define $\delta = \epsilon/M$. Suppose that $\|\vec{x} - \vec{y}\| < \delta$ and suppose that $k = f_{i_1} f_{i_2} ... f_{i_l}$; write $l = qp + r$ where $r \in \{0, 1, ..., p-1\}$. Then we can combine terms: $k = k_1 k_2 ... k_q k_{q+1}$ where $k_j = f_{i_{(j-1)p+1}} f_{i_{(j-1)p+2}} ... f_{i_{jp}}$ for $j = 1, 2, ..., q$ and $k_{q+1} = f_{i_{qp+1}} f_{i_{qp+2}} ... f_{i_l}$. Note that, in particular, for $j = 1, 2, ..., q$, $\|k_j\| \leq 1$.

Consider $\|k(\vec{x}) - k(\vec{y})\| = \|k(\vec{x} - \vec{y})\| = \|k_1 k_2 ... k_q k_{q+1}(\vec{x} - \vec{y})\| \leq \|k_2 k_3 ... k_q k_{q+1}(\vec{x} - \vec{y})\| \leq$ $... \leq \|k_{q+1}(\vec{x} - \vec{y})\| \leq \|k_{q+1}\| \cdot \|\vec{x} - \vec{y}\| \leq M\delta = \epsilon$. □

Example 3.2 Let r be a positive real number and define the linear map g_r of $\mathbf{R}^2$ to itself by $g_r(x, y) = (rx, ry)$. Consider the semigroup, $\mathcal{S}_r$ generated by g_r and the linear map,

$f(x, y) = (2y, x/4)$. Note that $\|f\| = 2$ but, as noted in Example 3.1, f is a topological contraction. Given a positive integer p, we can choose $r \in (0, 1)$ such that there exists $h_0 = fg_r^{p-1} \in \mathcal{S}^{(p)}$ with $\|h_0\| = 2r^{p-1} > 1$ but for all $h \in \mathcal{S}^{(p+1)}$, $\|h\| \leq 1$. Therefore, the level p in Proposition 3.4 could be large. $\square$

Therefore, given some specific collection of linear maps which are topological contractions it can be very difficult to decide whether the semigroup generated by these maps is equicontinuous.

Recall that a function f of $\mathbf{R}^m$ defined by $f(\vec{x}) = A\vec{x} + \vec{b}$, where A is an $m \times m$ matrix and $\vec{b}$ is some fixed vector, is called an *affine transformation*. We shall refer to A as the *derivative* of f. The following can be proved by induction.

Lemma 3.1 *Suppose that $f_1, f_2, ..., f_n$ are affine mappings of $\mathbf{R}^m$ with derivatives $A_1, A_2, ...,$ A_n, respectively. Then, for each finite sequence, $f_{i_1}, f_{i_2}, ..., f_{i_p}$, from $\{f_1, f_2, ..., f_n\}$ and for all $\vec{x}, \vec{y} \in \mathbf{R}^m$, we have*

$$\|f_{i_1} f_{i_2} ... f_{i_p}(\vec{x}) - f_{i_1} f_{i_2} ... f_{i_p}(\vec{y})\| = \|A_{i_1} A_{i_2} ... A_{i_p}(\vec{x}) - A_{i_1} A_{i_2} ... A_{i_p}(\vec{y})\|.$$

Using this lemma, the proof of the following is straightforward.

Proposition 3.5 *Suppose that $f_1, f_2, ..., f_n$ are affine mappings of $\mathbf{R}^m$ with derivatives $A_1,$ $A_2, ..., A_n$, respectively. The semigroup generated by $f_1, f_2, ..., f_n$ is equicontinuous if and only if the semigroup generated by $A_1, A_2, ..., A_n$ is equicontinuous.*

Lemma 3.2 *An affine mapping f is a (weak topological) (strong topological) contraction if and only if its derivative is a (weak topological) (strong topological) contraction.*

Proof. Suppose that $f(\vec{x}) = A\vec{x} + \vec{b}$ is a weak topological contraction. Let $\vec{z}$ be the fixed point of f and consider the map $h(\vec{x}) = \vec{x} + \vec{z}$. Note that $h^{-1}fh = A$ and, hence, f and the linear map defined by A are topologically equivalent. As noted in the introduction, it follows that A is a weak topological contraction.

Suppose that A is a weak topological contraction; by Proposition 3.1, all of the eigenvalues of A have magnitude < 1. If I is the identity $m \times m$ matrix then $A - I$ is an invertible matrix.

It follows that f has a fixed point $\vec{z} = -(A - I)^{-1}\vec{b}$. We can now use the argument in the first paragraph to show that f and the linear map defined by A are topologically equivalent and, hence, f is a weak topological contraction.

The proof that f is a strong topological contraction if and only if A is a strong topological contraction is similar. The fact that f is a contraction if and only if A is a contraction follows immediately from Lemma 3.1. $\square$

Corollary 3.2 *An affine transformation of $\mathbf{R}^m$ is a weak topological contraction if and only if it is a strong topological contraction if and only if the spectral radius of its derivative is less than 1.*

Again, as a consequence of the latter, we will also drop the adjectives "strong" and "weak" when referring to affine mappings, f, which are strong topological contractions and we will call f an *affine topological contraction*.

Corollary 3.3 *An affine transformation of $\mathbf{R}^m$ is a contraction if and only if the norm of its derivative is less than 1.*

The following example shows that the analogue of Proposition 3.3 for affine mappings is not true even in the case when the mappings are topological contractions.

Example 3.3 Let $\mathcal{S}$ be the semigroup generated by the following affine maps of the plane

$$f(\vec{x}) = A\vec{x} + \vec{d} \quad \text{and} \quad g(\vec{x}) = B\vec{x}$$

where

$$A = \begin{pmatrix} 0 & 1/2 \\ 1 & 0 \end{pmatrix}, \quad \vec{d} = \begin{pmatrix} 0 \\ 1 \end{pmatrix} \quad \text{and} \quad B = \begin{pmatrix} 0 & 1 \\ 1/2 & 0 \end{pmatrix}.$$

Note that

$$(BA)^n = \begin{pmatrix} 1 & 0 \\ 0 & 1/4^n \end{pmatrix}$$

and

$$(gf)^n(\vec{x}) = (BA)^n\vec{x} + \sum_{i=0}^{n-1}(BA)^i B\vec{d} = (BA)^n\vec{x} + \sum_{i=0}^{n-1} B\vec{d} = (BA)^n\vec{x} + nB\vec{d}$$

for each positive integer n. Therefore, for each non-zero $\vec{x} \in \mathbf{R}^2$, $\mathcal{S}(\vec{x})$ is unbounded; from this it follows that the closure of $\mathcal{S}$ is not compact. Since $\|A\| = \|B\| = 1$, Propositions 3.4 and 3.5 imply that $\mathcal{S}$ is equicontinuous and Proposition 3.1 implies that f and g are affine topological contractions. $\square$

Suppose that we have another norm $\| \cdot \|^*$ on $\mathbf{R}^m$. From linear algebra [12, p.170], this norm is equivalent to the usual norm; *i.e.*, there exist positive numbers K and L such that for all $\vec{x} \in \mathbf{R}^m$, $K \cdot \|\vec{x}\| \leq \|\vec{x}\|^* \leq L \cdot \|\vec{x}\|$. Note that this implies that the metrics induced from different norms are then equivalent. If g is a linear map defined on $\mathbf{R}^m$ then, as usual, we define its norm with respect to the norm $\| \cdot \|^*$ by

$$\|g\|^* = \sup_{\|\vec{x}\|^* = 1} \|g(\vec{x})\|^*.$$

The proof of Proposition 3.4 can be used to prove the following.

Corollary 3.4 *Suppose that $\mathcal{S}$ is a semigroup generated by a finite number of linear maps of $\mathbf{R}^m$ to itself such that, for some positive integer p and for each $g \in \mathcal{S}^{(p)}$, $\|g\|^* \leq 1$ for some norm $\| \cdot \|^*$. Then $\mathcal{S}$ is equicontinuous and the closure $\overline{\mathcal{S}}$ is compact.*

Corollary 3.5 *Suppose that $\mathcal{S}$ is a semigroup generated by a finite number of linear maps of $\mathbf{R}^m$ to itself. The closure $\overline{\mathcal{S}}$ is compact if and only if there exists a norm $\| \cdot \|^*$ such that, for each $g \in \mathcal{S}$, $\|g\|^* \leq 1$.*

Proof. Suppose that $\overline{\mathcal{S}}$ is compact. For each $\vec{x} \in \mathbf{R}^m$, define

$$\|\vec{x}\|^* = \sup\{\|g(\vec{x})\| : g \in \mathcal{S}\}.$$

One of the important properties of the compact–open topology [9, p.287] is the fact that the evaluation mapping $\mathcal{C}(\mathbf{R}^m) \times \mathbf{R}^m \to \mathbf{R}^m$ defined by $(f, \vec{x}) \to f(\vec{x})$ is continuous. It follows that $\|\vec{x}\|^*$ is finite; we leave to the reader to verify that $\| \cdot \|^*$ is the desired norm. $\square$

V.I. Opoitsev [10, Theorem 7.1] proved the following theorem.

Theorem 3.1 *Suppose that $\mathcal{S}$ is a semigroup generated by a finite number of linear maps of $\mathbf{R}^m$ to itself such that, for some positive integer p, for some $q \in (0, 1)$, and for each $g \in \mathcal{S}^{(p)}$, $\|g\| \leq q$. Then there exists a norm $\| \cdot \|^*$ and $q' \in (0, 1)$ such that for each $g \in \mathcal{S}$, $\|g\|^* \leq q'$.*

We say that a semigroup $\mathcal{S}$ generated by a finite number of continuous maps of a metric space (X, d) to itself is a *Z-semigroup* if, the closure of $\mathcal{S}$ in $\mathcal{C}(X)$ with the compact–open topology is compact and the only limit point of $\overline{\mathcal{S}}$ is a constant map.

Corollary 3.6 *Suppose that $\mathcal{S}$ is a semigroup generated by a finite number of linear maps, $f_1, f_2, ..., f_n$, of $\mathbf{R}^m$ to itself. $\mathcal{S}$ is a Z-semigroup if and only if there exists a norm $\|\cdot\|^*$ on $\mathbf{R}^m$ such that, for each $i = 1, 2, ..., n$, $\|f_i\|^* < 1$.*

Conjecture 3.1 *Let $\mathcal{S}$ be a semigroup generated by a finite number of linear maps of $\mathbf{R}^m$ to itself and suppose that each element of $\mathcal{S}$ is a topological contraction; then $\mathcal{S}$ is a Z–semigroup.*

We are able to prove this conjecture when we have commuting linear maps.

Proposition 3.6 *Suppose that $\mathcal{S}$ is a commutative semigroup generated by a finite number of linear topological contractions, $f_1, f_2, ..., f_n$, of $\mathbf{R}^m$ to itself; then $\mathcal{S}$ is a Z–semigroup.*

Proof. It suffices to show that if $\epsilon > 0$ is given then there exists P such that if $p > P$ and if $g \in \mathcal{S}^{(p)}$ then $\|g\| < \epsilon$. Since each f_i is a topological contraction there exists a positive integer t_i such that $\|f_i^{t_i}\| < 1/2$; let t be the maximum of the t_i's. For each i, let $M_i = \max\{\|f_i^j\| : j = 1, 2, ..., t_i - 1\}$ and define $M = M_1 M_2 ... M_n$. Choose σ so that if $s \geq \sigma$ then $1/2^s < \epsilon/M$. Finally, let $P = t(\sigma + n)$.

Suppose that $g \in \mathcal{S}^{(p)}$ for some $p > P$; then since the f_i's commute we can express $g = f_1^{s_1} f_2^{s_2} ... f_n^{s_n}$ where $s_1 + s_2 + ... + s_n = p$. For each i express $s_i = q_i t_i + r_i$ where $0 \leq r_i < t_i$. Consider

$$
\begin{aligned}
q_1 + q_2 + ... + q_n &= \frac{s_1 - r_1}{t_1} + \frac{s_2 - r_2}{t_2} + ... + \frac{s_n - r_n}{t_n} \\
&\geq \frac{s_1 + s_2 + ... + s_n - (r_1 + r_2 + ... + r_n)}{t} \\
&\geq \frac{p - nt}{t} > \frac{P - nt}{t} = \sigma.
\end{aligned}
$$

Hence

$$
\begin{aligned}
\|g\| &= \|f_1^{s_1} f_2^{s_2} ... f_n^{s_n}\| \\
&\leq \|f_1^{t_1}\|^{q_1} \cdot \|f_2^{t_2}\|^{q_2} \cdot ... \cdot \|f_n^{t_n}\|^{q_n} \cdot \|f_1^{r_1}\| \cdot \|f_2^{r_2}\| \cdot ... \cdot \|f_1^{r_n}\| \\
&\leq \frac{1}{2^{q_1}} \frac{1}{2^{q_2}} ... \frac{1}{2^{q_n}} M_1 M_2 ... M_n \leq \frac{1}{2^{\sigma}} M < \epsilon. \square
\end{aligned}
$$

Proposition 3.7 *Suppose that $f_1, f_2, ..., f_n$ are affine maps of $\mathbf{R}^m$ to itself with derivatives $A_1, A_2, ..., A_n$, respectively. If the semigroup $\mathcal{T}$ generated by $A_1, A_2, ..., A_n$ is a Z-semigroup then the semigroup $\mathcal{S}$ generated by $f_1, f_2, ..., f_n$ has compact closure.*

Proof. By Corollary 3.6, there exists a norm $\| \cdot \|^*$ such that $\|A_i\|^* < 1$ for each i. From Lemma 3.1, it follows that each f_i is a contraction on $\mathbf{R}^m$ using the metric induced by the norm $\| \cdot \|^*$. It follows that $\{\mathbf{R}^m; f_1, ..., f_n\}$ is a hyperbolic *IFS* and, by Corollary 2.2, the semigroup $\mathcal{S}$ has compact closure. $\square$

Corollary 3.7 *Suppose that $f_1, f_2, ..., f_n$ are affine topological contractions of $\mathbf{R}^m$ with derivatives $A_1, A_2, ..., A_n$, respectively. If the semigroup generated by $A_1, A_2, ..., A_n$ is a commutative semigroup then the semigroup generated by $f_1, f_2, ..., f_n$ has compact closure.*

The following example shows that the condition that $\mathcal{T}$ be a Z–semigroup in the previous Proposition is not necessary in order for $\overline{\mathcal{S}}$ to be compact.

Example 3.4 Consider the semigroup, $\mathcal{S}$, generated by the affine maps

$$f(x, y) = (x, y/2) \quad \text{and} \quad g(x, y) = (x/2, 1 - y/2)$$

of $\mathbf{R}^2$. Since the norms of the derivatives are ≤ 1, $\mathcal{S}$ is equicontinuous. One can show by induction on the length of elements of $\mathcal{S}$ that $\mathcal{S}(\vec{0}) \subset \{0\} \times [0, 1] \subset \mathbf{R}^2$ is bounded; by Corollary 2.1, the closure $\overline{\mathcal{S}}$ is compact. Since $\lim_{n \to \infty} f^n$ is not the zero map, the semigroup generated by the derivatives of f and g is not a Z–semigroup. $\square$

Suppose that f is an affine topological contraction of $\mathbf{R}^m$; by Theorem 1.1, f is topologically equivalent to the contraction $\vec{x} \to \vec{x}/2$. One of the consequences of Theorem 3.1 is that there exists a norm on $\mathbf{R}^m$ with respect to which f is a contraction. For application in later work, we wish to show that f is affinely equivalent to a contraction. From the proof of Lemma 3.2, we know that f is affinely equivalent to its derivative; therefore, it suffices to show that if f is a linear topological contraction of $\mathbf{R}^n$ to itself, then f is linearly conjugate to a contraction. In the light of our previous remarks, this is equivalent to

Theorem 3.2 *If A is an $n \times n$ real matrix whose eigenvalues are less than 1 in absolute value, then A is similar (over the reals) to a matrix whose singular values are less than 1.*

Proof. The important ideas of the proof are revealed in the 2×2 case. For the time being, we assume that A is a 2×2 real matrix whose eigenvalues have length less than one in magnitude. We also assume that we are not in the trivial situation where A is a multiple of the identity. There are three possibilities to consider.

1. A has distinct real eigenvalues. In this case, A is diagonalizable, and the proof is immediate.

2. A has a repeated eigenvalue p. Here, A is similar to the Jordan block

$$J = \begin{pmatrix} p & 1 \\ 0 & p \end{pmatrix}.$$

If J is conjugated by the matrix

$$\Delta = \begin{pmatrix} 1 & 0 \\ 0 & \delta \end{pmatrix},$$

the result is

$$T = \begin{pmatrix} p & \delta \\ 0 & p \end{pmatrix}.$$

By making δ small, we can make T close to $p * I$, which is a contraction. Since the largest singular value of a square matrix is just its l_2 norm, continuity implies that any matrix sufficiently close to a contraction is itself a contraction. Thus A is conjugate to a contraction.

3. A has complex eigenvalues. Suppose that the eigenvalues of A are z and $\bar{z}$, where $z = a + ib$ with $b > 0$. Then A is similar to the companion matrix of its characteristic polynomial,

$$C = \begin{pmatrix} 0 & -(a^2 + b^2) \\ 1 & 2a \end{pmatrix}.$$

If we conjugate C by

$$\begin{pmatrix} \sqrt{b} & -a/\sqrt{b} \\ 0 & 1/\sqrt{b} \end{pmatrix},$$

we obtain

$$T = \begin{pmatrix} a & -b \\ b & a \end{pmatrix},$$

which is easily seen to be a contraction.

In the general case, we may assume that A is in diagonal block form, where each block is either a standard $k \times k$ Jordan block or a $2k \times 2k$ matrix of the form

$$
S = \begin{pmatrix} C & I & & & \\ & C & I & & \\ & & \ddots & \ddots & \\ & & & \ddots & I \\ & & & & C \end{pmatrix},
$$

where I is the 2×2 identity matrix, and C is a 2×2 block in the form of T in the complex eigenvalue case above. (For a nice presentation of this version of the real Jordan form, see [2].) Therefore, it suffices to show that these forms are conjugate to contractions. If we conjugate S by

$$
\Delta = \begin{pmatrix} I & & & \\ & \delta I & & \\ & & \ddots & \\ & & & \delta^k I \end{pmatrix},
$$

we obtain the matrix

$$
\begin{pmatrix} C & \delta I & & & \\ & C & \delta I & & \\ & & \ddots & \ddots & \\ & & & \ddots & \delta I \\ & & & & C \end{pmatrix},
$$

which, by the continuity argument above, is a contraction if δ is sufficiently small. The case of a real Jordan block is handled similarly. $\square$

We thank O. Karakashian for showing us [12, p.284] which helped in the proof of the above theorem.

4 Equicontinuity and the Space of Compacta.

Let X be a locally compact metric space with metric d and let $\mathcal{H}(X)$ denote the space of non–empty compact subsets of X with the Hausdorff metric, h. Note that if $f : X \to X$ is a continuous map then f induces a continuous map, f_*, from $\mathcal{H}(X)$ to itself, defined by $f_*(A) = f(A)$; continuing with standard usage, we will designate f_* by f. Similarly, if $\mathcal{S}$

is a semigroup of continuous functions of X to itself, we can consider S as a semigroup of functions from $\mathcal{H}(X)$ to itself.

Lemma 4.1 *Suppose that $f_1, f_2, ..., f_n$ generates an equicontinuous semigroup, S, of maps of X to itself. The induced action of S on $\mathcal{H}(X)$ is equicontinuous at each element of $\mathcal{H}(X)$.*

Proof. Let $A \in \mathcal{H}(X)$ and let $\epsilon > 0$ be given. Choose $\delta_1 > 0$ such that the closed δ_1-neighborhood of A, U, in X is compact. Note that since U is a compact subset of X, then the restriction of S to U is uniformly equicontinuous; choose $\delta_2 > 0$ so that if $x, y \in U$ and if $d(x, y) < \delta_2$ then $d(g(x), g(y)) < \epsilon$ for all $g \in S$. Let δ be the minimum of δ_1 and δ_2 and suppose that $h(A, B) < \delta$. Note that $B \subseteq U$. Suppose that $b \in B$, then there exists $a \in A$ such that $d(a, b) < \delta$. Hence, if $g \in S$, then $d(g(a), g(b)) < \epsilon$; it follows that $g(B)$ is contained in the ϵ-neighborhood of $g(A)$. Similarly, $g(A)$ is contained in the ϵ-neighborhood of $g(B)$; hence, $h(g(A), g(B)) < \epsilon$. $\square$

Recall the following result from Hutchinson [5].

Lemma 4.2 *Let $\{A_1, A_2, ..., A_n, B_1, B_2, ..., B_n\} \subset \mathcal{H}(X)$; then*

$$h(\bigcup_{i=1}^{n} A_i, \bigcup_{i=1}^{n} B_i) = \max_{i}\{h(A_i, B_i)\}.$$

Definition 4.1. Suppose that $f_1, f_2, ..., f_n$ are continuous maps of a locally compact metric space X to itself; we define (as Hutchinson [5]) $F : \mathcal{H}(X) \rightarrow \mathcal{H}(X)$ by

$$F(A) = f_1(A) \cup f_2(A) \cup ... \cup f_n(A).$$

We say that F is the *union* of the maps $f_1, f_2, ..., f_n$.

Let S be the semigroup generated by the maps $f_1, f_2, ..., f_n$; we leave to the reader to check the following calculation of the iterates of F.

Lemma 4.3 *For $A \in \mathcal{H}(X)$ and for each positive integer i, $F^i(A) = \bigcup\{g(A) : g \in S^{(i)}\}$.*

Hypotheses 4.1 Suppose that $f_1, f_2, ..., f_n$ generates an equicontinuous semigroup of maps of a locally compact metric space X to itself and let F be the union of f_i's.

Proposition 4.1 *Assume Hypotheses 4.1; then the collection, $\mathcal{F}$, of positive iterates of F is equicontinuous at each point of $\mathcal{H}(X)$.*

Proof. Let $A \in \mathcal{H}(X)$ and let $\epsilon > 0$ be given. By Lemma 4.1, there exists $\delta > 0$ such that $h(A, B) < \delta$ implies that $h(g(A), g(B)) < \epsilon$ for all $g \in \mathcal{S}$. Let i be a positive integer; if $h(A, B) < \delta$ then, by Lemmas 4.2 and 4.3,

$$h(F^i(A), F^i(B)) = h(\bigcup_{g \in \mathcal{S}^{(i)}} g(A), \bigcup_{g \in \mathcal{S}^{(i)}} g(B)) = \max\{h(g(A), g(B)) : g \in \mathcal{S}^{(i)}\} < \epsilon. \quad \Box$$

The proof of the following lemma follows from the already noted fact that the evaluation map $\mathcal{C}(X) \times X \to X$ is continuous when one uses the compact–open topology on $\mathcal{C}(X)$.

Lemma 4.4 *Suppose that $\mathcal{S}$ is a collection of functions of X to itself such that its closure $\overline{\mathcal{S}}$ is compact; then, for each compact subset $A \subseteq X$, $\overline{\mathcal{S}}(A) = \cup\{g(A) \mid g \in \overline{\mathcal{S}}\} = \overline{\mathcal{S}(A)}$, the closure of $\mathcal{S}(A)$ in X, is compact.*

The following is a well-known property of the Hausdorff metric [8].

Proposition 4.2 *If Y is a compact metric space, then $\mathcal{H}(Y)$ with the Hausdorff metric is also compact.*

It follows that if X is a locally compact metric space, then $\mathcal{H}(X)$ is also locally compact.

Proposition 4.3 *Assume Hypotheses 4.1; then the closure, $\overline{\mathcal{F}}$, of $\mathcal{F}$ in $\mathcal{C}(\mathcal{H}(X))$, the space of all continuous mappings of $\mathcal{H}(X)$ to itself with the compact–open topology, is a compact semigroup.*

Proof. By Proposition 4.1 and the Arzela-Ascoli's Theorem, it suffices to show that, for each $A \in \mathcal{H}(X)$, $\mathcal{F}(A) = \{F^i(A) : i = 0, 1, 2, ...\}$ has compact closure in $\mathcal{H}(X)$. Note that for each i, $F^i(A) \subset \overline{\mathcal{S}(A)}$ which is a compact subset of X by Lemma 4.4. Note that the closure of $\mathcal{F}(A)$ in $\mathcal{H}(\overline{\mathcal{S}(A)})$ is the same as its closure in $\mathcal{H}(X)$; by Proposition 4.2, the former is compact. $\Box$

Suppose that $\overline{\mathcal{F}}$ is compact. Consider

$$\mathcal{K} = \bigcap_{n=1}^{\infty} \overline{\{F^i \mid i \geq n\}}$$

which is called in semigroup theory [11, p.27] the *kernel* or *minimal ideal* of $\overline{\mathcal{F}}$; consequently, we shall refer to $\mathcal{K}$ as the *kernel* of the *IFS* $\{X; f_1, ..., f_n\}$ or the *kernel* of F. The following is Theorem 3.1.1 from [11, p.109].

Proposition 4.4 *The kernel of the IFS* $\{X; f_1, ..., f_n\}$, $\mathcal{K}$, *is a compact monothetic group; i.e.,* $\mathcal{K}$ *is a compact Abelian group such that* $\{F^i \iota\}_{i=1}^{\infty}$ *is dense in* $\mathcal{K}$ *where* ι *is the identity element of* $\mathcal{K}$.

In the case that $\{X, f_1, ..., f_n\}$ is a hyperbolic *IFS* where X is a complete metric space then Hutchinson [5] showed that F is a contraction map (with respect to the Hausdorff metric) with a unique fixed point, A (which Hutchinson called an *attractor*). It follows that $\mathcal{K}$ is the trivial group where ι is the constant map which sends each element of $\mathcal{H}(X)$ to A. We refer to a fixed point of F as an *invariant set* of the *IFS* $\{X; f_1, ..., f_n\}$ or an *invariant set* of F.

Proposition 4.5 *If* $\mathcal{K}$ *is the kernel of the IFS* $\{X; f_1, ..., f_n\}$ *then* $A \in \mathcal{H}(X)$ *is an invariant set of the IFS* $\{X; f_1, ..., f_n\}$ *if and only if there exists* $B \in \mathcal{H}(X)$ *such that* $\mathcal{K}(B) = A$, *where* $\mathcal{K}(B) = \cup_{k \in \mathcal{K}} k(B)$.

Proof. Suppose that $F(A) = A$; since $\iota = \lim_{j \to \infty} F^{i_j}$ for some subsequence, then we have $\iota(A) = A$. Since $F^i \iota(A) = \iota F^i(A) = A$, it follows from Proposition 4.4 that $\mathcal{K}(A) = A$.

Conversely, suppose that $A = \mathcal{K}(B)$. Since $\mathcal{K}$ is a group containing $F\iota$ then $F\mathcal{K} = F(\iota\mathcal{K}) = (F\iota)\mathcal{K} = \mathcal{K}$ which implies that $F(A) = F(\mathcal{K}(B)) = \mathcal{K}(B) = A$. $\square$

Although ι is the identity element in $\mathcal{K}$, ι need not be the identity map of $\mathcal{H}(X)$; in the next section we show the consequences of assuming that ι is the identity map on $\mathcal{H}(X)$. In general, since $\iota^2 = \iota$, we have the following.

Proposition 4.6 ι *is a retraction of* $\mathcal{H}(X)$ *onto some subspace.*

Problem 4.1 What can be said about this subspace? Mogilski and Dijkstra tell us that $\mathcal{H}(\mathbf{R}^m)$, with $m > 1$, is homeomorphic to $\mathbf{Q} \times [0, 1)$ where $\mathbf{Q}$ is the Hilbert cube. It follows that $\iota(\mathcal{H}(\mathbf{R}^m))$ is an ANR (absolute neighborhood retract); note that $\mathcal{K}$ is a compact group acting on $\iota(\mathcal{H}(\mathbf{R}^m))$. Is $\iota(\mathcal{H}(\mathbf{R}^m))$ homeomorphic to $\mathbf{Q} \times [0, 1)$?

We leave to the reader the verification the following property of Hausdorff metrics.

Lemma 4.5 *Suppose that $A_1 \subseteq A_2 \subseteq A_3 \subseteq \ldots$ is an infinite sequence of nonempty nested compact subsets of X such that, for some compact subset $C \subseteq X$, $\cup_i A_i \subseteq C$. Then*

$$\lim_{i \to \infty} A_i = \overline{\bigcup_{i=1}^{\infty} A_i}.$$

Suppose that $A_1 \supseteq A_2 \supseteq A_3 \supseteq \ldots$ is an infinite sequence of nonempty nested compact subsets of X. Then

$$\lim_{i \to \infty} A_i = \bigcap_{i=1}^{\infty} A_i.$$

Hypotheses 4.2 Suppose that $\{X; f_1, \ldots, f_n\}$ is an *IFS* such that X is a locally compact metric space and such that the closure of the semigroup, $\mathcal{S}$, generated by the f_i's is compact; suppose that $\mathcal{K}$ is the kernel of this *IFS*.

Proposition 4.7 *Assume Hypotheses 4.2 and suppose that there exists $g \in \mathcal{S}^{(k)}$ which is periodic with period p; i.e. $g^p(x) = x$ for all $x \in X$. Then $F^{kp}\iota = \iota$ and the kernel, $\mathcal{K}$, is a cyclic group with order dividing kp.*

Proof. Suppose that $A \in \mathcal{H}(X)$; since $F^{kp}(A) = \cup\{q(A) : q \in \mathcal{S}^{(kp)}\}$, $A \subseteq F^{kp}(A)$. Applying F^{kp} to the latter, we obtain $F^{kp}(A) \subseteq F^{2kp}(A)$ and it follows that we have an increasing sequence

$$A \subseteq F^{kp}(A) \subseteq F^{2kp}(A) \subseteq F^{3kp}(A) \subseteq \ldots .$$

By Lemma 4.5, $\lim_{i \to \infty} F^{ikp}(A) = \overline{\cup_{i=1}^{\infty} F^{ikp}(A)}$. Hence, since A is arbitrary, $\tau = \lim_{i \to \infty} F^{ikp} \in \mathcal{K}$. Since $F^{kp}\iota\tau = F^{kp}\tau = \tau$ and $\mathcal{K}$ is a group, it follows that $(F\iota)^{kp} = F^{kp}\iota = \iota$. $\square$

Corollary 4.1 *Suppose that there exist periodic $g_1 \in \mathcal{S}^{(k)}$ and $g_2 \in \mathcal{S}^{(l)}$ with periods p and q, respectively, such that kp and lq are relatively prime; then $\mathcal{K}$ is the trivial group. In particular, $\lim_{i \to \infty} F^i = \iota$.*

Proposition 4.8 *Assume Hypotheses 4.2 and suppose that for some i and some $A \in \mathcal{H}(X)$, $f_i(A) \supseteq A$. Then, for all $k \in \mathcal{K}$, $k(A) = \overline{\cup_{i=1}^{\infty} F^i(A)}$.*

Proof. Since $A \subseteq f_i(A)$ implies that $A \subseteq F(A)$ then, as in the proof of Proposition 4.7, $\lim_{i \to \infty} F^i(A) = \overline{\cup_{i=1}^{\infty} F^i(A)}$. If $k \in \mathcal{K}$, then $k = \lim_{j \to \infty} F^{i_j}$ for some subsequence $\{i_j\}$ of the positive integers. The Proposition follows. $\square$

Corollary 4.2 *Suppose that x_0 is a fixed point of some f_i; then $\iota(\{x_0\}) = \mathcal{K}(\{x_0\})$.*

Now, let us just suppose that the semigroup $\mathcal{S}$ is generated by the continuous maps $f_1, f_2, ..., f_n$ of a locally compact metric space X to itself. If $x \in X$ then we define

$$\Omega(x) = \bigcap_{i=1}^{\infty} \overline{\{g(x) : g \in \mathcal{S}^{(p)}, p \geq i\}}.$$

Proposition 4.9 *If $\overline{\mathcal{S}(x)}$ is compact or if each f_i is one–to–one then $\Omega(x)$ is an invariant set of F.*

Proof. We need to show that

$$\bigcup_{i=1}^{n} f_i(\Omega(x)) = \Omega(x).$$

Clearly, we have $\cup_{i=1}^{n} f_i(\Omega(x)) \subseteq \Omega(x)$. Now suppose that $y \in \Omega(x)$; then, for each i,

$$y \in \overline{\{k(x) \mid k \in \mathcal{S}^{(p)}, p \geq i\}}$$

and, hence, we can find $k_i \in \mathcal{S}^{(p_i)}$ such that $p_i \geq i$ and $d(y, k_i(x)) < 1/i$. There exists some fixed j and a subsequence $\{k_{i_l}(x)\}$ such that for each l, $k_{i_l} = f_j k'_l$.

If we assume that f_j is one-to-one, then the sequence $\{k'_l(x)\}$ converges to some $y' \in X$. Clearly $y' \in \Omega(x)$ and, hence, $y = f_j(y') \in f_j(\Omega(x))$.

If $\overline{\mathcal{S}(x)}$ is compact then we can find a subsequence of $\{k'_l(x)\}$ which converges to some $y' \in X$. Again, we have $y' \in \Omega(x)$ and, hence, $y = f_j(y') \in f_j(\Omega(x))$. $\square$

An analogous proof obtained by expressing $h_{i_l} = h'_l f_j$ gives the following.

Proposition 4.10 $\Omega(x) = \bigcup_{i=1}^{n} \Omega(f_i(x))$.

By induction, we obtain

Corollary 4.3 *Let j be a positive integer; then*

$$\Omega(x) = \bigcup_{h \in \mathcal{S}^{(j)}} \Omega(h(x)).$$

Proposition 4.11 *Assume Hypotheses 4.2. If x_0 is a fixed point of some f_i then*

$$\Omega(x_0) = \mathcal{K}(\{x_0\}) = \overline{\mathcal{S}}(x_0).$$

Proof. By Corollary 4.2, $\mathcal{K}(\{x_0\}) = \iota(\{x_0\})$. By Proposition 4.8, $\iota(\{x_0\}) = \overline{\cup_{p=1}^{\infty} F^p(\{x_0\})}$. Clearly, $\overline{\cup_{p=1}^{\infty} F^p(\{x_0\})} = \overline{\mathcal{S}(x_0)} = \overline{\mathcal{S}}(x_0)$. Therefore, we have $\Omega(x_0) \subseteq \overline{\mathcal{S}}(x_0) = \mathcal{K}(\{x_0\})$.

Now suppose that $z \in \mathcal{K}(\{x_0\}) = \overline{\cup_{p=1}^{\infty} F^p(\{x_0\})}$; hence, $z = \lim_{i \to \infty} z_i$ where $z_i \in F^{p_i}(\{x_0\})$ for some positive integer p_i. Suppose that $z_i = g_i(x_0)$ where $g_i \in \mathcal{S}^{(p_i)}$. By the proof of Proposition 4.8, we can assume that the sequence $\{p_i\}$ is an increasing sequence. It follows that $z \in \Omega(x_0)$. $\square$

5 Stochastic Matrices - An Example

In this section, we apply the theory developed in the previous sections to study semigroups generated by column stochastic matrices. Hyperplanes in $\mathbf{R}^n$ of the form $\sum_{i=1}^{n} x_i = $ constant are invariant under linear mappings determined by column stochastic matrices. We will show that the restriction of the action to each such hyperplane gives rise to a hyperbolic *IFS* and that the invariant sets of such a stochastic *IFS* is the "cone" over the attractor of the induced *IFS*. First, we need to do some preliminary work; we leave the proof of the following lemma to the reader.

Lemma 5.1 *If $\{L_i\}$ is a sequence of compacta such that $\lim_{i \to \infty} L_i = L$ then the union of L and $\cup_{i=1}^{\infty} L_i$ is compact.*

Consider $\mathbf{R}^n \times \mathbf{R}$ with the metric $d((x, p), (y, q)) = d_0(x, y) + |p - q|$ where d_0 is a metric on $\mathbf{R}^n$; let $\pi : \mathbf{R}^n \times \mathbf{R} \to \mathbf{R}$ be the projection map onto the last factor. Suppose that $f_1, f_2, ..., f_r$ are continuous maps of $\mathbf{R}^n \times \mathbf{R}$ to itself such that $\pi f_i = \pi$ and such that (*Condition (*)*) there exists $s \in (0, 1)$ for which $d(f_i(x, p), f_i(y, p)) \leq s \cdot d((x, p), (y, p))$ for all $(x, p), (y, p) \in \mathbf{R}^n \times \mathbf{R}$ and $i = 1, 2, ..., r$. Therefore the restriction of the *IFS* $\{\mathbf{R}^n \times \mathbf{R}; f_1, f_2, ..., f_r\}$ to each hyperplane $\pi^{-1}(p) = \mathbf{R}^n \times \{p\}$ is a hyperbolic *IFS*. Let F be the union of the f_i's. From comments above and Hutchinson [5], we have the following.

Proposition 5.1 *If C is a compactum which lies in the hyperplane $\pi^{-1}(p)$ then the limit* $\lim_{i \to \infty} F^i(C)$ *exists and is a compactum which also lies in the hyperplane $\pi^{-1}(p)$.*

Corollary 5.1 *If $(x, p) \in \mathbf{R}^n \times \mathbf{R}$ then $L_p = \lim_{i \to \infty} F^i(\{(x, p)\})$ exists and is a compactum which lies in the hyperplane $\pi^{-1}(p)$.*

Let us assume that all elements of the semigroup $\mathcal{S}$ generated by the f_i's have the same Lipschitz constant; hence, $\mathcal{S}$ is equicontinuous and, consequently, has compact closure in the space of all continuous maps of $\mathbf{R}^n \times \mathbf{R}$ to itself. Then, all of the theory we have developed so far applies to $\mathcal{S}$.

Proposition 5.2 *If C is a compact subset of $\mathbf{R}^n \times \mathbf{R}$ then the limit $\lim_{i \to \infty} F^i(C)$ exists.*

Proof. Let $L = \bigcup_{p \in C} L_{\pi(p)}$. First, we want to show that L is compact. Let $\{y_i\} \subset L$ be an infinite sequence. Therefore, for each i, $y_i \in L_{\pi(p_i)}$ for a unique $p_i \in C$. By choosing subsequences, we can assume that the sequence $\{p_i\}$ converges to $p_0 \in C$. By Barnsley ([1, Theorem 1, p.113]), the sequence $\{L_{\pi(p_i)}\}$ converges to $L_{\pi(p_0)}$. It follows from Lemma 5.1 above that $\{y_i\}$ has a convergent subsequence in the union of $L_{\pi(p_0)}$ and $\bigcup_{i=1}^{\infty} L_{\pi(p_i)}$. Therefore L is compact.

We will now show that if the subsequence $\{F^{i_k}(C)\}$ converges to the compactum M then $M = L$. Once we have shown this, then it follows from Proposition 4.3 that the limit $\lim_{i \to \infty} F^i(C)$ exists. For $p \in \mathbf{R}$, let $C_p = C \cap \pi^{-1}(p)$. Since $C_p \subseteq C$, it follows that $L_p \subseteq M$ and, hence, $L \subseteq M$.

Suppose that there exists $y \in M \setminus L$; let $\epsilon = d(y, L)/3$. There exists N such that $k > N$ implies that $h(M, F^{i_k}(C)) < \epsilon$; in particular, for $k > N$, there exists $x_k \in C$ and $g_k \in \mathcal{S}^{(i_k)}$ such that $d(y, g_k(x_k)) < \epsilon$. By choosing a subsequence, we can assume that the sequence $\{x_k\}$ converges to some $x_0 \in C$. Since $\mathcal{S}$ is equicontinuous at x_0 there exists $\delta > 0$ such that $d(x_0, z) < \delta$ implies $d(g(x_0), g(z)) < \epsilon$ for all $g \in \mathcal{S}$. Note that

$$d(y, g_k(x_0)) \leq d(y, g_k(x_k)) + d(g_k(x_k), g_k(x_0)) < 2\epsilon$$

for k sufficiently large. Again, by choosing a subsequence, we can assume that the sequence $\{g_k(x_0)\}$ converges; by Proposition 5.1, it follows that the limit of the latter sequence lies in

$L_{\pi(x_0)}$. Hence, $d(y, L) \le d(y, L_{\pi(x_0)}) \le 2\epsilon$ and we have a contradiction. $\square$

Corollary 5.2 *The kernel $\mathcal{K}$ of the IFS $\{\mathbf{R}^n \times \mathbf{R}; f_1, f_2, ..., f_r\}$ is the trivial group.*

An $s \times s$ matrix (a_{ij}) is a *column stochastic matrix* if, for each i, $\sum_{j=1}^{s} a_{ij} = 1$ and if each entry $a_{ij} \ge 0$. (a_{ij}) is a *positive matrix* if each $a_{ij} > 0$.

Proposition 5.3 *If S is the semigroup generated by $s \times s$ column stochastic matrices, A_1, $A_2, ..., A_r$, then the closure of S is compact.*

Suppose that $A = (a_{ij})$ is an $(n+1) \times (n+1)$ column stochastic matrix. Let $B = (b_{ij})$ be the $(n+1) \times (n+1)$ matrix such that

$$b_{ij} = \begin{cases} a_{ij} - a_{i,n+1} & \text{if } i \le n \text{ and } j \le n \\ a_{ij} & \text{if } i \le n \text{ and } j = n+1 \\ 0 & \text{if } i = n+1 \text{ and } j \le n \\ 1 & \text{if } i = j = n+1. \end{cases}$$

Define ϕ on $\mathbf{R}^{n+1}$ by

$$\phi(x_1, x_2, ..., x_n, t) = (x_1, x_2, ..., x_n, t - x_1 - x_2 - ... - x_n).$$

Lemma 5.2 $A\phi = \phi B$.

We shall call B the ϕ–conjugate of A. Let B' be the $n \times n$ matrix obtained from B obtained by deleting the last row and last column of B. Consider the norm $\| \cdot \|$ on $\mathbf{R}^n$ defined by

$$\|(x_1, x_2, ..., x_n)\| = \sum_{i=1}^{n} |x_i|.$$

Note that, for an $n \times n$ matrix $D = (d_{ij})$, the induced matrix norm is given by

$$\|D\| = \max_{j=1,2,...,n} \{\sum_{i=1}^{n} |d_{ij}|\}.$$

Proposition 5.4 $\|B'\| \le 1$ *and if A is a positive matrix then $\|B'\| < 1$.*

Proof. Let $Q = \{\vec{x} \in \mathbf{R}^n : \|\vec{x}\| \le 1\}$, let $Q_+ = \{(x_1, x_2, ..., x_n) \in Q : x_i \ge 0\}$ and let $\{\vec{e}_i\}$ be the standard basis for $\mathbf{R}^n$. Note that Q_+ is an n–dimensional simplex the set of whose edges contains the standard basis. We need to show that $B'(Q) \subseteq Q$. Consider the affine map $f(\vec{x}) = B'(\vec{x}) + \vec{\alpha}$ where $\vec{\alpha} = (a_{1,n+1}, a_{2,n+1}, ..., a_{n,n+1})$. Since $f(\vec{e_i}) = (a_{1i}, a_{2i}, ..., a_{ni})$, it follows that $f(Q_+) \subseteq Q_+$. Note that $f(Q_+)$ is a simplex each of whose edges has length ≤ 1. Since $B'(Q_+)$ is a simplex which is the translate of $f(Q_+)$, it follows that, for each i, $\|B'(\vec{e_i})\| \leq 1$ and, hence, $B'(Q) \subseteq Q$. We leave to the reader to check that if A is a positive matrix then $\|B'\| < 1$. $\square$

Corollary 5.3

$$\|B(x_1, x_2, ..., x_n, p) - B(y_1, y_2, ..., y_n, p)\| \leq \|B'\| \cdot \|(x_1, x_2, ..., x_n, p) - (y_1, y_2, ..., y_n, p)\|.$$

Theorem 5.1 *Suppose that $f_1, f_2, ..., f_r$ are linear maps of $\mathbf{R}^{n+1}$ to itself such that, with respect to some basis, the matrices A_i associated to f_i are positive column stochastic matrices. Let ρ be the orthogonal projection of $\mathbf{R}^{n+1}$ onto the line λ given by $x_1 = x_2 = ... = x_{n+1}$ (using coordinates with respect to the above–mentioned basis).*

Then the semigroup S generated by the f_i's has compact closure and the kernel K is the trivial group. For each $\vec{x} \in \mathbf{R}^{n+1}$, $K(\{\vec{x}\})$ is the unique attractor of the restriction of $\overline{S}$ to the hyperplane $\rho^{-1}(\rho(\vec{x}))$. The invariant sets of the IFS $\{\mathbf{R}^{n+1}; f_1, f_2, ..., f_r\}$ are sets of the form

$$\bigcup_{\vec{x} \in C} K(\{\vec{x}\})$$

where C is a compact subset of λ.

$K(\{\vec{0}\}) = \{\vec{0}\}$ and for all $\vec{x}, \vec{y} \in \lambda \setminus \{\vec{0}\}$, $K(\{\vec{x}\})$ and $K(\{\vec{y}\})$ are (affinely) homeomorphic.

Proof. For each i, let B_i be the ϕ–conjugate of A_i and let B_i' be the submatrix obtained from B_i by deleting the last row and last column. Let $\pi : \mathbf{R}^{n+1} \to \mathbf{R}$ be the projection onto the last coordinate (using the coordinates mentioned in the hypotheses). By Corollary 5.3, the IFS $\{\mathbf{R}^{n+1}; B_1, B_2, ..., B_r\}$ satisfies Condition $(*)$. The reader may check the details of the rest of the proof. $\square$

Example 5.1. Consider the matrices

$$\begin{pmatrix} \frac{3}{5} & \frac{1}{10} & \frac{1}{10} \\ \frac{3}{10} & \frac{4}{5} & \frac{3}{10} \\ \frac{1}{10} & \frac{1}{10} & \frac{3}{5} \end{pmatrix}, \begin{pmatrix} \frac{4}{5} & \frac{3}{10} & \frac{3}{10} \\ \frac{1}{10} & \frac{3}{5} & \frac{1}{10} \\ \frac{1}{10} & \frac{1}{10} & \frac{3}{5} \end{pmatrix}, \text{ and } \begin{pmatrix} \frac{3}{5} & \frac{1}{10} & \frac{1}{10} \\ \frac{1}{10} & \frac{3}{5} & \frac{1}{10} \\ \frac{3}{10} & \frac{3}{10} & \frac{4}{5} \end{pmatrix}.$$

It is easily seen that $\mathcal{K}(\{(x_1, x_2, x_3)\})$ is homeomorphic to the Sierpinski triangle provided that $x_1 + x_2 + x_3 \neq 0$; otherwise, it is a single point. $\square$

Problem 5.1 For what hyperbolic *IFS*'s on $\mathbf{R}^n$ does there exist an *IFS* on $\mathbf{R}^{n+1}$ consisting of linear maps such that, with respect to some basis, the associated matrices are positive column stochastic matrices and such that the restriction of the latter *IFS* to the hyperplane $x_1 + x_2 + \ldots + x_{n+1} = 1$ is topologically equivalent to the given hyperbolic *IFS*?

6 A Special Case.

In this section, $\mathcal{S}$ is a semigroup generated by a finite number of continuous maps, $f_1, f_2, \ldots, f_n$, of a locally compact metric space X to itself. We shall assume that $\overline{\mathcal{S}}$ is compact, F is the union of the f_i's and $\mathcal{K}$ is the kernel of the *IFS* $\{X; f_1, \ldots, f_n\}$ with identity element ι.

Theorem 6.1 *Suppose that ι is the identity map of $\mathcal{H}(X)$; then $\overline{\mathcal{S}}$ is a compact monothetic group generated by f_1. The invariant sets of $\{X; f_1, \ldots, f_n\}$ are obtained by taking, for any $A \in \mathcal{H}(X)$, the set*

$$\overline{\mathcal{S}}(A) = \bigcup_{s \in \overline{\mathcal{S}}} s(A) = \overline{\bigcup_{i=1}^{\infty} f_1^i(A)}.$$

The kernel $\mathcal{K}$ is isomorphic to $\overline{\mathcal{S}}$.

Proof. Suppose that $\iota = \lim_{j \to \infty} F^{i_j}$. If $x_0 \in X$ then $F^{i_j}(\{x_0\}) \to \{x_0\}$ and it follows from the definition of F that $f_k^{i_j}(\{x_0\}) \to \{x_0\}$ for $k = 1, 2, \ldots, n$. Therefore, for each k, the sequence $f_k^{i_j}$ converges pointwise to the identity map 1_X of X.

Note that this convergence also implies that each f_k is one–to–one. For, suppose that $f_k(x) = f_k(z)$; then, for each i, $f_k^i(x) = f_k^i(z)$ and we have

$$x = \lim_{j \to \infty} f_k^{i_j}(x) = \lim_{j \to \infty} f_k^{i_j}(z) = z.$$

By Proposition 4.4, $\mathcal{K}$ is a group which contains $F\iota$; in particular, $F \in \mathcal{K}$ and, hence, F is invertible. Let $x \in X$ and suppose that $F^{-1}(\{x\}) = A$. Since $F(A) = \{x\}$ and since each f_i is one-to-one, there exists a unique $y \in X$ such that, for all i, $f_i(y) = x$. This implies that

$f_i = f_j$ for all i and j. In particular, for each $x \in X$, $F(\{x\}) = \{f_1(x)\}$ and it follows that f_1 is an onto map. The continuity of F^{-1} implies the continuity of f_1. $\square$

We note that if one starts with a homeomorphism f of X to itself such that the semigroup generated by f has compact closure in $\mathcal{C}(X)$ then the identity element of the kernel of the IFS $\{X; f\}$ is the identity map of $\mathcal{H}(X)$. As an example, if we let X be the m–dimensional torus (*i.e.*, the product of m one–dimensional spheres) and if we define $f : X \to X$ by

$$f(x_1, x_2, ..., x_m) = (x_1 e^{\alpha_1 \pi}, x_2 e^{\alpha_2 \pi}, ..., x_m e^{\alpha_m \pi})$$

where $\alpha_1, \alpha_2, ..., \alpha_m$ are irrational numbers which are linearly independent over the rationals, then the semigroup generated by f has compact closure in $\mathcal{C}(X)$.

7   Compact Groups.

In this section, we assume

Hypotheses 7.1 $\mathcal{S}$ is a semigroup generated by a finite number of homeomorphisms, $f_1, f_2, ...,$ f_n, of a locally compact metric space X to itself such that the closure $\overline{\mathcal{S}}$ is a compact group. Again we let $\mathcal{F}$ be the semigroup generated by F, the union of the f_i's, and $\mathcal{K}$ is the kernel of the IFS $\{X; f_1, ..., f_n\}$ with identity element ι.

Theorem 7.1 *Assume Hypotheses 7.1; then A is an invariant set of the IFS $\{X; f_1, ..., f_n\}$ if and only if $\overline{\mathcal{S}}(A) = A$.*

Proof. Suppose that $F(A) = A$, then, for each i, $f_i(A) \subseteq A$. It follows that for each $g \in \mathcal{S}$ that $g(A) \subseteq A$ and, hence, the same is true for each $g \in \overline{\mathcal{S}}$. Since the identity map is an element of $\overline{\mathcal{S}}$, we obtain $\overline{\mathcal{S}}(A) = A$.

Suppose that $\overline{\mathcal{S}}(A) = A$; since $\overline{\mathcal{S}}$ is a group, $f_i \overline{\mathcal{S}} = \overline{\mathcal{S}}$ for each i. Hence, for each i, $f_i(A) = f_i(\overline{\mathcal{S}}(A)) = (f_i \overline{\mathcal{S}})(A)) = \overline{\mathcal{S}}(A) = A$. $\square$

Theorem 7.2 *Assume Hypotheses 7.1 and suppose that $\overline{\mathcal{S}}$ is a compact Abelian group. Then there exists a continuous epimorphism $\phi : \overline{\mathcal{S}} \to \mathcal{K}$ such that, for each $i = 1, 2, ..., n$, $\phi(f_i) = F\iota$.*

Proof. If $A \in \mathcal{H}(X)$, consider the sequence $\{f_1^{-j} F^j(A)\}_{j=0}^{\infty}$. Note that, for each j, $f_1^{-j} F^j(A) \subseteq f_1^{-(j+1)} F^{j+1}(A)$ and, hence, by Lemma 4.5,

$$\widehat{A} = \lim_{j \to \infty} f_1^{-j} F^j(A)$$

exists. Let $\overline{S_1}$ be the closed subgroup generated by f_1; we want to define a function $\phi_1 : \overline{S_1} \to \mathcal{K}$. Suppose that $g \in \overline{S_1}$ and $A \in \mathcal{H}(X)$; we define

$$\phi_1(g)(A) = g(\widehat{A}).$$

First, we show that if $g = \lim_{p \to \infty} f_1^{i_p}$ for some subsequence $\{i_p\}$ of the positive integers then

$$\phi_1(g)(A) = \lim_{p \to \infty} F^{i_p}(A).$$

Let $\epsilon > 0$ and choose N such that $p > N$ implies that $d(g(x), f_1^{i_p}(x)) < \epsilon/2$ for all $x \in \overline{S}(A)$; since g induces a continuous map on $\mathcal{H}(X)$, we can also assume that $p > N$ implies that $h(g(\widehat{A}), g(f_1^{-i_p} F^{i_p}(A))) < \epsilon/2$. Then, for $p > N$, we have

$$h(g(\widehat{A}), F^{i_p}(A)) \leq h(g(\widehat{A}), g(f_1^{-i_p} F^{i_p}(A))) + h(g(f_1^{-i_p} F^{i_p}(A)), f_1^{i_p}(f_1^{-i_p} F^{i_p}(A))) < \epsilon.$$

Note that, in particular, we have $\phi_1(g) \in \mathcal{K}$. Suppose that $\kappa \in \mathcal{K}$; then $\kappa = \lim_{p \to \infty} F^{j_p}$ for some subsequence of $\mathcal{F}$. Note that the sequence $f_1^{j_p}$ has a subsequence which converges to some $g' \in \overline{S_1}$. Using this subsequence, it is straightforward to check that $\phi_1(g') = \kappa$ and, therefore, ϕ_1 is an onto map.

Suppose that $g_1 = \lim_{p \to \infty} f_1^{i_p}$ and $g_2 = \lim_{p \to \infty} f_1^{j_p}$ are elements of $\overline{S_1}$; then their product $g_1 g_2 = \lim_{p \to \infty} f_1^{i_p + j_p}$. From this, one can show that $\phi_1(g_1 g_2) = \phi_1(g_1)\phi_1(g_2)$; i.e., ϕ_1 is a homomorphism.

To show that ϕ_1 is continuous at g, let $\epsilon > 0$ be given. Suppose that $g' \in \overline{S_1}$ such that $d(g(x), g'(x)) < \epsilon$ for all $x \in \overline{S}(A)$. Clearly, we have $h(g(\widehat{A}), g'(\widehat{A})) < \epsilon$. Suppose that $I = \lim_{p \to \infty} f_1^{i_p}$ is the identity element of $\overline{S}$. Note that

$$\iota(A) = \phi_1(I)(A) = I(\widehat{A}) = \widehat{A}$$

and

$$\phi_1(f_1)(A) = \phi_1(f_1 I)(A) = \phi_1(\lim_{p \to \infty} f_1^{i_p + 1})(A) = \lim_{p \to \infty} F^{i_p + 1}(A) = F\iota(A).$$

Note that this implies that

$$\widehat{\widehat{A}} = \iota(\widehat{A}) = \iota(\iota(A)) = \iota(A) = \widehat{A}.$$

To summarize what we have done so far, we have defined a continuous epimorphism $\phi_1 : \overline{S_1} \to \mathcal{K}$. Similarly, for each $i = 2, 3, ..., n$, we can define $\phi_i : \overline{S_i} \to \mathcal{K}$ where $\overline{S_i}$ is the smallest closed subgroup of $\overline{S}$ containing f_i. If $g = g_1 g_2 ... g_n \in \overline{S}$, where each $g_i \in \overline{S_i}$, then we would like to define $\phi(g) = \phi_1(g_1) \phi_2(g_2) ... \phi_n(g_n)$. First, note that g can be expressed in this form but the expansion is not necessarily unique. To show that g is well-defined it suffices to show that if $g_1 g_2 ... g_n = I$ is the identity element of $\overline{S}$ then $\phi_1(g_1)\phi_2(g_2)...\phi_n(g_n) = \iota$, the identity element of $\mathcal{K}$. Note that if $A \in \mathcal{H}(X)$, then

$$\widehat{g_i(A)} = \lim_{j \to \infty} f_1^{-j} F^j(g_i(A)) = \lim_{j \to \infty} g_i f_1^{-j} F^j(A) = g_i(\widehat{A}).$$

Hence,

$$\begin{aligned}
\phi_1(g_1)\phi_2(g_2)...\phi_n(g_n)(A) &= \phi_1(g_1)\phi_2(g_2)...\phi_{n-1}(g_{n-1})g_n(\widehat{A}) \\
&= \phi_1(g_1)\phi_2(g_2)...\phi_{n-1}(g_{n-1})\widehat{g_n(A)} \\
&= ... \\
&= \widehat{g_1 g_2 ... g_n(A)} = \widehat{I(A)} = \widehat{A} = \iota(A).
\end{aligned}$$

It is straightforward now to check that ϕ is a continuous epimorphism. $\square$

Corollary 7.1 *If $x \in X$ then*

$$\iota(\{x\}) = K_\phi(x) = \{g(x) : g \in K_\phi\}$$

where K_ϕ is the kernel of the homomorphism ϕ.

Example 7.1 Let $X = \mathcal{S}^1$ be the one-dimensional sphere and let f and g be rotations of X through angles α and β, respectively, which are irrational multiples of π. Let S be the semigroup generated by f and g; it is well–known that $\overline{S}$ is a compact Abelian group. If $\alpha = \beta p/q$ where p and $q > 0$ are relatively prime integers, then S and $\mathcal{K}$ are isomorphic to the circle group or $SO(2)$ and ϕ is the q–fold covering homomorphism. For each $x \in X$, $\iota(\{x\})$ consists of precisely q–points, $x, xe^{\pi/q}, xe^{2\pi/q}, ..., xe^{(q-1)\pi/q}$. The only invariant set is the entire space X.

Now suppose that α is not a rational multiple of β; then S is still isomorphic to the circle group or $SO(2)$ but K is the trivial group and, hence, ϕ is the trivial homomorphism. Again, the only invariant set is the entire space X. $\square$

By extending this construction to higher dimensions, one can generate a plethora of examples. In the next section, we will discuss the case when S is a finite group.

Problem 7.1 To what extent can Theorem 7.2 be generalized to cover the case when S is not commutative?

8 Finite Groups.

In this section, we restrict our attention to the case when the semigroup S is a finite group. In this case we can obtain a calculation of both the kernel, K, and its identity element, ι.

Hypotheses 8.1 S is a semigroup generated by a finite number of homeomorphisms, $f_1, f_2, ...,$ f_n, of a locally compact metric space X to itself such that the closure $\overline{S}$ is a finite group. Again we assume that $\mathcal{F}$ is the semigroup generated by F, the union of the f_i's, and K is the kernel of the IFS $\{X; f_1, ..., f_n\}$ with identity element ι.

Theorem 8.1 *Suppose that $\overline{S}$ is a finite group such that for some point $x_0 \in X$ the function $\overline{S} \to \overline{S}(x_0)$ defined by $g \to g(x_0)$ is a bijection. Then*

1. *there exists k such that $S^{(k)}$ is a normal subgroup of $\overline{S}$.*

2. *Choose the smallest k such that $S^{(k)}$ is a normal subgroup of $\overline{S}$; then $\iota = F^k$.*

3. *Choose the smallest $j > k$ such that $F^j = F^k$; then j is the smallest integer greater than k such that $S^{(j)} = S^{(k)}$. The kernel K is isomorphic to $\mathbf{Z}_{j-k}$. $j - k$ is a factor of k.*

4. *If $S^{(k)}$ and $S^{(l)}$ are subgroups, then $S^{(k)} = S^{(l)}$.*

5. *The kernel K is isomorphic to the quotient group $\overline{S}/S^{(k)}$.*

Proof. Suppose that $\iota = \lim_{j \to \infty} F^{i_j}$ for some subsequence $\{i_j\}$. Since $\overline{S}$ is finite, for some subsequence $\{i_{j_l}\}$ of $\{i_j\}$, each $S^{(i_{j_l})}$ must be some $S^{(k)}$ for some fixed k. Choose the smallest

k with this property. Suppose that $S^{(k)} = \{g_i\}_{i=1}^p$, then from the comment above, it follows that $\iota(A) = F^k(A) = \cup_{i=1}^p g_i(A)$ for each $A \in \mathcal{H}(X)$.

Since $\iota^2(\{x_0\}) = \iota(\{x_0\})$,

$$\bigcup_{i,j=1}^{p} g_j g_i(\{x_0\}) = \bigcup_{i=1}^{p} g_i(\{x_0\}).$$

It follows from the choice of x_0 that $S^{(k)}S^{(k)} = \{g_j g_i\}_{i,j=1}^p = \{g_i\}_{i=1}^p = S^{(k)}$ and, from group theory, it follows that $S^{(k)}$ is a subgroup of $\overline{S}$.

To see that $S^{(k)}$ is a normal subgroup of $\overline{S}$, it suffices to show that $f_i S^{(k)} \subseteq S^{(k)} f_i$ for all i. Let $f_i g_j \in f_i S^{(k)}$; then its length is $k+1$. Consider $m = f_i^k \in S^{(k)}$ which has an inverse m^{-1} both of which have length k. Then $f_i g_j = f_i g_j m^{-1} m = q f_i$ where $q = f_i g_j m^{-1} f_i^{k-1}$ has length $3k$ and, hence, lies in $S^{(3k)} = S^{(k)}S^{(k)}S^{(k)} = S^{(k)}$. This completes the proof of 8.1.1 and 8.1.2.

By Proposition 4.4, $F\iota$ generates the kernel. Choose the smallest s such that $(F\iota)^s = \iota$. Consider $F^k = \iota = (F\iota)^s = F^s \iota^s = F^s \iota = F^{s+k}$. Therefore, if j is the smallest integer greater than k such that $F^j = F^k$ then it follows that $s = j - k$ and, hence, the kernel is isomorphic to $\mathbf{Z}_{j-k}$. Since $(F\iota)^k = F^k \iota^k = \iota\iota = \iota$, $j - k$ must divide k.

Since $F_j = F_k$ and, in particular, $F_j(\{x_0\}) = F_k(\{x_0\})$, it follows that $S^{(j)} = S^{(k)}$. By using this argument again and induction, it follows that $S^{(k+\alpha(j-k))} = S^{(k)}$ for all positive integers α. Conversely, if $S^{(k)} = S^{(p)}$ then $F_k = F_p$ and we have shown 8.1.3.

Suppose that $S^{(l)} = \{h_i\}_{i=1}^q$ is a group for some $l > k$. Then, for each $A \in \mathcal{H}(X)$, $F^l(A) = \cup_{i=1}^q h_i(A)$.

Since $S^{(l)}S^{(l)} = S^{(l)}$, it follows that

$$F^{2l}(A) = F^l(F^l(A)) = \cup_{i,j=1}^q h_j h_i(A) = \cup_{i=1}^q h_i(A) = F^l(A).$$

Consider $((F\iota)^l)^2 = F^{2l}\iota = F^l\iota = (F\iota)^l$; it follows that $(F\iota)^l = \iota$. Therefore, $j - k$ divides l and, hence, $j - k$ divides $l - k$. If $l - k = a(j-k)$, then $l = k + a(j-k)$ and the equality $S^{(l)} = S^{(k)}$ follows from the previous paragraph and we have shown 8.1.4.

Suppose that $f_1 g \in f_1 S^{(k)}$. For any $i = 2, 3, ..., n$, $f_i^k \in S^{(k)}$; let $g' \in S^{(k)}$ be the inverse of f_i^k. Then $f_1 g = f_i^k g' f_1 g = f_i g''$ where $g'' = f_i^{k-1} g' f_1 g \in S^{(k)}$ since it has length $3k$. Hence, we have that the cosets $f_1 S^{(k)}$ and $f_i S^{(k)}$ are the same. Since $f_1, f_2, ..., f_n$ generate $\overline{S}$ it follows that $f_1 S^{(k)}$ generates the quotient group $\overline{S}/S^{(k)}$.

One can use the same argument used in showing that the cosets $f_1 S^{(k)}$ and $f_i S^{(k)}$ are the same to show that if $l \leq k$ is a positive integer and $g \in S^{(l)}$ then the cosets $f_1^l S^{(k)}$ and $g S^{(k)}$ are the same. Choose the smallest l such that $f_1^l S^{(k)} = S^{(k)}$; it follows that $S^{(k+l)} = S^{(k)}$ and that l is the smallest positive integer for which $S^{(k+l)} = S^{(k)}$. Therefore, l is the smallest positive integer for which $F^{k+l} = F^k$ and we must have that $l = j - k$. $\square$

Let us refer to the condition that for some point $x_0 \in X$ the function $\overline{S} \to \overline{S}(x_0)$ defined by $g \to g(x_0)$ is a bijection as *condition* (*). While condition (*) may seem quite restrictive, in some circumstances, it really isn't. The following proposition gives a condition which implies condition (*). If X is Euclidean space then, by [13], the hypothesis of the following Proposition is satisfied and, hence, condition (*) is satisfied.

Proposition 8.1 *Suppose that, for each g in the finite group $\overline{S}$ other than the identity, the collection of fixed points of g, $Fixed(g)$, is nowhere dense in X. Then there exists a dense set of points in X for which the function $\overline{S} \to \overline{S}(x_0)$ defined by $g \to g(x_0)$ is a bijection.*

Proof. Let $V = \cap_{g \neq \iota} X \setminus Fixed(g)$ and let $x_0 \in V$. Suppose that $g(x_0) = h(x_0)$; then $h^{-1}g(x_0) = x_0$ implies that $h = g$. $\square$

The following example shows the necessity of Condition (*) in Theorem 8.1.

Example 8.1 Let $X = \{1, 2, 3, 4\}$ and consider the following two cycles: $f_1 = (12)$ and $f_2 = (34)$ in the permutation group of X. They generate a group S which is isomorphic to the direct sum $\mathbf{Z}_2 \oplus \mathbf{Z}_2$. If $A \subseteq X$ is non–empty then note that $F(A)$ is one of the three sets $\{1, 2\}, \{3, 4\}$ or X. It is straightforward to check that $F^2(A) = F(A)$ from which it follows that the kernel $\mathcal{K}$ is the trivial group. Note that $S^{(2)}$ is the subgroup of S which is generated by $f_1 f_2$ and, hence, if Theorem 8.1 applied, then the kernel $\mathcal{K}$ would be isomorphic to $\mathbf{Z}_2$. $\square$

Proposition 8.2 *Assume Hypotheses 8.1 and suppose that $\overline{S} = \mathbf{Z}_{p_1} \oplus ... \oplus \mathbf{Z}_{p_n}$, where f_i generates $\mathbf{Z}_{p_i}$. Further suppose that $\gcd\{p_i\} = p_1$, $t \geq \sum_{k=2}^{n}(p_k - 1)$ and is divisible by p_1.*

Then $S^{(t)}$ is a group.

Proof. First note that the identity element $f_1^{p_1(t/p_1)} \in S^{(t)}$. Hence, $S^{(t)} \subseteq S^{(t)}S^{(t)}$. Now, consider any two elements of $S^{(t)}$, $\prod_{k=1}^{n} f_k^{i_k}$ and $\prod_{k=1}^{n} f_k^{j_k}$, whose product is $\prod_{k=1}^{n} f_k^{i_k+j_k}$. Choose nonnegative integers a_k and r_k, where $r_k < p_k$, so that, for all k, $i_k + j_k = a_k p_k + r_k$. Then note that

$$2t = \sum_{k=1}^{n}(i_k + j_k) = \sum_{k=1}^{n} r_k + \sum_{k=1}^{n}(a_k p_k) \leq r_1 + \sum_{k=2}^{n}(p_k - 1) + \sum_{k=1}^{n}(a_k p_k) \leq r_1 + t + \sum_{k=1}^{n}(a_k p_k),$$

and hence, $\sum_{k=1}^{n}(a_k p_k) + r_1 \geq t$. However, since $\sum_{k=1}^{n}(a_k p_k)$ and t are divisible by p_1, and $r_1 < p_1$, $\sum_{k=1}^{n}(a_k p_k) \geq t$.

Further, $\sum_{k=1}^{n}(a_k p_k) = t + Ap_1$, for some nonnegative integer A, implies that $t = \sum_{k=1}^{n} r_k + Ap_1$ and, hence,

$$\prod_{k=1}^{n} f_k^{i_k+j_k} = \prod_{k=1}^{n} f_k^{r_k} = \prod_{k=1}^{n} f_k^{r_k} \cdot f_1^{Ap_1} \in S^{(t)},$$

that is, $S^{(t)}S^{(t)} \subseteq S^{(t)}$. From group theory, since $S^{(t)}S^{(t)} = S^{(t)}$, $S^{(t)}$ is a subgroup. $\square$

Proposition 8.3 *Assume Hypotheses 8.1 and suppose that $\overline{S} = \mathbf{Z}_{p_1} \oplus \ldots \oplus \mathbf{Z}_{p_n}$ where f_i is a generator of $\mathbf{Z}_{p_i}$ and suppose that p_1 is the minimum of $\{p_1, \ldots p_n\}$. Suppose further that $S^{(t)}$ is a group. Then t is divisible by $\gcd\{p_i\}$, and $t \geq \sum_{k=2}^{n}(p_k - 1)$.*

Proof. First, note that $S^{(t)}$ contains the identity exactly when t is some sum $\sum_{k=1}^{n} \alpha_k p_k$, which is divisible by $\gcd\{p_i\}$.

Suppose that $t < \sum_{k=2}^{n}(p_k - 1)$; then we can write $\sum_{k=2}^{n}(p_k - 1) = qt + r$ for integers q and r where $0 \leq r < t$ and $q \geq 1$. Since $\sum_{k=2}^{n}(p_k - 1) + (t - r) = (q + 1)t$,

$$f_1^{t-r} \cdot \prod_{k=2}^{n} f_k^{p_k-1} \in S^{(t)}.$$

Therefore, we can find non–negative integers

$$\alpha_1 = t - r + \beta_1 p_1 \quad \text{and} \quad \alpha_i = p_i - 1 + \beta_i p_i$$

for $i = 2, 3, \ldots, n$ such that $\alpha_1 + \ldots + \alpha_n = t$. Note that for $i \geq 2$, $\alpha_i \geq 0$ implies that $\beta_i \geq 0$. Substituting, we get

$$t = \alpha_1 + \ldots + \alpha_n = t - r + \beta_1 p_1 + \sum_{i=2}^{n}(p_i - 1 + \beta_i p_i) = (t - r) + qt + r + \beta_1 p_1 + \sum_{i=2}^{n} \beta_i p_i.$$

Cancelling, we get

$$0 = qt + \beta_1 p_1 + \sum_{i=2}^n \beta_i p_i.$$

Since $\alpha_1 = t - r + \beta_1 p_1 \geq 0$,

$$qt + \sum_{i=2}^n \beta_i p_i = -\beta_1 p_1 \leq t - r.$$

Since $q \geq 1$ and $\beta_i \geq 0$ for $i \geq 2$, this implies that $q = 1$ and $r = 0$ which, in turn, implies that $t = \sum_{k=2}^n (p_k - 1)$, a contradiction. $\square$

Theorem 8.2 *Assume Hypotheses 8.1 and suppose that $\overline{S} = \mathbf{Z}_{p_1} \oplus \ldots \oplus \mathbf{Z}_{p_n}$ where f_i is a generator of $\mathbf{Z}_{p_i}$. Suppose that p_1 is the gcd of $\{p_1, \ldots p_n\}$. Then the minimal t for which $\mathcal{S}^{(t)}$ is a group is the (unique) multiple of p_1 satisfying*

$$\sum_{k=2}^n (p_k - 1) \leq t \leq \sum_{k=1}^n (p_k - 1).$$

$\mathcal{K}$ *is isomorphic to $\mathbf{Z}_{p_1}$.*

Corollary 8.1 *Assume Hypotheses 8.1 and suppose that $\overline{S} = \mathbf{Z}_{p^{i_1}} \oplus \ldots \oplus \mathbf{Z}_{p^{i_n}}$, for some fixed integer p, suppose that f_j generates $\mathbf{Z}_{p^{i_j}}$ and suppose i_1 is the minimum of $\{i_1, \ldots i_n\}$. Then the minimal t for which $\mathcal{S}^{(t)}$ is a group is the (unique) multiple of p^{i_1} satisfying*

$$\sum_{k=2}^n (p^{i_k} - 1) \leq t \leq \sum_{k=1}^n (p^{i_k} - 1).$$

$\mathcal{K}$ *is isomorphic to $\mathbf{Z}_{p^{i_1}}$.*

Corollary 8.2 *Assume Hypotheses 8.1 and suppose that the group $\overline{S}$ is the direct sum of n copies of $\mathbf{Z}_2$ such that each f_i is a generator. The minimal t for which $\mathcal{S}^{(t)}$ is a group is n when n is even and is $n - 1$ when n is odd. In both cases, the kernel is $\mathbf{Z}_2$.*

Theorem 8.3 *Assume Hypotheses 8.1; if $\overline{S}$ is the direct sum of $\mathbf{Z}_p$ and $\mathbf{Z}_q$ with generators f_1 and f_2, respectively, and if s is the lcm of p and q then the smallest k such that $\mathcal{S}^{(k)}$ is a group is*

    1. $s - 1$, when p and q are relatively prime

2. s, otherwise.

In both cases, the kernel $\mathcal{K}$ is a cyclic group whose order is the gcd of p and q.

Proof. Note that for any k, $\mathcal{S}^{(k)} = \{f_1^i f_2^{k-i}\}_{i=0}^k$. First, we will show that $\mathcal{S}^{(s)}$ is a subgroup. Let $f_1^i f_2^{s-i}, f_1^j f_2^{s-j} \in \mathcal{S}^{(s)}$; then their product is $f_1^{i+j} f_2^{2s-i-j}$. If $i+j \le s$, then $f_1^{i+j} f_2^{2s-i-j} = f_1^{i+j} f_2^{s-i-j} f_2^s = f_1^{i+j} f_2^{s-i-j} \in \mathcal{S}^{(s)}$. If $i+j > s$ then $f_1^{i+j} f_2^{2s-i-j} = f_1^s f_1^{i+j-s} f_2^{s-(i+j-s)} = f_1^{s-i-j} f_2^{s-(i+j-s)} \in \mathcal{S}^{(s)}$. Thus $\mathcal{S}^{(s)} \mathcal{S}^{(s)} \subseteq \mathcal{S}^{(s)}$.

Since $f_1^i f_2^{s-i} = f_1^s f_1^i f_2^{s-i} \in \mathcal{S}^{(s)} \mathcal{S}^{(s)}$, we have $\mathcal{S}^{(s)} \subseteq \mathcal{S}^{(s)} \mathcal{S}^{(s)}$ and, hence, equality of these sets. As above this implies that $\mathcal{S}^{(s)}$ is a subgroup.

Now suppose that there exists $k < s$ such that $\mathcal{S}^{(k)}$ is a subgroup. We will show that if $f_1^i f_2^{k-i} = f_1^j f_2^{k-j}$ for some $i, j \le k$ then $i = j$. Note that $i \equiv j \pmod{p}$ and $k - i \equiv k - j \pmod{q}$; the latter implies that $i \equiv j \pmod{q}$. Hence, if $i > j$, the difference $i - j = as$ for some positive integer a which implies that $i = j + as > k$, a contradiction. Since we have unique representation for the elements of $\mathcal{S}^{(k)}$ as $f_1^i f_2^{k-i}$, the order of $\mathcal{S}^{(k)}$ must be $k + 1$.

If we repeat the above argument to count the number of elements of $\mathcal{S}^{(s)}$ we get to the point $i = j + as \ge s$. Therefore $j = 0$ and $i = s$; this implies that the only element of the set $\{f_1^i f_2^{s-i}\}_{i=0}^s$ which is repeated is the identity element $= f_1^s = f_2^s$. Hence, the order of $\mathcal{S}^{(s)}$ must be s.

Therefore, if we assume that $\mathcal{S}^{(k)}$ is a group with $k < s$ then, by Theorem 8.1, $\mathcal{S}^{(k)} = \mathcal{S}^{(s)}$ and, hence, $k + 1 = s$.

Suppose that p and q are relatively prime; we will show that, when $k = s - 1$, $\mathcal{S}^{(k)}$ is a group. By Corollary 4.1 and Theorem 8.1, $\mathcal{S}^{(s)} = \overline{S}$. Note that $s = pq$ and $\overline{S}$ is a cyclic group generated by $f_1 f_2$. Let $i \in \{0, 1, 2, ..., pq - 1\}$; we want to express $(f_1 f_2)^i$ in the form $f_1^j f_2^{k-j}$. By the Chinese Remainder Theorem, there exists a unique j modulo pq such that $j \equiv i \pmod{p}$ and $j \equiv k - i \pmod{q}$. Hence, $(f_1 f_2)^i = f_1^i f_2^i = f_1^{j+ap} f_2^{k-j+bq} = f_1^j f_2^{k-j}$. Therefore $\mathcal{S}^{(k)} = \mathcal{S}^{(s)}$ when $k + 1 = s = pq$.

Suppose that p and q are not relatively prime and suppose that $\mathcal{S}^{(k)}$ is a group for some $k < s$. From above, we have that $k + 1 = s$. By Theorem 8.1.3, the kernel $\mathcal{K}$ is the trivial group. By Theorem 8.1.5, the kernel $\mathcal{K}$ is isomorphic to the quotient group $\overline{S}/\mathcal{S}^{(k)} = \overline{S}/\mathcal{S}^{(s)}$

which has order $\frac{pq}{s} = gcd(p,q) \neq 1$. This contradiction implies that s is the smallest positive integer such that $\mathcal{S}^{(s)}$ is a group.

Note that this theorem covers the previous case when $p = q$. $\square$

To a certain extent, the space X plays an insignificant role in this section. Note that if we start with a finite group $\mathcal{S}$ with generators $f_1, ..., f_n$, then there is as faithful representation of $\mathcal{S}$ into the special orthogonal group $SO(k)$ for some k. Consequently, we have an action of $\mathcal{S}$ on $\mathbf{R}^k$ which satisfies condition (*) and, hence, Theorem 8.1 applies. This suggests that the theory developed in this paper for finitely generated compact semigroups may be explored independently of the space X. We pursue this approach in another paper.

9 Topological Contractions.

Theorem 9.1 *Suppose that $f_1, f_2, ..., f_n$ are maps of a complete locally compact metric space X to itself such that the closure, $\overline{S}$, of the semigroup generated by these maps is compact and suppose that some f_i is a topological contraction. Then there exists a unique minimal invariant set for F, the union of the f_i's; i.e., there exists an invariant set A for F such that for any other invariant set B for F, $A \subseteq B$. If x_0 is the fixed point of the topological contraction f_i, then $A = \mathcal{K}(\{x_0\})$.*

Proof. It follows from Proposition 4.5 that F has an invariant set $B = \mathcal{K}(C)$ for any $C \in \mathcal{H}(X)$. Let $x \in B$; then $x \in g(C)$ for some $g \in \mathcal{K}$. Suppose that f_1 is a topological contraction with fixed point x_0; then $\lim_{i \to \infty} f_1^i(x) = x_0$. Note that $f_1^i(x) \in f_1^i g(C) \subseteq F^i g(C) \subseteq \mathcal{K}(C) = B$ for all i. It follows that $x_0 \in B$. Define A to be the intersection of all invariant sets of F; since $x_0 \in A$, A is non–empty. Clearly A is the unique minimal invariant set for F. By Proposition 4.5, $\mathcal{K}(\{x_0\})$ is an invariant set for F; since $x_0 \in A$, $\mathcal{K}(\{x_0\}) \subseteq A$. From the definition of A, we must have equality $\mathcal{K}(\{x_0\}) = A$. $\square$

Let (X, d) be a complete locally compact metric space. Suppose that $\mathcal{G} = \{g_1, g_2, ..., g_n\}$ is a finite group of isometries of (X, d) and that $\{f_1, f_2, ..., f_m\}$ is a collection of contractions of (X, d). We shall construct three related *IFS*'s on X and study the relationship between their kernels and invariant sets.

First, define

$$h_{i,j} = g_i f_j g_i^{-1}$$

for $i = 1, 2, ..., n$ and $j = 1, 2, ..., m$. Using the fact that the g_i's are isometries, it is easily seen that $\{X; h_{1,1}, ..., h_{n,m}\}$ is a hyperbolic *IFS*. Let F_1 be the union of the $h_{i,j}$'s and let $\mathcal{S}$ be the semigroup generated by the latter; by Corollary 2.2, $\mathcal{S}$ has compact closure and, by comments after Proposition 4.4, this *IFS* has a unique invariant set A. In this situation, the kernel, $\mathcal{K}_1$, is the trivial group.

Lemma 9.1 *For each* $i = 1, 2, ..., n$, $g_i(A) = A$.

Proof. Consider

$$
\begin{aligned}
g_i(A) &= g_i(\textstyle\bigcup_{j,l} h_{j,l}(A)) \\
&= \textstyle\bigcup_{j,l} g_i g_j f_l g_j^{-1}(A) \\
&= \textstyle\bigcup_{j,l} (g_i g_j) f_l (g_i g_j)^{-1}(g_i(A)) \\
&= \textstyle\bigcup_{j,l} h_{j,l}(g_i(A)).
\end{aligned}
$$

Hence, $g_i(A)$ is an invariant set for the hyperbolic *IFS* $\{X; h_{j,l}\}$ and the conclusion of the lemma follows from the uniqueness of the invariant set A. $\square$

The second *IFS* which we shall consider is $\{X; f_1, f_2, ..., f_m, k_1, k_2, ..., k_p\}$ where $\{k_1, k_2, ..., k_p\}$ is a collection of generators of $\mathcal{G}$. Now, let F_2 be the union of the f_i's and k_i's and let $\mathcal{T}$ be the semigroup generated by the f_i's and k_i's Again, since the k_i's are isometries, it is easily seen that $\mathcal{T}$ is an equicontinuous semigroup.

Lemma 9.2 *If* $\alpha \in \mathcal{T}$ *then either* $\alpha \in \mathcal{G}$ *or* $\alpha = gh$ *where* $g \in \mathcal{G}$ *and* $h \in \mathcal{S}$.

Proof. The proof is by induction on the length of α. If $\alpha \in \mathcal{T}^{(1)}$ then we have two cases: if $\alpha = f_j$ then let g_i be the identity element of $\mathcal{G}$ and we have $\alpha = h_{i,j}$. The second case is $\alpha = k_j \in \mathcal{G}$.

Suppose that $\alpha \in \mathcal{T}^{(p+1)}$; we can express $\alpha = f_j \alpha_0$ or $\alpha = k_j \alpha_0$ where $\alpha_0 \in \mathcal{T}^{(p)}$. By induction hyspostheses, $\alpha_0 \in \mathcal{G}$ or $\alpha_0 = gh$ where $g \in \mathcal{G}$ and $h \in \mathcal{S}$. Assume the latter; then we have

$$\alpha = f_j \alpha_0 = f_j gh = g((g^{-1} f_j g)h)$$

or

$$\alpha = k_j \alpha_0 = k_j gh = (k_j g)h.$$

The proof in the other case is similar. □

Lemma 9.3 *For each $x \in X$, $T(x)$ has compact closure.*

Proof. Let $\{\alpha_i(x)\} \subseteq T(x)$ be an infinite sequence; we need to find a convergent subsequence. By the previous lemma, for each i, either $\alpha_i \in \mathcal{G}$ or $\alpha_i = \beta_i \gamma_i$ for some $\beta_i \in \mathcal{G}$ and $\gamma_i \in \mathcal{S}$. For infinitely many of the i's we must have either the former or the latter. Suppose the former; since $\mathcal{G}$ is finite then we must have a constant subsequence. Suppose the latter; then, again, since $\mathcal{G}$ is finite, we can find a subsequence α_{i_j} where each $\beta_{i_j} = g$, a fixed element of $\mathcal{G}$. Since $\mathcal{S}$ has compact closure, the sequence $\{\gamma_{i_j}\}$ has a convergent subsequence $\{\gamma_{i_{j_k}}\}$; $\{\alpha_{i_{j_k}}(x)\}$ is the desired subsequence. □

By the Ascoli–Arzela Theorem, we have the following.

Corollary 9.1 *T has compact closure in the space of continuous maps of X to itself with the compact–open topology.*

Since each f_i is a contraction and, hence, has a fixed point, then, by Theorem 9.1, there exists a unique minimal invariant set B for F_2, the union of the maps f_i's and k_i's. Let $\mathcal{K}_2$ be the associated kernel.

Lemma 9.4 *The invariant sets, A and B, of F_1 and F_2, respectively, are equal.*

Proof. Since B is an invariant set for F_2, it follows that, for each i and j, that $h_{i,j}(B) \subseteq B$. Since F_1 is the union of the $h_{i,j}$'s, we have $F_1(B) \subseteq B$. It follows from the comments after Proposition 4.4 that $A = \lim_{n \to \infty} F_1^n(B) \subseteq B$.

By Lemma 9.1, for each i, $k_i(A) = A$ and since, for each i, $f_i(A) \subseteq A$, we have $F_2(A) = A$. Since B is the unique minimal invariant set for F_2, $A = B$. □

For each i, let r_i be the contractivity constant for f_i; i.e.,

$$r_i = \inf\{r : d(f_i(x), f_i(y)) \le r \cdot d(x,y) \text{ for all } x,y \in X\}.$$

For $x \in X$, let $d(x, A)$ denote the distance from x to the set A. We leave the verification of the following lemma to the reader.

Lemma 9.5 *1. If $x \in X$ and $g \in G$ then $d(g(x), A) = d(x, A)$.*

        2. For each i, $d(f_i(x), A) \le r_i d(x, A)$.

The third *IFS* which we consider is $\{X; k_1, k_2, ..., k_p\}$; let F_3 be the union of the k_i's and let $\mathcal{K}_3$ be the kernel of this *IFS*.

Lemma 9.6 *Suppose that there exists $x_0 \in X$ such that the function $\mathcal{G} \to \mathcal{G}(x_0)$ given by $g \to g(x_0)$ is a bijection, then the kernels, $\mathcal{K}_2$ and $\mathcal{K}_3$, are isomorphic.*

Proof. By Theorem 8.1, if we choose the smallest k and smallest $l > k$ such that $\mathcal{G}^{(k)}$ and $\mathcal{G}^{(l)}$ are subgroups of $\mathcal{G}$ then $\mathcal{K}_2$ is isomorphic to the cyclic group of order $l - k$. Let $C \in \mathcal{H}(X)$; by Lemma 4.3, we have $C \subseteq F_3^k(C) \subseteq F_2^k(C)$ and it follows that, for each positive integer i, $F_2^{ik}(C) \subseteq F_2^{(i+1)k}(C)$. By Lemma 4.5, the function τ of $\mathcal{H}(X)$ to itself defined by

$$\tau(C) = \lim_{i \to \infty} F_2^{ik}(C) = \overline{\bigcup_{i=0}^{\infty} F_2^{ik}(C)}$$

exists. Since $\tau^2 = \tau$ it follows that τ is the identity element of $\mathcal{K}_2$. Note also that $F_2^k \tau = \tau$. Similarly, we obtain $F_2^l \tau = \tau$; since $F_2 \tau$ generates $\mathcal{K}_2$, it follows that $\mathcal{K}_2$ is a cyclic group of order q where q divides $l - k$. We now want to show that $q = l - k$.

Let t be a non–negative integer and consider $g(x_0) \in F_2^t \iota(\{x_0\}) \cap \mathcal{G}(x_0)$; from above, we have $g(x_0) \in \overline{\bigcup_{i=0}^{\infty} F_2^{t+ik}(\{x_0\})}$. Suppose that for some i, $g(x_0) \in F_2^{t+ik}(\{x_0\})$; by Lemma 4.3, $g(x_0) = \alpha(x_0)$ for some $\alpha \in \mathcal{T}^{(t+ik)}$. If some f_j appears in α then it follows from Lemma 9.5 that $d(\alpha(x_0), A) < d(x_0, A) = d(g(x_0), A)$, a contradiction. Therefore no f_j appears in α and we must have $\alpha \in \mathcal{G}$; the hypotheses on x_0 implies that $g = \alpha \in \mathcal{G}^{(t+ik)} = \mathcal{G}^{(t+k)}$. Suppose that $g(x_0) = \lim_{j \to \infty} \alpha_j(x_0)$ where $\alpha_j \in \mathcal{T}^{(t+i_j k)}$. Let $r = \max\{r_i\}$; since $r < 1$ we can find $\epsilon > 0$ such that $(r \cdot d(x_0, A) + \epsilon)$–neighborhood of A and the ϵ–neighborhood of $g(x_0)$ do not intersect. For all j's such that $d(\alpha_j(x_0), g(x_0)) < \epsilon$ it follows as above that $\alpha_j \in \mathcal{G}$ and, hence, $\alpha_j \in \mathcal{G}^{(t+k)}$. Again, we obtain that $g \in \mathcal{G}^{(t+k)}$. Since $\mathcal{G}^{(t+k)}(x_0) \subseteq \mathcal{T}^{(t+k)}(x_0) = F_2^{t+k}(x_0) = F_2^t \tau(x_0)$, we have

$$F_2^t \tau(\{x_0\}) \cap \mathcal{G}(x_0) = \mathcal{G}^{(t+k)}(x_0).$$

Since $F_2^q \tau = \tau$ it follows that

$$\mathcal{G}^{(q+k)}(x_0) = F_2^q \tau(\{x_0\}) \cap \mathcal{G}(x_0) = \tau(\{x_0\}) \cap \mathcal{G}(x_0) = \mathcal{G}^{(k)}(x_0)$$

and, by hypotheses on x_0,

$$\mathcal{G}^{(q+k)} = \mathcal{G}^{(k)}.$$

Since q divides $l - k$ it follows from Theorem 8.1.3 that $q = l - k$. $\square$

Combining these lemmas together we get the following.

Summary. Let (X, d) be a complete locally compact metric space and suppose that $\mathcal{G} = \{g_1, g_2, ..., g_n\}$ is a finite group of isometries of (X, d) with generators $\{k_1, k_2, ..., k_p\}$. Suppose that $\{f_1, f_2, ..., f_m\}$ is a collection of contractions of (X, d). The *IFS* $\{X; g_1 f_1 g_1^{-1}, ..., g_n f_1 g_n^{-1}, ..., g_n f_m g_n^{-1}\}$ is a hyperbolic *IFS* with a unique invariant set A. The *IFS* $\{X; f_1, ..., f_m, k_1, ..., k_p\}$ has a unique minimal invariant set which is also A and its kernel is isomorphic to the kernel of the *IFS* $\{X; k_1, ..., k_p\}$. $\square$

We will now apply these techniques to study several classes of examples.

Example 9.1. Let X be the plane with its elements considered as complex numbers and define $k(x) = e^{2\pi i p/q} x$ and $f(x) = ax + b$ where $p < q$ are relatively prime positive integers and a and b are non–zero complex numbers with $|a| \in (0, 1)$. Consider the *IFS* $\{X; f, k\}$; note that f is a contraction and k is an isometry which generates a cyclic group of order q. By the Summary (and using the notation from above), the kernel $\mathcal{K}_2$ exists; furthermore, if c is the fixed point of f then the unique minimal invariant set is $B = \mathcal{K}_2(\{c\})$, which is the unique invariant set of the hyperbolic *IFS* $\{X; f, kfk^{-1}, ..., k^{q-1} f k^{-(q-1)}\}$. $\mathcal{K}_2$ is isomorphic to the kernel of the *IFS* $\{X; k\}$ which, by Theorem 6.1, is isomorphic to the cyclic group $\mathbf{Z}_q$. For example, as a special case, if $k(x) = e^{2\pi i/3} x$ and $f(x) = \frac{1}{2}x + \frac{1}{2}$, then the minimal invariant set, $\mathcal{K}_2(\{1\})$, of the *IFS* $\{X; f, k\}$ is Sierpinski's triangle [7, p.142]. We want to now investigate the action of the kernel $\mathcal{K}_2 = \{\tau, F_2 \tau, ..., F_2^{q-1} \tau\}$. For each $x \in B$, it is straightforward to check that $\tau(\{x\}) = B$ and, hence, for all i, $F_2^i \tau(\{x\}) = B$. Similarly, if x lies in the fixed set of k then, by Corollary 4.2 and Proposition 4.11, we have that, for all

$i \geq 0$, $F_2^i \tau(\{x\}) = \mathcal{K}(\{x\}) = \overline{S}(x)$. In the special case above, $F_2^i \tau(\{0\})$ is the union of a Sierpinski's triangle (which is $\mathcal{K}_2(\{1\})$) and the set of centers of each of the triangles whose interiors are removed in the alternate method of constructing Sierpinski's triangle.

Suppose that there exists $x \in X$ which satisfies the hypotheses of Lemma 9.6. From the proof of this lemma, $\tau(\{x\}) = \overline{\{g(x) : g \in \mathcal{T}^{(iq)} : i = 1, 2, ...\}}$ and, for all $i = 0, 1, 2, ..., q - 1$, $F_2^i \tau(\{x\}) = k^i(\tau(\{x\}))$. In the special case above, for all non–zero $x \in X \setminus B$, $\tau(\{x\})$ is the union of a Sierpinski's triangle and a sequence of points which converges to the Sierpinski's triangle. $F_2 \tau(\{x\})$ and $F_2^2 \tau(\{x\})$ are the images of $\tau(\{x\})$ under the rotations by $2\pi/3$ and $4\pi/3$, respectively; their intersections with $\tau(\{x\})$ and each other is Sierpinski's triangle. $\square$

Example 9.2 Let X again be the plane with its elements considered as complex numbers and define

$$f_1(x) = \frac{1}{3}x + 2 + 2i, \quad f_2(x) = \frac{1}{3}x + \frac{2}{3} + 2i \quad \text{and} \quad k(x) = e^{\pi/2}x.$$

Proceeding with the analyses as in Example 9.1, the *IFS* $\{X; f_1, f_2, k\}$ has a unique minimal invariant set which is Sierpinki's carpet [7, p.144]. The kernel is isomorphic to $\mathbf{Z}_4$. Note that $\mathcal{K}_2(\{0\})$ is the union of a Sierpinski's carpet (which is $\mathcal{K}_2(\{3 + 3i\})$) and the set of centers of each of the rectangles whose interiors are removed in the alternate way to obtain Sierpinski's carpet. $\square$

Example 9.3 Let X again be the plane with its elements considered as complex numbers and define

$$k(x) = e^{\sqrt{2}\pi}x \quad \text{and} \quad f(x) = \frac{1}{2}x.$$

The semigroup, $\mathcal{T}$, generated by k and f has compact closure in the space of continuous maps of X to itself. The minimal invariant set, $\mathcal{K}(\{0\})$, is the singleton $\{0\}$. If $x \in X \setminus \{0\}$, then $\mathcal{K}(\{x\})$ is the union of a countable collection of circles whose center is 0 and whose radii are $|x|/2^i$ for $i = 0, 1, 2, ...$. We wish to calculate $\gamma(\{x\})$ for elements γ of the kernel, $\mathcal{K}$; it was this calculation which led to the proof of Theorem 7.2.

Since f and k commute, note that $\mathcal{T}^{(n)} = \{k^i f^{n-i}\}_{i=0}^n$.

Suppose that the sequence $\{k^{i_n}\}$ converges to $g \in \overline{\mathcal{T}}$. We want to show that the sequence $\{F_2^{i_n}\}$ converges to some $\gamma \in \mathcal{K}$. Fix $A \in \mathcal{H}(X)$ and let j be a nonnegative integer. Note that $k^{i_n-j} f^j \to gk^{-j} f^j$. Hence, the sequence $\{k^{i_n-j} f^j(A)\}_{n=1}^{\infty}$ converges to the compact set $A_j = gk^{-j} f^j(A) \in \mathcal{H}(X)$. Note that $A_j \to \{0\}$ in $\mathcal{H}(X)$. Let $Y = \overline{\{A_j\}_{j=0}^{\infty}} = \{0\} \cup \{A_j\}_{j=0}^{\infty} = \overline{\cup_{j=0}^{\infty} A_j}$.

We will now show that $\lim_{n\to\infty} F_2^{i_n}(A) = Y$.

Let $\epsilon > 0$ be given; suppose that N is an integer such that the Hausdorff distance $h(A_j, \{0\}) < \epsilon/2$ for $j > N$. Choose $N_1 \geq N$ such that if $n > N_1$ then $h(g(A), k^{i_n}(A)) < \epsilon$; note that this implies that for each j that $h(A_j, k^{i_n-j} f^j(A)) < \epsilon$.

Suppose that $n > N_1$ and let $y \in Y$; we have two cases to consider. If $y = 0$ or if $y \in A_j$ for some $j > N$ then $i_n > N_1 \geq N$ implies that if $z \in k^{i_n-N-1} f^{N+1}(A)$ then both z and y lie in the $\epsilon/2$ ball about 0. If $y \in A_j$ for some $j \leq N$ then from the previous paragraph, $h(A_j, k^{i_n-j} f^j(A)) < \epsilon$; choose $z \in k^{i_n-j} f^j(A)$ with $|z - y| < \epsilon$. In both cases, we have found $z \in k^{i_n-j} f^j(A)$ such that $|z - y| < \epsilon$; it follows that Y is contained in the ϵ-neighborhood of $F_2^{i_n}(A)$. Similarly, it can be shown that $F_2^{i_n}(A)$ is contained in the ϵ-neighborhood of Y and, hence, $h(Y, F_2^{i_n}(A)) < \epsilon$. Define $\gamma(A) = Y$; the above argument shows that $F_2^{i_n} \to \gamma$. The correspondence $g \to \gamma$ determines a homomorphism of $SO(2)$ to the kernel; we leave to the reader to check that this is, in fact, an isomorphism. If $A = \{x\}$ is a singleton with $x \neq 0$ and if g is the identity map, then the set $Y(= \tau(A))$ defined above consists of 0 and the infinite sequence $\{r^{-j}x/2^j : j = 0, 1, 2, ...\}$. For each γ in the kernel, $\gamma(\{x\}) = g(\tau(\{x\}))$ where $g \to \gamma$ under the above-mentioned isomorphism of $SO(2)$ to the kernel. $\square$

Example 9.4. Let X and k be as in Example 9.3 but redefine $f(x) = ax + b$ where $a \in (0, 1)$ and $b > 0$. Let c be the fixed point of f. Again, the semigroup, $\mathcal{T}$, generated by k and f has compact closure in the space of continuous maps of X to itself. If $a \geq 1/2$ then the minimal invariant set of this IFS is the disk $\{x \in X : |x| \leq |c|\}$. If $a < 1/2$ then the minimal invariant set is the annulus $\{x \in X : b - ac \leq |x| \leq c\}$.

In the special case when $a = b = \frac{1}{2}$, the invariant set $\mathcal{K}(\{2\})$ is the union of the circle of radius 2 with center at the origin and the disk of radius $\frac{3}{2}$ with center at the origin.

We leave to the reader to check that the kernel is isomorphic to the special orthogonal group $SO(2)$. $\square$

Finally, we remark that the Julia sets of the quadratic functions, $z \to z^2 + c$, and of the function given by Newton's method to solve the equations $z^n = c$ are the invariant sets of *IFS*'s on the Riemann sphere with kernels isomorphic to $\mathbf{Z}_2$ and $\mathbf{Z}_n$, respectively.

References

[1] M. Barnsley, *Fractals Everywhere*, Academic Press, Inc., Boston (1988).

[2] W. C. Brown, *A Second Course in Linear Algebra*, John Wiley, (1988).

[3] T. Homma and S. Kinoshita, *On a topological characterization of the dilatation in E^3*, Osaka Math. J. 6 (1954), 135-144.

[4] L. S. Husch, *A topological characterization of the dilation in E^n*, Proc. Amer. Math. Soc. 28 (1971), 234-5.

[5] J. E. Hutchinson, *Fractals and Self Similarity*, Indiana Univ. Math. J. 30 (1981), 713-747.

[6] B.v.Kerékjártó, *Topologische Charakterisierung der linearen Abbildungen*, Acta Litt. Acad. Sci. Szeged. 6 (1934), 235- 262.

[7] B. B. Mandelbrot, *The Fractal Geometry of Nature*, W. H. Freeman and Co., New York (1983).

[8] E. Michael, *Topologies on spaces of subsets*, Trans. Amer. Math. Soc. 71 (1951), 152-182.

[9] J. R. Munkres, *Topology, A First Course*, Prentice-Hall, Inc., New Jersey (1975).

[10] V. I. Opoitsev, *A Converse to the Principle of Contracting Maps*. Russian Math. Surveys 31:4 (1976), 175–204.

[11] A. B. Paalman-DeMiranda, *Topological Semigroups* Mathematical Centre Tracts No. 11, Mathematisch Centrum Amsterdam (1964).

[12] G. W. Stewart, *Introduction to Matrix Computations*, Academic Press (New York) 1973.

[13] C. T. Yang, *Transformation groups on a homological manifold*, Trans. Amer. Math. Soc. 87 (1958), 261-283.

The Forced Damped Pendulum and the Wada Property

JUDY KENNEDY   University of Delaware, Newark, Delaware

JAMES A. YORKE   University of Maryland, College Park, Maryland

ABSTRACT: We prove that an idealized model of the Poincaré return map of the forced damped pendulum has basins of attraction that satisfy the Wada property and contain densely in their common boundary the stable manifold of a saddle fixed point. We also investigate in some detail other dynamical and topological properties of the system.

I. Introduction. One of the objectives of science is prediction. For a dynamical system with several attractors, the main prediction problem often is to determine which basin a given point is in, that is, which attractor the trajectory through that point will be attracted to.

This research was supported in part by NSF grant DMS–9006931 and AFOSR grant F49620–J–0033

When boundaries are complicated in a specified bounded region of initial points [McDGOY], a small uncertainty in the position of the initial position of the initial point may yield a large uncertainty as to which attractor the trajectory will go to. Consider a possible, but incorrect analogy: in the United States, one might not know which state a point is in, if there is some uncertainty in the position, but the list of possible states would be small. We, however, report on situations where if there is any uncertainty as to which basin the point is in, then in fact the point might be in any of the basins. None can be ruled out because any point on the boundary is arbitrarily close to points in all the basins.

Suppose that Γ is a finite or countably infinite set with at least three members, and that X is a 2–manifold. If for each $i \in \Gamma$, B_i is an open, connected, nonempty subset of X, then the sets B_i satisfy the **Wada property** if each boundary point of any B_i is in the boundary of all the B_i. The first explicit example in which this property occurs was reported in a paper by Yoneyama [Y] in 1917. This beautiful example is known as the Lakes of Wada continuum. In his paper Yoneyama states that the example was reported to him by a Mr. Wada, but no one seems to know anything else about Mr. Wada. In 1924, C. Kuratowski [K] proved that if K is a continuum (a compact, connected metric space) in the plane, and the complement of K consists of three or more disjoint, open, connected, nonempty sets B_i that satisfy the Wada property, then K is either an indecomposable continuum or the union of two indecomposable continua. (An **indecomposable continuum** is a continuum with the property that each of its proper subcontinua is nowhere dense in the continuum.) Thus, when the Wada property occurs in a 2–manifold, and the common boundary ∂B_i is a continuum, then that continuum is complicated. Since

Kuratowski gave an example in which ∂B_i is an indecomposable continuum, and another example in which ∂B_i is the union of two distinct indecomposable continua, his theorem cannot be improved. However, Kuratowski's theorem does not hold in higher–dimensional spaces. In $I\!R^3$, for example, there is a continuum that is an ANR, and the complement of that continuum consists of disjoint open sets that satisfy the a three–dimensional version of the Wada property. (This is the reason for restricting the Wada property to 2–manifolds.)

We believe that the Wada property may be quite common in dynamics. When this property arises, it translates into real problems with prediction. Recall that the closed set A is an attractor for a dynamical system (X, f) if $A \subseteq X$, and there is a closed neighborhood O of A in X such that $\cap_{i=0}^{\infty} f^i(O) = A$. We say that each point in O is **attracted** to the set A. If $O = X$, then A is a **global attractor** for f. Further, the **basin of attraction** of either an attractor or a global attractor A is the set of all points that are attracted to the set A. In "Basins of Wada" [KY], we gave an apparent example using the Poincaré return map of the forced damped pendulum, choosing parameters so that there were four attracting fixed points for the map. We argued that their basins of attraction satisfied the Wada property. However, the evidence for this example is mostly numerical. (The specific equation investigated was $\frac{d^2\theta}{dt^2} + \frac{1}{10}\frac{d\theta}{dt} + \sin\theta = \frac{7}{4}\cos t$.)

We wanted to prove that the example had the Wada property. Since we were not able to do this completely, we constructed an idealized model of the Poincaré return map. We can prove that this model has the Wada property. The construction and proof for this idealized model were outlined in "Basins of Wada," but since the paper appeared in a physics journal, we did not give all the mathematical details. Those details are given here, and at least one error is corrected. Also, we changed the set–up slightly: this has been done because the proofs are a little easier and the number of cases to be considered are reduced with the changes. None of the essential properties has been changed.

II. The example. Consider the space which is the product of the circle S^1 and the real line $\mathbb{R}^1$, $S^1 \times \mathbb{R}^1$. We define a diffeomorphism W on $S^1 \times \mathbb{R}^1$. The map W is the composition of three simpler maps s, σ, and α. For $(x,y) \in S^1 \times \mathbb{R}^1$, define $s(x,y) = (x, 2^{-7}y + (127)2^{-7}\sin(2\pi x))$. Note that s fixes the sine curve, the points (x,y) satisfying $y = \sin(2\pi x)$, and s maps the annulus $A = S^1 \times [-2,2]$ into a band of width $1/32$ about the sine curve. The map s ripples and contracts A. The reader will note that we have chosen to use explicit values in the map (and so the proof requires these values to be carried throughout). Explicit values make it possible for the reader to make reliable pictures of the attractor.

The map σ is a shear. For (x,y) in $S^1 \times \mathbb{R}^1$, define $\sigma(x,y) = ((x+y)(\mathrm{mod}\,1), y)$. The composite map $\sigma \circ s$ contracts, ripples and shears A. It also maps A into itself and both s and σ have fixed points $(0,0), (1/4,1), ((1/2,0)$, and $(3/4,-1)$.

Write $(1/4,1) = P_R, (1/2,0) = P_B, (3/4,-1) = P_G$. We define the third map α as follows. Suppose (x,y) is a point of $S^1 \times \mathbb{R}^1$ which is at a distance of $1/4$ or more from each of P_R, P_B, and P_G. Then $\alpha(x,y) = (x,y)$. If (x,y) is within $1/4$ unit from one of P_R, P_B, and P_G, then it is within $1/4$ unit of only one of these points. If $(x,y) \neq P_R$ is inside the circle of radius $1/4$ centered at P_R, then it lies on exactly one radial line L of the circle. The map α maps each radial line L to

itself, leaving fixed P_R and the unique point on both the radial line L and the circle.

If a point is on L at a distance $0 < d \leq 3/16$ from P_R, then that point maps to

the unique point on L at a distance $d(d^3 + 2^{-7})$ from P_R. If a point is on L at a

distance $3/16 < d < 1/4$ from P_R, then that point maps to the unique point on L

at a distance

$$(72423/4096) - (1026136/4096)d + (4765552/4096)d^2 - (7213568/4096)d^3$$

from P_R. (This is just a cubic polynomial fitted so that both the cubic and its

derivative agree at the ends of its domain with the other maps making up α. There

are many other ways to define α, and then to extend α so that it will be a diffeo-

morphism and most of these would work for the proof given here. In particular, it is

possible to define and extend α so that it is a C^∞ diffeomorphism.) Inside circles of

radius $1/4$ about P_B and P_G, we define α analogously. Thus, α is a diffeomorphism.

Now $\alpha(A) = A$ and $\sigma \circ s(A) \subseteq A$. Let $W = \alpha \circ \sigma \circ s$. Therefore, $W(A) = \alpha \circ \sigma \circ$

$s(A) \subseteq A$, which implies $W^{n+1}(A) \subseteq W^n(A)$ for all n. Then $\{A, W(A), W^2(A), \ldots\}$

is a nested collection of continua, and $Y = \cap_{n=0}^{\infty} W^n(A)$ is a continuum. Further,

since for each n, the set $W^n(A)$ separates $S^1 \times \mathbb{R}^1$ into two disjoint, open sets, each

homeomorphic to $S^1 \times \mathbb{R}^1$, Y also separates $S^1 \times \mathbb{R}^1$ into two disjoint, open sets,

each homeomorphic to $S^1 \times \mathbb{R}^1$. Let U denote those points of $S^1 \times \mathbb{R}^1$ which lie

in the open set of points above Y and V denote those points of $S^1 \times \mathbb{R}^1$ which lie

in the open set below Y. The continuum Y is the global attractor for the map W

on $S^1 \times \mathbb{R}^1$.

In this paper $I\!N$ denotes the positive integers, $\tilde{I\!N}$ denotes the nonnegative integers, and $Z\!\!\!Z$ denotes the negative integers. We use d to denote the usual metric on $S^1 \times I\!R^1$, and for $x \in S^1 \times I\!R^1$ and $\epsilon > 0, D_\epsilon(x) = \{y \in S^1 \times I\!R^1 | d(x,y) < \epsilon\}$.

Let $P(x) =$

$$(72423/4096) - (1026136/4096)x + (4765552/4096)x^2 - (7213568/4096)x^3$$

for $x \in [3/16, 1/4]$. (This is the cubic polynomial used in the definition of α.) Then

$P'(x) =$

$$-(128267/512) + (1191388/512)x - (2705088/512)x^2$$

and $P'(x) \geq 1$ for $x \in [128779/676272, 1/4]$. Further, $128779/67627$ (approximately 0.190424) is larger than $3/16 = .1875$, and $P(128779/676272) =$ 1487590632415/351240052211712 (approximately 0.004235) is smaller than $1/64 = 0.015625$. Let $128779/676272 = a_0$. Computation also reveals that the maximum value for a derivative of P is less than 6.

III. The proofs.

Lemma 1. The points P_R, P_B, and P_G are attracting fixed points for W. The strip $[3/16, 5/16] \times [-2, 2]$ is a subset of the basin of attraction of P_R, the strip $[31/64, 33/64] \times [-2, 2]$ is a subset of the basin of attraction of P_B, and the strip $[11/16, 13/16] \times [-2, 2]$ is a subset of the basin of attraction of P_G. The fixed point $(0,0)$ is a saddle point for W.

Proof. Since $\alpha(x,y) = (x,y)$ in some neighborhood of $(0,0)$, the Jacobian DW

evaluated at $(0,0)$ is the 2×2 matrix

$$\begin{bmatrix} 1 + (127\pi/64) & 1/128 \\ (127\pi/64) & 1/128 \end{bmatrix}$$

and the eigenvalues of this matrix are, with two significant digits, approximately

7.2 and 1.1×10^{-3}. Thus, $(0,0)$ is a saddle point.

Suppose $(x,y) \in [3/16, 5/16] \times [-2,2]$. Then $\sigma \circ s(x,y) = (x + 2^{-7}y + (127)2^{-7} \sin(2\pi x), 2^{-7}y + (127)2^{-7} \sin(2\pi x))$ and $d(\sigma \circ s(x,y),(1/4,1)) = ((x + 2^{-7}y + (127)2^{-7} \sin(2\pi x) - 5/4)^2 + (2^{-7}y + (127)2^{-7} \sin(2\pi x) - 1)^2)^{1/2} = ((x - 1/4)^2 + 2|x - 1/4| |2^{-7}y + (127)2^{-7} \sin(2\pi x) - 1| + 2(2^{-7}y + (127)2^{-7} \sin(2\pi x) - 1)^2)^{1/2} \leq \sqrt{5}(1/16) < 3/16$, since $|2^{-7}y + (127)2^{-7} \sin(2\pi x) - 1| \leq 1/64 + |127/128 - 1|$ for

$x \in [3/16, 5/16]$. Thus, $\sigma \circ s(x,y) \in D_{3/16}(P_R)$ and $W(x,y) = \alpha \circ \sigma \circ s(x,y) \in D_{1/64}(P_R)$.

For $\epsilon < 1/8, (1 - \cos(2\pi\epsilon))^2 + (\sin(2\pi\epsilon))^2 < 2\pi\epsilon$ (consider the unit circle) and

this leads to the inequality $1 - \pi\epsilon < \cos(2\pi\epsilon) = \sin(\pi/2 \pm 2\pi\epsilon) = \sin 2\pi(1/4 \pm \epsilon)$.

It follows that if $(a,b) \in D_\epsilon(P_R)$, then $(a,b) \in [\frac{1}{4} - \epsilon, \frac{1}{4} + \epsilon] \times [1 - \pi\epsilon, 1 + \pi\epsilon]$, and

$s(a,b) \in [1/4 - \epsilon, 1/4 + \epsilon] \times [1 - \pi\epsilon, 1 + \pi\epsilon]$, since s takes vertical intervals to shorter

vertical intervals about the $\sin(2\pi x)$ curve graph. If $s \circ W(x,y) = (q_1, q_2)$, then

$d(\sigma \circ s \circ W(x,y), P_R) = ((q_1 + q_2 - 1 - 1/4)^2 + (q_2 - 1)^2)^{1/2} \leq (5(\pi/64)^2)^{1/2} < 1/8$,

and $d(W^2(x,y), P_R) < 3(2^{-12}) < 2^{-9}$.

Then, similarly, $W^3(x,y) \in D_{2^{-12}}(P_R)$, and, continuing, for $n > 3, W^n(x,y) \in D_{2^{-(3n+3)}}(P_R)$. Thus, $[3/16, 5/16] \times [-2,2]$ is contained in the basin of attraction

of P_R. A similar argument proves that $[11/16, 13/16] \times [-2,2]$ is contained in the

basin of attraction of P_G.

Suppose that $(x,y) \in [31/64, 33/64] \times [-2,2]$. Then $d(\sigma \circ s(x,y), (1/2,0)) =$
$((x + 2^{-7}y + (127)2^{-7}\sin(2\pi x) - 1/2)^2 + (2^{-7}y + (127)2^{-7}\sin(2\pi x))^2)^{1/2} = ((x -$
$1/2)^2 + 2|x - 1/2|\,|2^{-7}y + (127)2^{-7}\sin(2\pi x)| + 2(2^{-7}y + (127)2^{-7}\sin(2\pi x))^2)^{1/2} <$
$((1/64)^2 + 2(1/64)(1/64 + .1) + 2(1/64 + .1)^2)^{1/2} < (0.0343)^{1/2} < 0.186 < 3/16,$
since $|\sin(2\pi x)| < \pi/32 < 0.1$ for $x \in [31/64, 33/64]$. Then $d(\alpha \circ \sigma \circ s(x,y), P_B) <$
$(3/16)^4 + (3/16)2^{-7} < 2^{-8}$.

For $(a,b) \in D_\epsilon(\frac{1}{2},0), \epsilon < 1/4, 1/2 - \epsilon < a < 1/2 + \epsilon$, and $-\epsilon < b < \epsilon$,
which implies that $-2\pi\epsilon < \sin(-2\pi\epsilon) = \sin(\pi + 2\pi\epsilon) < \sin(2\pi a) < \sin(\pi - 2\pi\epsilon) =$
$\sin(2\pi\epsilon) < 2\pi\epsilon$. Then $s(a,b) \in [1/2 - \epsilon, 1/2 + \epsilon] \times [-2\pi\epsilon, 2\pi\epsilon]$. It follows that
$(q_1, q_2) = s \circ W(x,y) \in [1/2 - 2^{-8}, 1/2 + 2^{-8}] \times [-2\pi(2^{-8}), 2\pi(2^{-8})]$, and
$d(s \circ W(x,y), (1/2,0)) < 1/32$. Thus, $d(W^2(x,y), (1/2,0)) < 3/32(2^{-6}) < 2^{-9}$.
Then likewise, $d(W^3(x,y), (1/2,0)) < 2^{-10}, d(W^4(x,y), (1/2,0)) < 2^{-11}$, etc. In
general, for $n \geq 2, d(W^n(x,y), (1/2,0)) < 2^{-(n+7)}$. Thus, $[31/64, 33/64] \times [-2,2]$ is
contained in the basin of attraction of P_B.   ∎

Remark:

In [KY] we state in Lemma 1 that $[7/16, 9/16] \times [-2,2]$ is contained in the
basin of attraction of P_B. This is incorrect since the graph of the $\sin(2\pi x)$ curve is
quite steep when it goes through this strip.

Let U_R, which we call the "red basin," U_B, and U_G (the "blue" and "green"
basins) denote the basins of attraction of P_R, P_B, and P_G, respectively. Each of
these sets can be shown to be homeomorphic to an open disk.

Lemma 2. Suppose that p is a point of $S^1 \times \mathbb{R}^1$ that is in the boundary of one of U_R, U_B, and U_G. Then p is in the boundary of each of U_R, U_B, and U_G, i.e., $\partial U_R \cup \partial U_B \cup \partial U_G = \partial U_R \cap \partial U_B \cap \partial U_G$, where ∂ denotes the boundary. Further, each point x in $S^1 \times I\!R^1$ is in $U_R \cup U_G \cup U_B \cup \partial U_R$.

Proof. Let A° denote the interior of A. The component C_R of $U_R \cap A^\circ$ that contains $[3/16, 5/16] \times (-2, 2)$ is homeomorphic to the interior of an open disk (that is, it is connected and has no holes in it), and its closure intersects both boundaries of A. Examination of $W(A)$ reveals that this component C_R intersects $W(A)$ in at least four components each of which runs from the top of $W(A)$ to the bottom of $W(A)$. Thus, $W^{-1}(C_R \cap W(A))$ has at least four components that stretch between both boundaries of A. Similar statements can be made about the corresponding components C_B and C_G of $U_B \cap A^\circ$ and $U_G \cap A^\circ$, respectively, except that $C_B \cap W(A)$ has at least three components that run from the top to the bottom boundaries of $W(A)$. (See Figure 1.)

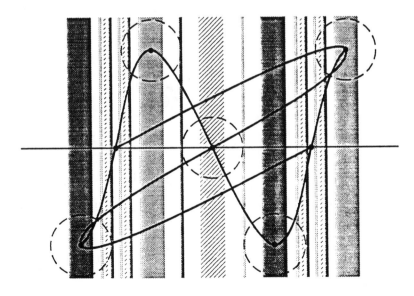

Figure 1.

Further, the components of the basins in A discussed so far alternate in a predictable, orderly way, and all run between both boundaries of A. If a connected open subset of A contained in U_R, U_B, or U_G extends from the top of A to the bottom of A, we call it a basin strip. We look at the inverse images of the basin strips, and notice that each inverse image of a basin strip contains several basin strips in A. Continuing the process generates ever more basin strips, all disjoint. There are infinitely many basin strips and between any two, there are red, blue, and green basin strips, that is, basin strips in the basins of P_R, P_B, and P_G. Showing that each boundary in A of each strip is in the boundary of each basin requires more work however. We must consider three cases.

(Case 1.) Let $D = Cl(\cup\{C | C$ is a component of U_B, U_G, or U_R in A° and $C \neq C_R\}$. Suppose there is some point x in $(\partial C_R - D) \cap A^\circ$. We may assume x is in the "left" boundary of C_R (that portion of ∂C_R in A° that has the property that each point contained in it has its first coordinate between 0 and 1/4), since eventually $W^n(x)$ is in the left boundary of C_R. Likewise, we may assume that the second coordinate of x is less than 1.

There is an open disk o containing x and contained in $(0, \frac{1}{4}) \times (-2, 2)$ such that $o \cap D = \emptyset$. Further, $\hat{o} = o - C_R \neq \emptyset$, and $\bar{\hat{o}} \cap C_R \neq \emptyset$. There is a horizontal arc B contained in o such that $B \cap C_R$ is the right endpoint r of B and $B - \{r\} \subseteq \hat{o}$. Let l denote the left endpoint of B. Choose B so that it is at a height less than 1.

What happens to B under iteration by W? Since the right endpoint r of B is in the left boundary of C_R, each point in B has first coordinate between 0 and 1/4, and no point of B is in D, it follows that for each $n, W^n(B)$ is an arc each point of which has first coordinate between 0 and 1/4 with one endpoint in $\bar{C}_R$ and each other point contained in $A° - (\bar{C}_R)$.

The set $s(B)$ in $A°$ is an increasing graph such that the length of $\pi_1 s(B)$ is the same as the length of B (since s leaves first coordinates unchanged). However, $\pi_2 s(B)$ is an arc, too, and the length of $\pi_1(\sigma \circ s(B))$ is the sum of the lengths of $\pi_1 s(B)$ and $\pi_2 s(B)$. Also, $\sigma \circ s(B)$ is an increasing graph. If $\sigma \circ s(B)$ intersects $D_{1/4}(P_R)$, then $Q = \sigma \circ s(B) \cap \overline{D_{1/4}(P_R)}$ is an arc that does not intersect $D_{a_0}(P_R)$, for otherwise $W(B)$ would contain a point of $D_{1/64}(P_R) \subseteq C_R$.

Recall that $P'(x) \geq 1$ for $x \in [a_0, 1/4]$. Because of geometrical considerations, $\pi_1 \alpha(Q)$ is longer than $\pi_1(Q)$: Suppose r_l denotes the radial line of $\overline{D_{1/4}(P_R)}$ containing the left endpoint of Q. See Figures 2 and 3.

(Case 1a.) If r_l is below r_h, the radial line containing the right endpoint of Q, then if any points on Q are on radial lines above r_h, then rotate those points counterclockwise, keeping distances to $(1/4, 1)$ constant, to points on the radial line of r_h. Likewise rotate any points on Q that are on radial lines below the radial line of r_l clockwise until they are on the radial line of r_l. Then the resulting arc $\hat{Q}$ is still increasing and $\pi_1 \hat{Q} = \pi_1 Q$. Project $\hat{Q}$ vertically onto r_l. Since $\hat{Q}$ is increasing this projection is one–to–one. Suppose Q' denotes the projected $\hat{Q}$ on r_l. In the figure A_1 and A_7 denote the left and right endpoints of Q', while A_1 and A_2 denote the left and right endpoints of $\hat{Q}$. Further, $\alpha(A_1) = A_8, \alpha(A_2) = A_3$, and $\alpha(A_7) = A_9$. The arc $\alpha(Q')$ extends from A_8 to A_9 on r_l.

The right endpoint A_2 of Q must be on the highest radius r_h involved. The point A_6 on r_h corresponds to A_7 on r_l (in the sense that they are the same distance

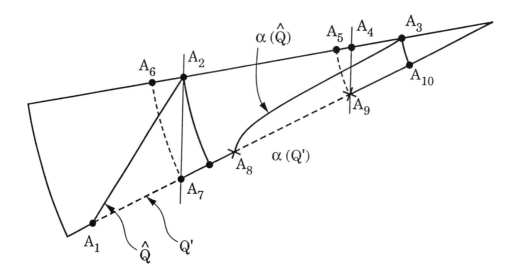

Figure 2.

from P_R), A_5 on r_h corresponds to A_9 on r_l, and A_3 on r_h corresponds to A_{10} on r_l. Because the triangles $\triangle A_6 A_7 A_2$ and $\triangle A_5 A_9 A_4$ are similar, the length of the interval from A_5 to A_4 on r_h is less than the length from A_6 to A_2 on r_h. Therefore, since $\alpha(Q')$ is longer than Q', and the interval $\overline{A_5 A_3}$ is longer than the interval $\overline{A_6 A_2}$, $A_3 = \alpha(A_2)$ must be closer to P_R than is A_4. (Remember that $P' \geq 1$ for the portions of r_l and r_h that intersect Q and $\hat{Q}$.) Hence $\pi_1 \alpha(\hat{Q})$ is longer than $\pi_1 \alpha(Q')$, which is longer than $\pi_1 Q' = \pi_1 Q = \pi_1 \hat{Q}$. Further, $\pi_1 \alpha(Q)$ is at least as long as $\pi_1 \alpha(\hat{Q})$. It follows that $\pi_1 W(B)$ is longer than $\pi_1 B$.

(Case 1b.) If the radial line containing r_h is below or the same as the radial line containing r_l, rotate points counterclockwise on Q that are on radial lines above the radial line of r_h until they are on the radial line of r_h. Let Q_1 denote the rotated Q. Then rotate points on Q clockwise until each is on the radial line of r_l. Let Q_2 denote the resulting rotated arc on the radial line of r_l. Again, because of derivative considerations, $\alpha(Q_1)$ and $\alpha(Q_2)$ are longer than Q_1 and Q_2, and $\pi_1 Q_1 \cup \pi_1 Q_2 = \pi_1 Q$, while $\pi_1 \alpha(Q_1) \cup \pi_1 \alpha(Q_2) = \pi_1 \alpha(Q)$. Then, in either case, $\pi_1 W(B)$ is longer than $\pi_1 \sigma \circ s(B)$, which is longer than $\pi_1 B$.

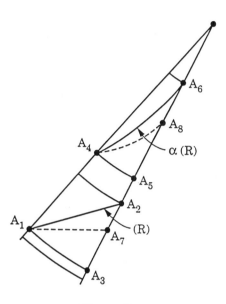

Figure 3.

Further, $W(B)$ is an increasing arc, in either case. This can be seen by combining the arguments just given. If some subarc R of Q is a case $1b$ type of arc, then $\alpha(R)$ is also increasing. To see why this is so, consider Figure 3. If the curve from A_1 to A_2 denotes R, then $\alpha(A_1) = A_4, \alpha(A_2) = A_6$, and $\alpha(R)$ is the curve running from A_4 to A_6. The point A_3 is the point on the lower radial line that corresponds to A_1, and similarly, A_5 corresponds to A_4. The line $\overline{A_1 A_7}$ is horizontal, so A_7 is between A_3 and A_2 on the lower radial line. The point A_8 is $\alpha(A_7)$, and it must be between A_5 and A_6 on the lower radial line. Recall that $P' \geq 1$ for the entire region concerned here. Thus, A_8 must lie above A_4, which means that A_6 is also above A_4. It follows that $\alpha(R)$ is increasing. Now if R is a subarc of Q that is a case 1a. type arc, then since the right endpoint of $\alpha(Q)$ lies to the right of the right endpoint of $\alpha(Q')$, as well as lying on a radial line above that of $\alpha(Q')$, it follows that $\alpha(R)$ is increasing.

Let B_1 denote the horizontal arc having as its left endpoint the lower left endpoint of $W(B)$ with $\pi_1 B_1 = \pi_1 W(B)$. Then $\pi_1 s \circ W(B) = \pi_1 s(B_1) = \pi_1 W(B), \pi_1 \sigma \circ s \circ W(B)$ is longer than $\pi_1 \sigma \circ s(B_1)$ which is longer than $\pi_1 s \circ W(B)$, and, using a geometrical argument similar to the one we used before, $\pi_1 \alpha \circ \sigma \circ s \circ W(B) = \pi_1 W^2(B)$ is longer than $\pi_1 \sigma \circ s \circ W(B)$ and $W^2(B)$ is an increasing graph. Continuing, the sequence $\pi_1 W(B), \pi_1 W^2(B), \ldots$ is a sequence of intervals in $[0, 3/16]$, and the lengths of these intervals are increasing. Some subsequence $W^{n_1}(B), W^{n_2}(B), \ldots$ of $W(B), W^2(B), \ldots$ converges to a continuum K, and $\pi_1 K$ is an interval in $[0, 3/16]$ with length longer than any $\pi_1 W^n(B)$. Further, K must be a nondecreasing arc in $[0, 3/16] \times [-2, 2]$. Then, as before, $\pi_1 \sigma \circ s(K)$ is longer than $\pi_1 s(K)$, and $\pi_1 W(K)$ is longer than $\pi_1 \sigma \circ s(K)$. But then $W^{n_1+1}(B), W^{n_2+1}(B), \ldots$ converges to $W(K)$ and some $\pi_1 W^{n_i+1}(B)$ is longer than $\pi_1 K$, which is impossible. Therefore, each point of the boundary of C_R is in D.

(Case 2.) The argument that each point in the boundary of C_G is also in $D_G = Cl(\cup\{C|C$ is a component of $(U_R \cup U_B \cup U_G) \cap A^\circ$ and $C \neq C_G\})$ is essentially the same as the one just given, but we need to prove that each point in ∂C_B is also in $D_B = Cl(\cup\{C|C$ is a component of $(U_R \cup U_B \cup U_G) \cap A^\circ$ and $C \neq C_B\})$.

(Case 3.) Now the left boundary of C_B is mapped into the right boundary of C_B, and the right boundary of C_B is mapped into the left boundary of C_B by W. Suppose there is some point x in $(\partial C_B - D_B) \cap A^\circ$. Then we may assume that

$W^n(x)$ is in the left boundary of C_B for n even, and $W^n(x)$ is in the right boundary

of C_B for n odd.

As before, it follows that we can find a horizontal arc B' such that the right

endpoint r' of B' is in $\bar{C}_B$ and $B' = \{r'\}$ is contained in $A^\circ - (D_B \cup \bar{C}_B)$ and is

contained in $[1/4, 1/2] \times [-2, 2]$. Further, we may assume that B' is below the x–axis

(since eventually some such arc will be). Then $\pi_1 s(B') = \pi_1 B'$. Since $\sigma \circ s(B')$

consists of points of the form

$$(x + 2^{-7}c + 127(2^{-7})\sin(2\pi x), 2^{-7}c + 127(2^{-7})\sin(2\pi x)),$$

it must be the case that $0.35 < x < 0.5$, for otherwise $\pi_1 \sigma \circ s(B')$ intersects $[0, 1/4]$

and $\sigma \circ s(B')$ intersects $\bar{C}_B$ and $\bar{C}_G$, which is impossible. For $x \in [0.35, 0.5]$,

$$d(x + 2^{-7}c + 127(2^{-7})\sin(2\pi x))/dx$$

takes on values between -1 and -6, so that $\pi_1 \sigma \circ s(B')$ is longer than $\pi_1 s(B')$. Let

$Q_B = \sigma \circ s(B') \cap \overline{D_{1/4}(P_B)}$. Now $\sigma \circ s(B') \cap \overline{D_{3/16}(P_B)} = \emptyset$, since otherwise $W(B')$

contains a point of U_B. For $\sigma \circ s(x, c) \in Q_B, d(\sigma \circ s(x, c), (1/2, 0)) = ((x - 1/2)^2 +$

$2|x - 1/2| \, |2^{-7}c + 127(2^{-7})\sin(2\pi x)| + 2(2^{-7}c + 127(2^{-7})\sin(2\pi x))^2)^{.5}$. Now Q_B is

an increasing curve in the first quadrant with respect to $1/2, 0)$. (It is once again

derivative considerations that allow us to conclude this, in this case the derivative

$$\frac{\frac{d(2^{-7}c+127(2^{-7})\sin(2\pi x))}{dx}}{\frac{d(x+2^{-7}c+127(2^{-7})\sin(2\pi x))}{dx}}$$

Also, Q_B has a lower left endpoint r_L and an upper right endpoint r_U.

First suppose that the radial line containing r_U is above the radial line contain-

ing r_L. (Rotate Figure 2 by π radians clockwise for an illustration of this case.) If

any points on Q_B are on radial lines above r_U, then rotate those points clockwise, keeping distances to $(1/2, 0)$ constant, to points on the radial line of r_U. Likewise rotate any points on Q that are on radial lines below the radial line of r_L counterclockwise until they are on the radial line of r_L. Then the resulting arc $\hat{Q}_B$ is still increasing and $\pi_1 \hat{Q}_B = \pi_1 Q_B$. Then it follows, using an argument similar to that for Q and Q' in $\overline{D_{1/4}(P_R)}$, that $\pi_1 \alpha(Q_B)$ and $\pi_1 W(B')$ are at least as long as $\pi_1 Q_B$ and $\pi_1 \sigma \circ s(B')$, respectively, and that $\pi_1 W(B')$ is an increasing arc longer than $\pi_1 B'$.

If the radial line containing r_U is below or the same as the radial line containing r_L, rotate points clockwise on Q_B that are on radial lines above the radial line of r_U until they are on the radial line of r_U. Let Q_{B1} denote the rotated Q_B. Then rotate points on Q_B counterclockwise until each is on the radial line of r_L. Let Q_{B2} denote the resulting rotated arc on the radial line of r_L. Again, because of derivative considerations, $\alpha(Q_{B1})$ and $\alpha(Q_{B2})$ are at least as long as Q_{B1} and Q_{B2}, and $\pi_1 Q_{B1} \cup \pi_1 Q_{B2} = \pi_1 Q_B$. Then, in either case, $\pi_1 W(B')$ is at least as long as $\pi_1 \sigma \circ s(B')$, which is longer than $\pi_1 B'$. Further, $W(B')$ is an increasing arc. Then we can parametrize $W(B')$ with x, getting an increasing function $y_1(x)$ from $\pi_1 W(B')$ onto $\pi_2 W(B')$. Now for x in $[0.35, 0.5]$,

$$\frac{d(2^{-7}c + 127(2^{-7})\sin 2\pi x)/dx}{d(x + 2^{-7}c + 127(2^{-7})\sin 2\pi x)/dx} = \frac{\frac{127\pi}{64}\cos 2\pi x}{1 + \frac{127\pi}{64}\cos 2\pi x}$$

is in $[0, 2]$. The map α applied to $\sigma \circ s(B')$ stretches the curve further. We need to know how large $y_1'(x)$ can be.

Each subarc of $\sigma \circ s(B')$ is increasing. Each $y_1'(x)$ can be approximated by considering the slopes of secant lines to y_1. By considering sufficiently small subarcs of $\sigma \circ s(B')$, we can approximate the inverse images of the secant lines under α with secant lines of appropriate subarcs of $\sigma \circ s(B')$. At this point it becomes important to consider whether those subarcs $\hat{B}$ of $\sigma \circ s(B')$ are of the type in case 1a (the lower left endpoint of $\hat{B}$ is on a radial line below the radial line of the upper right endpoint of $\hat{B}$), or they are of the type of case 1b (the lower left endpoint of $\hat{B}$ is on a radial line above the radial line of the upper right endpoint). For those arcs $\hat{B}$ of the type in case 1a, α flattens $\hat{B}$ and $y_1'(x)$ is less than $\frac{d(\pi_2 \sigma \circ s(x))/dx}{d(\pi_1 \sigma \circ s(x))/dx}$. For those arcs $\hat{B}$ of the type in case 1b, $y_1'(x)$ must be less than the slope of the steepest radial line involved. Careful computations of the derivatives and distances involved reveal that the slopes of these possible radial lines are less than 10.

It follows that $\frac{2^{-7} y_1'(x) + \frac{127x}{64} \cos 2\pi x}{1 + 2^{-7} y_1'(x) + \frac{127x}{64} \cos 2\pi x}$, the value of the slope of the tangent line to the curve $\sigma \circ s \circ W(B')$ at the point corresponding to x, is in $[0, 2]$. Then, considering the possible effects of α again, we see that $W^2(B')$ has tangent lines with positive slopes less than 10.

Arguments similar to those given before then prove that $\pi_1 W(B')$ $= \pi_1 s \circ W(B')$, and the length of $\pi_1 s \circ W(B')$ is less than the length of $\pi_1 \sigma \circ s \circ W(B')$, which is less than the length of $\pi_1 W^2(B')$. And we can continue: $W^2(B')$ can be parametrized differentiably by x, obtaining a function $y_2 : [.35, .5] \rightarrow S^1 \times \mathbb{R}^1$ with $y_2'(x) \in [0, 10]$. Then $\pi_1 W^3(B')$ is longer than $\pi_1 W^2(B')$ and each tangent to the curve $W^3(B')$ has slope in $[0, 10]$, etc.

Continuing this process, $B', W^2(B'), \ldots$ gives us a sequence of arcs in $[1/4, 1/2] \times [-2, 2]$ such that the lengths $|\pi_1 W^{2i}(B')|$ are increasing. Then some subsequence $W^{2n_1}(B'), W^{2n_2}(B')$ converges with respect to the Hausdorff metric to an arc K' in $[1/4, 1/2] \times [-2, 2]$. But $\pi_1 W^2(K')$ must be longer than each member of

$\pi_1 W^{2n_1+2}(B')$,

$\pi_1 W^{2n_2+2}(B'), \ldots$, which is a contradiction. Finally, we can conclude that $\partial C_B \cap A^\circ \subseteq D_B$. Hence, each point in some basin boundary is in each basin boundary.

We need finally to show that no open set o fails to intersect each of the basins. Suppose such an open set o exists. We may as well assume that o is an open disk contained in A. Suppose that B'' is a horizontal arc in o. Now B'' intersects no basin or basin boundary, so, in particular, it does not intersect $C_R \cup C_B \cup C_G$. Then the length of $\pi_1 B''$ is the same as the length of $\pi_1 s(B'')$. Taking into account values of $d(x + 2^{-7}c + 127(2^{-7}\sin(2\pi x)))/dx$ on the allowed regions, it follows that the length of $\pi_1 \sigma \circ s(B'')$ is longer than the length of $\pi_1 s(B'')$, and that $\sigma \circ s(B'')$ is an increasing arc. If $\sigma \circ s(B'')$ intersects $D_{1/4}(P_R) \cup D_{1/4}(P_B) \cup D_{1/4}(P_G)$, then, reasoning as before in cases 1, 2, and 3, the length of $\pi_1 W(B'')$ is longer than the length of $\pi_1 \sigma \circ s(B'')$. Then the length of $\pi_1 W^2(B'')$ is longer than the length of $\pi_1 W(B'')$, and so on. As before, this leads us eventually to a contradiction, so no such arc exists.   ∎

A continuum X is said to be **arclike** if for each $\epsilon > 0$, there is a map f_ϵ from X onto $[0,1]$ such that for each $x \in [0,1]$, diam $f_\epsilon^{-1}(x) < \epsilon$. A continuum X is **circlelike** if for each $\epsilon > 0$, there is a map g_ϵ from X onto the circle S^1 such that for each $x \in S^1$, diam $g_\epsilon^{-1}(x) < \epsilon$. Roughly, an arclike (circlelike) continuum is one that can be approximated by arcs (circles). Arclike and circlelike continua have been widely studied. Arcs, the topologist's $\sin(1/x)$ curve, and the Knaster buckethandle continuum are examples of arclike continua. The circles S^1 and the solenoids are circlelike. As we see next, Y is both circlelike and indecomposable.

If x is a point in the continuum X, then the **composant** of x in X is the union of all proper subcontinua of X that contain x. If X is an indecomposable continuum, then the composants of X form an uncountably infinite collection of mutually disjoint, dense sets.

In Lemma 1 we proved that $(0,0)$ is a saddle fixed point for our C^1 diffeomorphism W. If $F : M \to M$ is a C^1 diffeomorphism from a 2–manifold M onto itself, and p is a saddle fixed point for F, then

$$W^s(p) = \{x \in M | x, F(x), F^2(x), \ldots \text{ converges to } p\}$$

is known as the **stable manifold of** p, and

$$W^u(p) = \{x \in M | x, F^{-1}(x), F^{-2}(x), \ldots \text{ converges to } p\}$$

is known as the **unstable manifold of** p.

For F a diffeomorphism on a 2–manifold with p a saddle fixed point for F, $W^s(p)$ and $W^u(p)$ are one–to–one continuous images of real lines. Further, by Hartmann's Theorem, there is a homeomorphism $\psi : \mathbb{B} \to M$, where $\mathbb{B} = \{(x_1, x_2) \in \mathbb{R}^2 | x_1^2 + x_2^2 \leq 1\}$, such that $W_{\mathrm{loc}}^u(p) = \psi(\{(x_1, 0) \in \mathbb{B} | |x_1| \leq 1\})$ is the component of $W^u(p)$ in $\psi(\mathbb{B})$, and $W_{\mathrm{loc}}^s(p) = \psi(\{(0, x_2) \in \mathbb{B} | |x_2| \leq 1\})$ is the component of $W^s(p)$ in $\psi(\mathbb{B})$.

Recall that since Y is the intersection of the nested collection of continua $W^n(A)$, Y is a continuum. Since if $x \in S^1 \times \mathbb{R}^1$, there is some integer N such that if $n \geq N$, then $W^n(x) \in A$, it also follows that Y is the global attractor for W. However, we need to prove more.

Lemma 3. The global attractor Y is circlelike and indecomposable with

(1) each composant of Y having the property that it is the one–to–one image of the line $\mathbb{R}^1$, or the one–to–one image of the ray $[0, \infty)$ of nonnegative numbers, and

(2) the composant that contains $(0,0)$ is the unstable manifold of $(0,0)$.

Proof. Each $W^n(A)$ is an annulus, so that Y is the nested intersection of increasingly thin annuli. There is more to proving that Y is circlelike, however. Let $A_t = S^1 \times \{t\}$ for $t \in [-2, 2]$. For closed sets P and Q in A, let $\nu(P, Q)$ denote the Hausdorff distance from P to Q. If $\lim_{n \to \infty} \nu(W^n(A_2), W^n(A_{-2})) = 0$, then it follows that Y is circlelike. Thus, we need to prove that $\lim_{n \to \infty} \nu(W^n(A_2), W^n(A_{-2})) = 0$.

Since $W(A) \subseteq S^1 \times [-5/4, 5/4]$, we need only start with the set $E = S^1 \times [-5/4, 5/4]$, rather than all of A. Each vertical line in E has length $5/2$, and each vertical line in $s(E)$ has length $5/256$. Applying σ to each of the vertical lines in $s(E)$ yields a straight line slanted at an angle of $\pi/4$ that has length $\sqrt{2}(5/256)$.

In order to know the slope of the tangent to the curve $W(A_t)$ for $t \in [-5/4, 5/4]$, we need to compute the derivatives, with respect to x, of $f_t(x) = 2^{-7}t + 127(2^{-7}) \sin(2\pi x)$ and of $g_t(x) = x + f_t(x)$, and then divide $df_t(x)/dx$ by $dg_t(x)/dx$. This yields the derivative (of a curve ϕ_t parameterized by x)

$$d\phi_t(x)/dx = (127\pi/64) \cos(2\pi x)/(1 + (127\pi/64) \cos(2\pi x)).$$

Now for $x \in [-1/4, 1/4]$ and for $x \in [\arccos(-64/(127\pi))/(2\pi), 1 - \arccos(-64/(127\pi))/(2\pi)]$ (approximately $[0.2756405, 0.724394]$), $d\phi_t/dx \geq 0$, so that the curves are increasing. Let $J_1 = [-1/4, 1/4]$, $J_2 = [\arccos(-64/(127\pi))/(2\pi), 1 - \arccos(-64/(127\pi)/(2\pi)]$, and $J = J_1 \cup J_2$. Note that the intervals where the curves are decreasing (have tangents with negative slopes) give curve points inside $D_{3/16}(P_R) \cup D_{3/16}(P_G)$. Further, no point of any $\phi_t|[-1/4, 1/4]$ is in $D_{1/4}(P_B)$, for the following reasons: For $x \in [0, 0.03]$, $1/2 - \pi_1\phi_t(x) > 1/4$, and for $x \in [0.05, 1/4]$, $\pi_2\phi_t(x) > 1/4$. Although for $x \in [0.03, 0.05]$, it may be the case that both $1/2 - \pi_1\phi_t(x) < 1/4$ and $\pi_2\phi_t(x) < 1/4$, $\phi_t(x)$ fails for this interval to be in $D_{1/4}(P_B)$, too, as careful computation reveals that $((\pi_1\phi_t(x) - 1/2)^2 + (\pi_2\phi_t(x) - 0)^2)^{.5} > 1/4$. Symmetry considerations mean that no point of $\phi_t|[-1/4, 0]$ is contained in $D_{1/4}(P_B)$, either.

Thus, $W(A_t)$ increases except for some portions that are contained in regions where α contracts strongly. Then no vertical line contained in $\sigma \circ s(E)$ and not contained in $D_{3/16}(P_R) \cup D_{3/16}(P_G)$ is longer than $\sqrt{2}(5/256)$. (We established that the "diagonals" are no longer than this. Since the curves are increasing on this region, the verticals cannot be longer than the diagonals.)

For convenience, let $D_{3/16}(P_R) \cup D_{3/16}(P_B) \cup D_{3/16}(P_G) = T$ and let $D_{1/64}(P_R) \cup D_{1/64}(P_B) \cup D_{1/64}(P_G) = T'$. Now each component of the graph of ϕ_t in $(S^1 \times I\!\!R^1) - T$ is increasing, and using arguments similar to those given with the last lemma, and referring to Figures 2 and 3, it can be shown that each component of the graph of $\alpha(\phi_t)$ in $(S^1 \times I\!\!R^1) - T'$ is increasing. For $x \in S^1 \times I\!\!R^1$ and $\epsilon > 0$, let $S_\epsilon(x)$ denote the square, with interior, of side 2ϵ and center x. Let $\tilde{S}_\epsilon = S_\epsilon(P_R) \cup S_\epsilon(P_B) \cup S_\epsilon(P_G)$. If L is a vertical line in $\sigma \circ s(E) - T$, then the distance from one endpoint of $\alpha(L)$ to the other is less than $7/32$. (Remember that the maximum value of a derivative of P is less than 6.) Thus, the distance from the top boundary of $W(E)$ to the bottom boundary of $W(E)$ in $(S^1 \times I\!\!R^1) - \tilde{S}_{1/64}$ is less than $7/32$. Further, each square in $\tilde{S}_{1/64}$ contains both top boundary and bottom boundary points of $W(E)$, and since the diameter of each square is $\sqrt{2}/32$, the Hausdorff distance from $W(A_{5/4})$ to $W(A_{-5/4})$ is less than $1/4$. Since each component of each $W(A_t) - \tilde{S}_{1/64}$ is an increasing open arc, no vertical line in $W(E)$ is longer than $1/4$.

Continuing, no vertical line in $s \circ W(E)$ is longer than 2^{-9}, and if L is some vertical line in this set, then $\sigma(L)$ is a straight line inclined at an angle of $\pi/4$ with length equal to $\sqrt{2}(2^{-9}) < 2^{-8}$. Then no vertical line in $\sigma \circ s \circ W(E)$ that is not in T has length longer than 2^{-8}. (Tangent lines to the curves $\sigma \circ s \circ \alpha(\phi_t)$ have positive slopes in this region.) As before, each component of the graph of $W \circ \alpha(\phi_t)$ increases on this region, and therefore, no vertical line in $W^2(E) - \tilde{S}_{1/64}$ has length greater than $7(2^{-8})$. Since the diameter of $\alpha(\tilde{S}_{1/64})$ is less than $(2^{-18} + 2^{-11}) < 2^{-10}$, the Hausdorff distance from the top boundary of $W^2(E)$ to the bottom boundary of $W^2(E)$ inside $\tilde{S}_{2^{-10}}$ is less than 2^{-9}. Points inside $\tilde{S}_{1/64}$ but outside $\tilde{S}_{2^{-10}}$ are mapped into $\tilde{S}_{1/64}$ by W. Since no vertical line in $s \circ W(E)$ has length longer than 2^{-9}, and the diagonal lines in $\sigma \circ s \circ W(E)$ may be divided into those parts not in T and those parts in T (where strong contraction occurs), the distance from the top boundary of $W^2(E)$ to the bottom boundary is less than $7(2^{-8}) + 2^{-8} + 2^{-14} + 2^{-9} < 2^{-4}$. Thus, the Hausdorff distance from the top boundary of $W^2(E)$ to the bottom boundary of $W^2(E)$ is less than 2^{-4}.

Now having reached the inductive step, we continue. No vertical line extending from the top boundary of $W^2(E)$ to the bottom boundary of $W^2(E)$, and not intersecting $\tilde{S}_{1/64}$ has length greater than 2^{-4}, and the distance from the top boundary of $W^2(E)$ to the bottom boundary of $W^2(E)$ inside $\tilde{S}_{1/64}$ is less than 2^{-4}. Then no vertical line in $W^3(E) - \tilde{S}_{1/64}$ has length greater than $7(2^{-10})$, $W(\tilde{S}_{2^{-10}}) \subseteq \tilde{S}_{2^{-14}}$ has diameter less than 2^{-13}, $W(\tilde{S}_{1/64} - \tilde{S}_{2^{-10}}) \subseteq \tilde{S}_{2^{-10}}$, and $W(T - \tilde{S}_{1/64}) \subseteq \tilde{S}_{1/64}$.

Then the distance from the top boundary of $W^3(E)$ to the bottom boundary of $W^3(E)$ in $\tilde{S}_{1/64}$ is less than $7(2^{-10}) + 2^{-10} + 2^{-16} + 2^{-11}$. Thus, the distance from the top boundary of $W^3(E)$ to the bottom boundary of $W^3(E)$ is less than 2^{-6}. In general, the maximum length of a vertical line in $W^n(E) - \tilde{S}_{1/64}$ is less than $7(2^{-2n-4})$, and the distance from the top boundary of $W^n(E)$ to the bottom boundary of $W^n(E)$ is less than $7(2^{-2n-4}) + 2^{-2n-4} + 2^{-2n-10} + 2^{-2n-5}$, so that the Hausdorff distance from the top of $W^n(E)$ to the bottom of $W^n(E)$ is less than 2^{-2n}. It follows that Y is circlelike.

The rest of the proof now follows easily. It is easy to check that no proper sub-continuum of Y contains two of the points in the set $\{(0,0), (1/4, 1), (1/2, 0), (3/4, -1)\}$, so Y must be indecomposable. Further, from the construction it follows that each composant of Y except for two, namely the composants containing $(1/4, 1)$ and $(3/4, -1)$, is the one–to–one image of a real line. The composant containing $(1/4, 1)$ is the one–to–one image of a ray (with endpoint $(1/4, 1)$), and so is the composant containing $(3/4, -1)$. Then the unstable manifold $W^u(0,0)$ is contained in the composant of $(0,0)$, and since the unstable manifold crosses the basins, it follows that it is the composant of $(0,0)$. ∎

Theorem 4. The unstable manifold $W^u(0,0)$ crosses all three of the basins U_R, U_B, and U_G. The stable manifold $W^s(0,0)$ intersects both boundaries of A and is dense in the common basin boundary. Every point in $S^1 \times I\!\!R^1$ is either in U_R, U_B, or U_G, or is in the common basin boundary.

Proof. The first statement is obvious. The stable manifold must cross the top and bottom boundaries of some $W^n(A)$, since Y contains no triods and the stable manifold cannot be in the same composant of Y as the unstable manifold. Then the stable manifold crosses both boundaries of A, as it is invariant under W.

Suppose O is an open disk that intersects the basin boundary. There is an arc P in O that intersects all three basins. Since P is compact, there is an n such that $W^n(P)$ is in A, and $W^n(P)$ comes within $1/16$ unit of each of P_R, P_G, and P_B. Then $W^{n+1}(P)$ intersects the stable manifold. (See Figure 1.) Therefore, O intersects the stable manifold, and the stable manifold is dense in the basin boundary. ■

Theorem 5. The diffeomorphism W is dissipative. If $S^1 \times \mathbb{R}^1$ is compactified with two points at $+\infty$ and $-\infty$ (to yield a copy of the sphere), then the closure of the stable manifold $W^s(0,0)$ in the compactification is an indecomposable continuum.

Proof. To prove the first statement, we need to prove that if θ is an open, bounded set in $S^1 \times \mathbb{R}^1$, then the area of $W(\theta)$ is less than the area of θ. We need only consider interiors of rectangles. A line integral computation then gives that if R is a rectangle with interior and area A, then the area of $\sigma \circ s(R)$ is $\frac{1}{128}A$. Since the map α can only multiply this by a maximum less than of 7, the area of $W(A)$ is less than $(7/128)A$.

The second statement follows from a theorem of Marcy Barge [B], since the stable manifold crosses the unstable manifold in a point other than $(0,0)$. ■

References.

[B] M. Barge, Homoclinic intersections and indecomposability, *Proc. AMS.* **101** (1987) 541.

[KY] J. Kennedy and J.A. Yorke, Basins of Wada, to appear, *Physica D..*

[K] C. Kuratowski, Sur les coupures irreducibles des plan, *Fund. Math.* **6** (1924) 130.

[McDGOY] S. McDonald, C. Grebogi, E. Ott, and J.A. Yorke, Fractal Basin Boundaries, *Physica 17D* (1985) 125.

[Y] K. Yoneyama, Theory of continuous sets of points, *Tohoku Math. J.* 11-12 (1917) 43.

Denjoy Meets Rotation on an Indecomposable Cofrontier

JOHN C. MAYER and LEX G. OVERSTEEGEN   University of Alabama at Birmingham, Birmingham, Alabama

ABSTRACT. We construct an example of an indecomposable cofrontier minimally invariant under a homeomorphism of the plane such that the prime end rotation number is the same from both sides. However, on one side the induced map on the circle of prime ends is conjugate to a rigid rotation, and on the other side the induced map on the circle of prime ends is a Denjoy map.

1. INTRODUCTION

Our construction of an indecomposable confrontier invariant under a homeomorphism of the plane which is rotation-like from one side and Denjoy-like from the other is based upon one of the family of cofrontiers admitting irrational pseudorotations constructed in [BGM] (these Proceedings, page 59). The reader is expected to refer to [BGM] for certain details of the construction herein.

Our interest in this example is motivated in part by the Siegel disk question from complex analytic dynamics: *Is the boundary of a Siegel disk always a simple closed curve?* A Siegel disk is a maximal domain on which a rational function of the Riemann sphere is analytically conjugate to a rigid rotation of the unit disk. The cofrontier we construct herein is a possible candidate for the boundary of a Siegel disk. There are a number of examples of what one might call "topological Siegel disks;" see [Ro1-2] for surveys (for further information about complex analytic dynamics in general, see [Mi]). Two recent examples which involve Denjoy dynamics are in [Br2,Wa].

In [Br2] (these Proceedings page 51), a cofrontier X is constructed (in fact, an uncountable family of such), invariant under a homeomorphism g of the plane, so that on Int(X), g is conjugate to a rigid irrational rotation, and the induced prime end map from both sides is the same Denjoy map. In [Wa], a cofrontier Y, the "split hairy circle," is constructed, invariant under a homeomorphism h of the plane, for which the prime end rotation number is the same from both sides; on one side the induced prime end map is

---

Parts of this paper were presented by the first author at the AMS/MAA Mathfest Special Session on Continuum Theory and Dynamical Systems, August 8-11, 1991.

a rigid rotation, and on the other side the induced prime end map is a Denjoy map. The cofrontiers in [Br2,Wa] are decomposable, and the homeomorphism is neither recurrent, nor minimal on the cofrontier. Our example improves on these particulars: the cofrontier Λ' we construct is indecomposable, and the homeomorphism h' of the plane which leaves Λ' invariant is minimal on Λ'.

1.1. Cofrontiers. A continuum Λ contained in the complex plane $\mathbf{C}$ is a *cofrontier* iff Λ irreducibly separates $\mathbf{C}$ into exactly two complementary domains. It follows that Λ is the common boundary of these two domains. We denote the bounded domain by $\mathrm{Int}(\Lambda)$ and the unbounded domain by $\mathrm{Ext}(\Lambda)$.

1.2. Minimally invariant and recurrent. Let $h : (\mathbf{C}, \Lambda) \to (\mathbf{C}, \Lambda)$ be a homeomorphism of the complex plane leaving the cofrontier $\Lambda \subset \mathbf{C}$ invariant. We say Λ is *minimally invariant* under h iff Λ is the only closed, nonempty subset of Λ invariant under h. In this case, we also say h is *minimal* on Λ. Equivalently, h is minimal on Λ iff the orbit of every point in Λ is dense in Λ. We say that h is *recurrent* on Λ iff some subsequence of $\{h^n\}_{n=1}^{\infty}$ converges uniformly to the identity on Λ.

1.3. Prime end uniformization. Let $\widehat{\mathbf{C}} = \mathbf{C} \cup \{\infty\}$ denote the Reimann sphere. Let $U \subset \widehat{\mathbf{C}}$ be a simply connected domain with nondegenerate boundary. Let $\phi : \mathbf{D} \to U$ be a conformal isomorphism, as guaranteed by the Riemann Mapping Theorem. We call ϕ a *(prime end) uniformization* of U. Basic definitions and further references concerning prime ends may be found in [BGM,Br1,P].

1.4. Induced mappings on prime ends. Let h and Λ be as in §1.2. Let $\phi_i : \mathbf{D} \to \mathrm{Int}(\Lambda)$ and $\phi_e : \widehat{\mathbf{C}} - \overline{\mathbf{D}} \to \mathrm{Ext}(\Lambda)$ be prime end uniformizations. Then $\phi_i^{-1} h \phi_i$ induces a homeomorphism H_i on $\overline{\mathbf{D}}$, and $\phi_e^{-1} h \phi_e$ induces a homeomorphism H_e on $\widehat{\mathbf{C}} - \mathbf{D}$. Hence, $H_i | \partial \mathbf{D}$ and $H_e | \partial \mathbf{D}$ are homeomorphisms of the circle. In general, there is no reason to suppose that there is any relationship between $H_i | \partial \mathbf{D}$ and $H_e | \partial \mathbf{D}$. See Figure 1.

If $H_i | \partial \mathbf{D}$ and $H_e | \partial \mathbf{D}$ are each topologically conjugate to the same rigid rotation R_α of the circle (where $\alpha \in \mathbf{R}/\mathbf{Z}$), then we say that h is a *pseudorotation* of Λ with rotation number α.

1.6. Conjecture. Our goal is to construct an indecomposable cofrontier Λ', minimally invariant under a homeomorphism h' of $\mathbf{C}$, so that $H_i' | \partial \mathbf{D}$ is conjugate to the rotation R_α, but $H_e' | \partial \mathbf{D}$ is a Denjoy homeomorphism of the circle with irrational rotation number α.

We conjecture that no homeomorphism h of a simply-connected plane domain U with nondegenerate boundary which has the properties that

 (1) h extends continuously to ∂U and
 (2) the induced mapping H on the circle of prime ends of U is a Denjoy map

can be recurrent on ∂U. However, we prove this only for the particular cofrontier Λ' that we construct in §3.

2. CONSTRUCTION AND PROPERTIES OF Λ

We first construct a cofrontier Λ which admits a pseudorotation h_α for an irrational rotation number α. We then modify Λ to produce the desired cofrontier Λ' on which, from

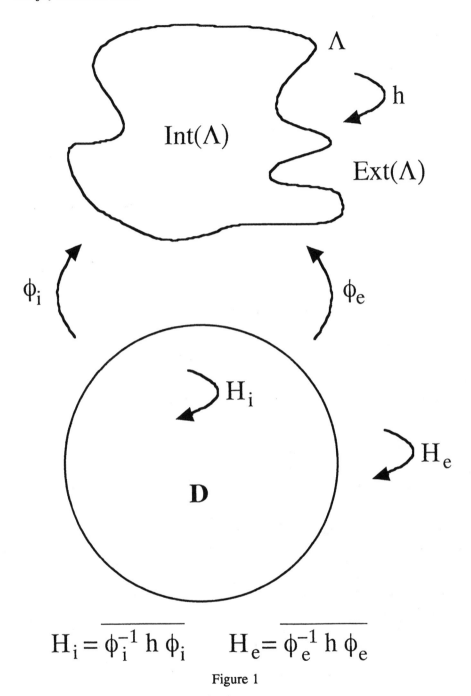

$$H_i = \overline{\phi_i^{-1} \, h \, \phi_i} \qquad H_e = \overline{\phi_e^{-1} \, h \, \phi_e}$$

Figure 1

the point of view of the prime end structures, a Denjoy homeomorphism meets a rotation. Refer to Figures 2 and 3 for an illustration of the first few stages of the construction.

2.1. Definition of Λ. We define

$$\Lambda = \bigcap_{i=1}^{\infty} A_i$$

where each A_i is a closed annulus such that

    (1) A_i is contained essentially in the interior of A_{i-1},
    (2) A_i has a rational rotational symmetry within the annular structure of A_{i-1},
    (3) A_i is long and thin, but has relatively few bends, with respect to A_{i-1}, and
    (4) each "loop" of A_i meets every link of A_{i-1}.

A general recipe, and more precise description, for the construction of cofrontiers such as Λ may be found in [BGM] to which the reader is referred. In the language of [BGM], for our construction here we have chosen $q_1 = 4$, $c_{i+1} = \frac{q_i-1}{q_i}$, and $r_{i+1} = q_i$, with minimal bending. (Figures 2 and 3 are harmlessly inaccurate as we have used smaller numbers for r_2 and r_3 in order to produce a reasonable drawing.) We use A_i, somewhat ambiguously, to denote an annulus, and to denote the circular chain of closed topological disk links whose union is that annulus.

The topological, dynamical, prime end, and accessiblity properties of Λ are summarized in the following four theorems:

2.2. Theorem (Topological properties of Λ). *Let Λ be constructed as described in §2.1. Then*

    (1) Λ *is a cofrontier.*
    (2) Λ *is an indecomposable continuum.*
    (3) *Every proper subcontinuum of Λ is chainable.*
    (4) *Every point $x \in \Lambda$ is either an endpoint or an interior point of an arc in Λ.*

Proof. Since Λ is circularly chainable, and A_i sits essentially in A_{i-1}, Λ is a cofrontier. It follows from condition 2.1(4) and the Cook-Ingram criterion of indecomposablility (compare [BGM] Theorems 3.3 and 4.1) that Λ is indecomposable.

An argument similar to that which follows can be used to establish (3) for the proper subcontinua of any circularly chainable continuum.

To establish (4), we will show that every point x of Λ is contained in a proper subcontinuum K_x of Λ that is chainable and itself has property (4). Let $x \in \Lambda$. Chose three links of A_1 forming a chain $\mathcal{C}_1 = \{C_{1,1}, C_{1,2}, C_{1,3}\}$ so that x lies in the interior of their union. (We choose a link containing x and the link joining it on either side.) Let $\mathcal{C}_i = \{C_{i,1}, \ldots, C_{i,k_i}\}$ be the maximal subchain of A_i that contains x and lies in $\cup \mathcal{C}_{i-1}$. Because of condition 2.1(3), $\mathcal{C}_i$ contains at most one "bend link;" that is, a link where $\mathcal{C}_i$ reverses direction in $\mathcal{C}_{i-1}$. Only two kinds of patterns are possible for $\mathcal{C}_i$ in $\mathcal{C}_{i-1}$: either $\mathcal{C}_i$ runs straight through $\mathcal{C}_{i-1}$ with one endlink of $\mathcal{C}_i$ in one endlink of $\mathcal{C}_{i-1}$ and the other endlink of $\mathcal{C}_i$ in the opposite endlink of $\mathcal{C}_{i-1}$, or both endlinks of $\mathcal{C}_i$ lie in the same endlink of $\mathcal{C}_{i-1}$, thus forming a "U"-chain. (Compare Example 3.1, Figure 4, following.)

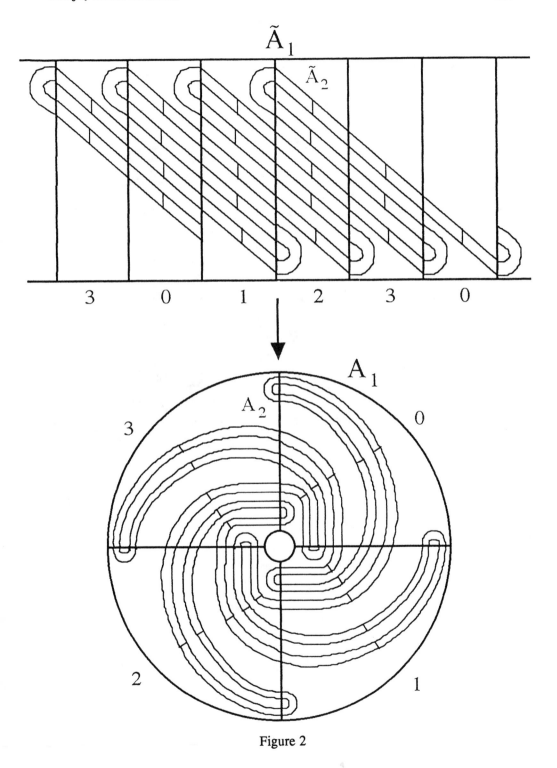

Figure 2

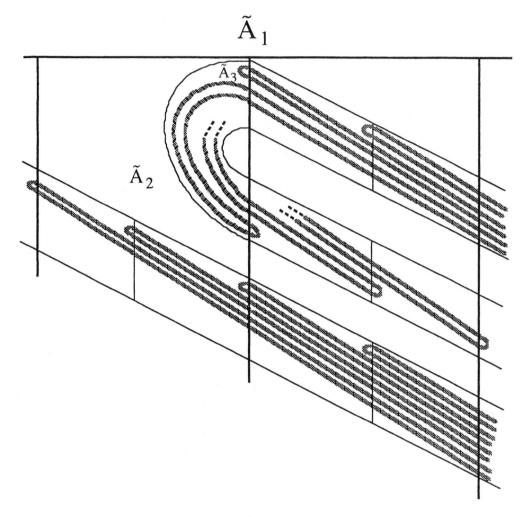

Figure 3

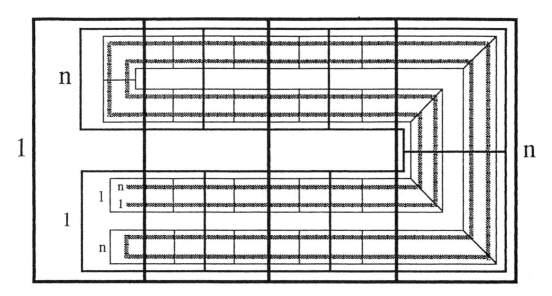

Figure 4

Let $K_x = \bigcap_{i=1}^{\infty} UC_i$. Then K_x is a chainable subcontinuum of Λ containing x. The continuum K_x might be decomposable or indecomposable; it might even be an arc. It is well-known that chainable continua which are the intersection of only straight chains and "U"-chains have property (4): each point of K_x is an endpoint or an interior point of an arc in K_x. □

2.3. Theorem (Dynamical properties of Λ). *Let Λ be constructed as described in §2.1. Then*

(5) *Λ admits a pseudorotation h_α with irrational prime end rotation number α.*
(6) *h_α is minimal and recurrent on Λ, though not almost periodic.*
(7) *α is not of constant type.*
(8) *The rotation number of every point in Λ is α.*

Proof. How an irrational pseudorotation may be induced on Λ is explained in §3.4 of [BGM]. Properties (5) through (8) follow from Theorem 5.8 of [BGM]. □

2.4. Theorem (Prime end properties of Λ). *Let Λ be constructed as described in §2.1. Then*

(9) *The impression of every prime end of $\mathrm{Int}(\Lambda)$ (respectively, $\mathrm{Ext}(\Lambda)$) is all of Λ.*
(10) *Points of $\partial\mathbf{D}$ corresponding to prime ends of $\mathrm{Int}(\Lambda)$ (respectively, $\mathrm{Ext}(\Lambda)$) which are simple dense canals (Lake-of-Wada channels) form a residual set in $\partial\mathbf{D}$.*

Proof. By Theorems 2.2 and 2.3, we have an indecomposable cofrontier Λ which admits an irrational pseudorotation h_α. Theorem 2.12 of [BGM] guarantees that properties (9) and (10) follow. □

2.5. Definitions. A *ray* is a one-to-one continuous image of the half-line $[0,\infty)$. A point $x \in \Lambda$ is *accessible* from $\mathrm{Int}(\Lambda)$ (respectively, $\mathrm{Ext}(\Lambda)$) iff there is a ray R in $\mathrm{Int}(\Lambda)$ (respectively, $\mathrm{Ext}(\Lambda)$) such that $\overline{R} - R = \{x\}$. The *composant* $C(x)$ of a point x in a nondegenerate continuum X is the union of all proper subcontinua of X containing x. A composant is a dense F_σ in X. If X is indecomposable, then the composants of X partition X into continuum-many pairwise disjoint sets. A composant C of an indecomposable continuum X is said to be a *ray-composant* iff C is a ray. The *endpoint* of C is the image of 0 under the one-to-one map from $[0,\infty)$ to C.

2.6. Theorem (Accessibility Properties of Λ). *Let Λ be constructed as described in §2.1. Then*

(11) *No composant of Λ is accessible at more than one point.*
(12) *Every accessible point $x \in \Lambda$ is the endpoint of a ray-composant $C(x)$ in Λ.*

Proof. Since Λ is an indecomposable cofrontier admitting an irrational pseudorotation, property (11) follows from Corollary 2.13 of [BGM].

However, property (12) depends upon our precise construction of Λ using circular chains with many long loops and (relatively) few bend links. Figure 3 will be helpful in visualizing the argument below. Suppose that x is a point of Λ accessible from $\mathrm{Ext}(\Lambda)$. Then x is the unique principal point of a prime end ξ of $\mathrm{Ext}(\Lambda)$ of the second kind. Let R be a ray from ∞ in $\mathrm{Ext}(\Lambda)$ whose endpoint in Λ is x. (R is the image under the uniformization of

Ext(Λ) of a radial ray in $\widehat{\mathbf{C}} - \overline{\mathbf{D}}$ ending at $\xi \in \partial\mathbf{D}$.) We may suppose that $x = \bigcap_{i=1}^{\infty} C_i$, where C_i is a link of A_i containing x and $C_i \subset C_{i-1}$. Without loss of generality, we may further suppose that R meets only the link C_i of A_i, since the limit set of R in Λ is a point. (A sequence of isotopies of the plane, each sucesssively closer to the identity, can change a ray hitting several links to one hitting only the one link at each stage.)

Since Ext(Λ) is the increasing union of Ext(A_i)'s, ray R approaches each C_i from the outside with respect to A_i. From the construction, the only links of A_i accessible from the outside within C_{i-1} are the one bend link on the outside of C_{i-1} and the links of the outermost straight subchain of A_i that runs from one side of C_i to the other (the top in Figure 3).

First, suppose that infinitely often, C_i is a bend link of A_i. Then, by taking the bend link and several links joining it on either side, forming a "U"-chain as in the proof of Theorem 2.2, we may construct an arc A in Λ as the intersection of these "U"-chains so that x is an endpoint of A. Moreover, any attempt to enlarge A will only lead to a longer arc one of whose endpoints is still x. An increasing union of longer and longer arcs in Λ, each terminating in the subarc A, is the ray-composant $C(x)$.

Now suppose that for every i, C_i is not a bend link, but a link of a straight chain running through A_{i-1}. The straight subchain of A_i that is closest to the outside of link C_{i-1} eventually runs into a bend link in the counterclockwise direction in A_i, but a bend link which is contained in a link C'_{i-1} of A_{i-1} adjacent to C_{i-1}. It follows that $x = \bigcap_{i=1}^{\infty}(C'_i \cup C_i)$. Let $\mathcal{C}_1 = \{C'_1, C_1\}$, and in general, let $\mathcal{C}_i$ be the maximal "U"-chain contained in $\cup\mathcal{C}_{i-1}$ that includes $\{C'_i, C_i\}$. Then $A = \bigcap_{i=1}^{\infty} \cup\mathcal{C}_i$ is an arc in Λ with endpoint x. Moreover, A can only be lengthened in one direction, to form the ray-composant $C(x)$. $\square$

3. Construction and Properties of Λ'

We will produce the desired cofrontier Λ' from Λ by "splitting" a countable collection of ray-composants of Λ.

3.1. Example: splitting a single ray-composant. As an illustration of the technique of splitting a ray-composant, we apply it to a certain embedding of the Knaster U-continuum ("bucket-handle") K. Figure 4 illustrates the construction of K as a nested intersection of "U"-chains of closed topological disks, where 1 indicates the first link and n indicates the last link of each chain. The intersection of the links numbered 1 is the unique endpoint p of K. Note how successive chains alternate between ascending and descending in their predecessors. As a consequence, the composant $C(p)$ containing p is accessible only at p.

Figure 5 shows the ray-composant $C(p)$, and the prime end uniformization of $\widehat{\mathbf{C}} - K$, with the unique prime end whose impression is all of K denoted by π. Prime end π corresponds to the endpoint p of K; that is, the radial limit set (principal set) of π is exactly the accessible point p.

Figure 6 illustrates the technique of "splitting" the ray-composant $C(p)$ by inserting in its place a tapered strip in the plane whose width decreases in such a way that its area is finite. The "side" boundaries of the strip become two ray-composants $C(p_1)$ and $C(p_2)$ in the new continuum K', bordering the simple dense canal that is thus created in K'. An

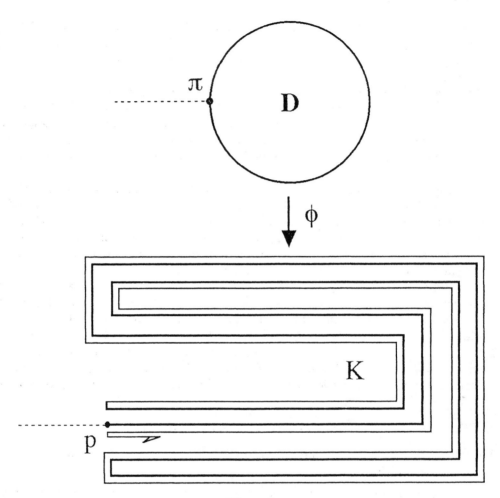

Figure 5

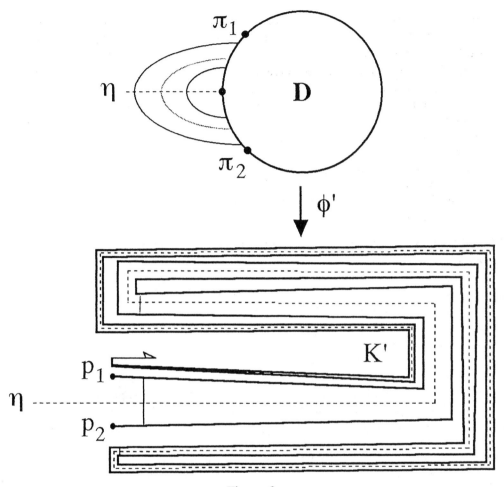

Figure 6

initial part of a chain of crosscuts defining a prime end η is pictured. (The lighter crosscut is not included in the chain because a chain of crosscuts must converge to a single point, say p_2, and not to both p_2 and p_1.)

Comparing Figures 5 and 6, the point π in the prime end circle for $\widehat{\mathbf{C}} - K$ has been replaced by an interval $[\pi_1, \pi_2]$ of prime ends in the prime end circle for $\widehat{\mathbf{C}} - K'$ with the center η of the interval corresponding to the unique simple dense canal in K'. However, there are now three prime ends whose impression is all of K': η, the unique prime end of the third kind, and the two prime ends π_1 and π_2 of the second kind corresponding, respectively, to the accessible points p_1 and p_2 which resulted from splitting p. All the remaining prime ends (both those originally in K and those added along the "sides" of the canal) are prime ends of the first kind. Clearly, K' is still indecomposable. It can also be shown that K' is still chainable. (Split the chains in Figure 4 all the way to the last link of stage i paralleling the "pocket" in chain $i - 1$ as a guide for the splitting line.)

3.2 Splitting an orbit of ray-composants. Choose a ray-composant $C(x_0)$ in Λ with endpoint x_0 accessible from $\mathrm{Ext}(\Lambda)$. Observe that the orbit of x_0 consists of accessible endpoints $x_n = h_\alpha^n(x_0)$ of pairwise disjoint ray-composants $C(x_n)$, one for each $n \in \mathbf{Z}$. See Figure 7 for a schematic of Λ from the point of view of $\mathrm{Ext}(\Lambda)$.

By ξ_n we denote the point in $\partial \mathbf{D}$ corresponding, with respect to the prime end uniformization $\phi_e : \widehat{\mathbf{C}} - \overline{\mathbf{D}} \to \mathrm{Ext}(\Lambda)$, to the accessible point $x_n \in \Lambda$. The orbit of ξ_0 under the induced prime end homeomorphism H_e is $\{\xi_n \mid n \in \mathbf{Z}\}$. We may take this to be the orbit of ξ_0 under the irrational rotation R_α.

Split each ray-composant $C(x_n)$ in a fashion similar to that in Example 3.1. The tapered strips inserted must each have finite area, and the sum of the areas of all of them must be finite. See Figure 8. The result is a continuum Λ' which is still an indecomposable cofrontier in $\mathbf{C}$. In fact, Λ' still has all the topological properties mentioned in Theorem 2.2.

The original pseudorotation h_α induces a homeomorphism $h' : (\mathbf{C}, \Lambda') \to (\mathbf{C}, \Lambda')$. If the modifications of h_α and Λ to create h' and Λ' are carried out in a straight-forward way, replacing each point $x \in C(x_n)$ by an arc A_x and defining $h'(A_x) = A_{h_\alpha(x)}$, then h' will induce

   (1) a homeomorphism $H_i'|\partial \mathbf{D}$ conjugate to the rigid rotation R_α, and
   (2) a homeomorphism $H_e'|\partial \mathbf{D}$ conjugate to the Denjoy homeomorphism obtained from R_α by replacing the orbit of ξ_0 by a doubly infinite null sequence of intervals.

For each n, we have split a prime end ξ_n of $\mathrm{Ext}(\Lambda)$ of the second kind into two prime ends ξ_n' and ξ_n'', each of the second kind, forming the endpoints of the wandering interval I_n referred to in item (2) above. The center of I_n will be a point η_n corresponding to a new simple dense canal in Λ bordered by the ray-composants $C(x_n')$ and $C(x_n'')$ which resulted from splitting $C(x_n)$. Thus, for each n, one prime end of the second kind of $\mathrm{Ext}(\Lambda)$ becomes two prime ends of the second kind, one of the third kind, and two intervals of prime ends of the first kind, of $\mathrm{Ext}(\Lambda')$.

It can be shown that h' still has most of the dynamical properties mentioned in Theorem 2.3, except that (5) must be replaced by

(5') h' *has the same irrational prime end rotation number α from both sides.*

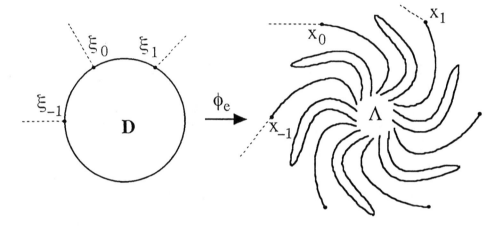

Figure 7

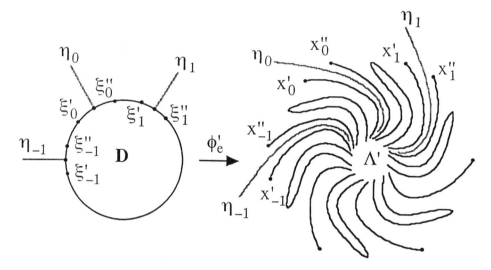

Figure 8

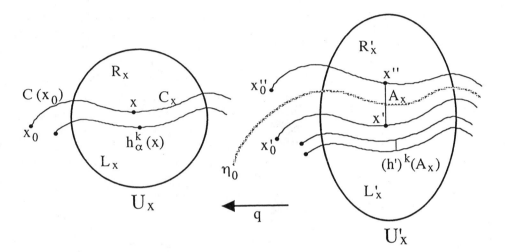

Figure 9

In particular, h' is minimal on Λ'. To see this consider the following: In forming Λ' from Λ, we replaced each point x in each composant $C(x_n)$ with a small interval $A_x = [x', x'']$ whose endpoints x' and x'' lie in the composants $C(x_n')$ and $C(x_n'')$, respectively, of Λ', but whose intrinsic interior (x', x'') lies in $\text{Ext}(\Lambda')$. Consequently, the quotient map $q : (\mathbf{C}, \Lambda') \to (\mathbf{C}, \Lambda)$ that carries A_x to x is *quasi-interior* on Λ' (that is, the image of an open set of Λ' has interior in Λ). Since h_α is minimal on Λ, and $h'(A_x) = A_{h_\alpha(x)}$, it follows that h' is minimal on Λ'.

However, essentially because of the Denjoy prime end dynamics, h' is not recurrent on Λ'. To see this, assign a counterclockwise orientation to the prime end circle $\partial \mathbf{D}$ of $\text{Ext}(\Lambda')$, and indicate left and right sides of each simple dense canal η_n corresponding to left and right halves, respectively, of the wandering interval I_n under $H_e'|\partial \mathbf{D}$. Because $H_e'|\partial \mathbf{D}$ is a Denjoy map, the sides of the canal η_n in Λ' can never map back to themselves under a power of h'. (If they did, the prime end rotation number would be rational, since the unique simple dense canal in I_n [not to mention the endpoints of the ray-composants forming the sides] would constitute a fixed prime end under a power of the map.)

Choose a point $x \neq x_0$ in the composant $C(x_0) \subset \Lambda$ and a topological disk neighborhood U_x of x in $\mathbf{C}$ so that the arc C_x of $C(x_0) \cap U_x$ containing x separates U_x into exactly two components R_x and L_x. Consider the neighborhood $U_x' = q^{-1}(U_x)$ of the arc A_x in $\mathbf{C} = q^{-1}(\mathbf{C})$. Then $U_x' \cap \Lambda'$ is a neighborhood of both endpoints x' and x'' of A_x. Note that we have defined the quotient map q so that it is *monotone* (that is, the preimage of each point is connected). See Figure 9. Let $C_x' = q^{-1}(C_x)$. Then C_x' is a "channel" in U_x' (down the "center" of which the simple dense canal η_0 goes) which separates U_x' into components $R_x' = q^{-1}(R_x)$ and $L_x' = q^{-1}(L_x)$. We may suppose $x' \in \overline{L_x'}$ and $x'' \in \overline{R_x'}$, on opposite sides of the channel. Let $\epsilon = \mathrm{d}(\overline{R_x'}, \overline{L_x'}) > 0$ which we may regard as the "width" of the channel.

By way of contradiction, suppose h' is recurrent on Λ'. Then there is an integer $k \in \mathbf{Z}^+$ such that $\mathrm{d}((h')^k(x'), x') < \frac{\epsilon}{2}$, $\mathrm{d}((h')^k(x''), x'') < \frac{\epsilon}{2}$, and $h_\alpha^k(x) \in U_x$. Without loss of generality, suppose $h_\alpha^k(x)$ is in component L_x of $U_x - C_x$. Then $(h')^k(A_x) = A_{h_\alpha^k(x)}$ lies in L_x'. Since $(h')^k(x'') \in (h')^k(A_x)$, $(h')^k(x'')$ lies in L_x' "across the channel" from $x'' \in \overline{R_x'}$; thus, $\mathrm{d}((h')^k(x''), x'') \geq \epsilon$, a contradiction.

The prime end and accessibility properties of Λ' differ considerably from those of Λ, since we have inserted two intervals of prime ends of the first kind flanking a prime end of the third kind for each of countably infinitely many prime ends of the second kind in Λ. We leave the investigation of details to the reader.

REFERENCES

[Br1] B. L. Brechner, *On stable homeomorphisms and imbeddings of the pseudo arc*, Illinois J. of Math. 22 (1978), 630–661.

[Br2] ———, *Irrational rotations on simply connected domains*, these Proceedings, 51–57.

[BGM] B. L. Brechner, M. D. Guay, and J. C. Mayer, *The rotational dynamics of cofrontiers*, these Proceedings, 59–82.

[FO] R. J. Fokkink and L. G. Oversteegen, *An example related to the Birkhoff conjecture*, Transactions AMS (to appear).

[Mi] J. Milnor, *Dynamics in one complex variable: introductory lectures*, Preprint #1990/5, Institute for Mathematical Sciences, SUNY-Stony Brook.

[P]   G. Piranian, *The boundary of a simply connected domain*, Bull. Amer. Math. Soc. **64** (1958), 45–55.

[Ro1] J. T. Rogers, Jr., *Intrinsic rotations of simply connected regions and their boundaries*, Complex Variable Theory Appl. (to appear).

[Ro2] _____, *Indecomposable continua, prime ends, and Julia sets*, these Proceedings, 263–276.

[Wa]   R. Walker, *Basin boundaries with irrational rotations on prime ends*, Transactions AMS **324** (1991), 303–317.

New Problems in Continuum Theory

SAM B. NADLER, JR.   West Virginia University, Morgantown, West Virginia

A <u>continuum</u> is a nonempty, compact, connected metric space. This paper contains a number of problems about continua and about mappings. The problems are my own and have not appeared in the literature before. Many are rather specific, some are related to known unsolved problems, and a few are of a general nature. They are accompanied by (hopefully) motivating discussions which include some new theorems. A number of these theorems give partial solutions to the problems or suggest techniques which may be useful and, thus, they are not conclusive.

The problems fall into four categories as described by the titles of the sections: I. HYPERSPACES, INDUCED MAPS, AND UNIVERSAL MAPS; II. FIXED POINT SETS; III. DISCONNECTION NUMBERS; IV. QUOTIENT MAPS.

Most of our terminology and notation is standard and is as in [5], [19], [23], [41], [44], and/or [66]. For completeness, some standard definitions are included. Non-standard ideas and notation will be explained as they come up. We shall assume the reader is familiar with basic ideas in continuum theory. We remark that the literature is referenced often, but our references are not intended to be comprehensive. We also note that throughout the paper

all spaces are metric and a <u>map</u> means a continuous function.

I.   HYPERSPACES, INDUCED MAPS, AND UNIVERSAL MAPS

We let 2^X and $C(X)$ denote the hyperspaces of a continuum X [44]. If X

and Y are continua and $f: X \to Y$ is a map (continuous), there are natural

induced maps $f^*: 2^X \to 2^Y$ and $\hat{f}: C(X) \to C(Y)$ defined by

$$f^*(A) = f(A) \text{ for all } A \in 2^X, \hat{f} = f^* | C(X).$$

These maps are important in the study of hyperspaces as well as continuum

theory in general. For example: they are the induced bonding maps used in

proving hyperspaces commute with inverse limits ([61], [62], or [44, p. 171]),

and the notion of Class (W) (e.g., [11]-[13]) may obviously be reformulated in

terms of the surjectivity of $\hat{f}$.

A map $f: X \to Y$ is said to be <u>universal</u> provided that for any map

$g: X \to Y$, $f(x) = g(x)$ for some $x \in X$ [17]. As we shall see in the next

theorem, universal maps onto n-cells I^n are related to maps onto I^n having a

stable value (a point $y \in f(X)$ is a <u>stable</u> <u>value</u> of $f: X \to Y$ provided that for

some $\delta > 0$, any map $g: X \to Y$ within δ of f has y as a value [19, p. 74]). The

existence of maps onto I^n with some stable value characterizes those spaces of

dim $\geq$ n [19, pp. 75-77]. We also remark that universal maps onto I^n are the

same as AH-essential maps (f: $X \to I^n$ is called <u>AH</u>-<u>essential</u> provided that

$$f | f^{-1}(\dot{I}^n): f^{-1}(\dot{I}^n) \to \dot{I}^n \text{ where } \dot{I}^n = \text{manifold boundary of } I^n$$

can not be extended to a continuous function mapping X to $\dot{I}^n$). This equiva-

lence has been noted by several people (e.g., 1.1 of [18] and [48, p. 222]).

We note that the existence of AH-essential maps onto I^n <u>also</u> characterizes spaces of dim $\geq$ n, as follows easily using the theorem in [19, p. 83] and Tietze's extension theorem. We now state the following result which is simple to prove using B of [19, pp. 78-79].

1.1 THEOREM. If f: X $\rightarrow$ I^n is AH-essential (or universal), then every point in I^n - $\overset{\bullet}{I}{}^n$ is a stable value of f.

The converse of this theorem is false even for maps from I^2 onto I^2, as the following example shows.

1.2 EXAMPLE. Let X and Y be the two 2-cells in the plane drawn below and defined analytically as follows: X = $X_1 \cup X_2 \cup X_3$ where

X_1 = [-7, -11/2] $\times$ [-1, 1]

X_2 = [-11/2, -9/2] $\times$ [-1/2, 1/2]

X_3 = [-9/2, -3] $\times$ [-1, 1]

and let Y = [-1, 1] $\times$ [-1, 1]

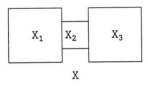

X   Y

Define a map f: X $\rightarrow$ Y as follows: f translates X_1 and X_3 to the right by six and four units respectively, and, after rotating X_2 in 3-space through 180° about the line x = -5, f translates X_2 to the right five units (i.e., f on X_2 is the natural "linear" extension of f on $X_1 \cup X_3$). It is easy to see that every point of Y - $\overset{\bullet}{Y}$ is a stable value of f but that f is not AH-essential.

1.3 PROBLEM. Determine conditions which are both necessary and sufficient in order that a map of a continuum (or a space) onto I^n, n $\geq$ 2, have all values in I^n - $\overset{\bullet}{I}{}^n$ stable.

1.4   PROBLEM.   Let us say that a continuum Y is in <u>Class (S)</u> provided

that every map of any continuum onto Y has a stable value.   What continua are

in Class (S)?   Some comments are in the paragraph below.

It is evident that any continuum with a cut point (i.e., separating

point) is in Class (S).   More generally, any continuum containing a separating

C-set is in Class (S) (a subcontinuum K of X is a <u>C-set</u> in X provided that any

subcontinuum of X intersecting K and X - K contains K).   Hence, e.g., all

hereditarily decomposable arc-like continua are in Class (S).   On the other

hand, the Buckethandle continuum is not in Class (S).   I do not know an in-

trinsic characterization of those arc-like continua which are in Class (S).

J. F. Davis and I are currently working on trying to determine what continua

are in Class (S).   The problem is especially interesting for Peano continua.

Among other results, we have shown regular continua are in Class (S).

Let us return to the induced hyperspace maps f^* and $\hat{f}$ defined above.   In

[57], it was shown that C(X) has the fixed point property when X is arc-like.

In [19], AH-essential maps were used to give a shorter, simpler proof of this

result and to prove that C(X) has the fixed point property when X is circle-

like.   The proofs of these facts as done in [22] may be summarized as follows:

If f: X → [0,1] or S^1 is continuous and onto, then $\hat{f}$: C(X) → C([0,1]) or $C(S^1)$

is AH-essential and, if f is an ϵ-map, $\hat{f}$ is an ϵ-map (same ϵ) - thus, by

Lokuciewski's theorem [27], C(X) has the fixed point property.   It is a well-

known problem to determine whether or not 2^X has the fixed point property

when X is arc-like or circle-like.   Since, as is easy to see, f^* is an ϵ-map

if f is an ϵ-map, an affirmative solution to this problem would follow from an

affirmative solution to the following problem.

1.5   PROBLEM.   Let Y = [0,1] or S^1.   If f maps a continuum X onto Y, then must f^*: $2^X \to 2^Y$ be universal? On considering the problem about the fixed point property for 2^X mentioned above, it may be assumed here that X is arc-like or circle-like and that, if X is proper circle-like, f: $X \to S^1$ is essential.

In connection with the fixed point problem, we also mention that 2^X and C(X) have the fixed point property when X is an inverse limit of dendrites with quasi-monotone (onto) bonding maps (3.3 of [45, p. 753]), a special case of which is open bonding maps. This was used in [45] to give an example of an indecomposable continuum X such that 2^X has the fixed point property.

Knowledge about when the induced map $\hat{f}$ is universal may lead to a solution of the dimension question for hyperspaces - if X is a 2-dimensional continuum, then is C(X) infinite-dimensional? The answer is yes if the rank of $\check{H}^1(X)$ is finite [21]. Background concerning this question is in Chapter II of [44], so we do not take the time to discuss it here. We state the following problem related to this question (a <u>simple n-odd</u> (n ≥ 3) is the union of n arcs emanating from a single point which is an end point of each arc).

1.6 PROBLEM. If Y is a simple n-odd and f is a universal map of a continuum X onto Y, then must $\hat{f}$: C(X) → C(Y) be universal? A positive answer would yield a positive answer to the dimension question for hyperspaces (by using the existence of a universal map of any 2-dimensional continuum onto I^2, noted near the beginning of this section, and applying 2.2 of [48]).

We have seen that it is of interest to know when $\hat{f}$ and f^* are universal. It is also of interest to know when f is universal assuming $\hat{f}$ and/or f^* is universal. Such a general inquiry is motivated in part by the theorem we shall prove next. First, we prove a lemma.

Recall that a <u>selection</u> for a subset S of $C(X)$ is a continuous function $\sigma: S \to X$ such that $\sigma(A) \in A$ for all $A \in S$. A <u>$C(X)$-selection continuum</u> is a continuum X such that there is a selection for $C(X)$.

1.7 LEMMA. Let X be a $C(X)$-selection continuum. If $f: X \to Y$ is a map such that $\hat{f}: C(X) \to C(Y)$ is universal, then f is universal.

PROOF. Let $\sigma: C(X) \to X$ be a selection. Let $i: Y \to C(Y)$ be given by $i(y) = \{y\}$ for each $y \in Y$. Now, to prove f is universal, let $g: X \to Y$ be a map. Since $\hat{f}$ is universal, there exists $A \in C(X)$ such that

$$\hat{f}(A) = i \circ g \circ \sigma(A).$$

In other words, $\hat{f}(A) = \{g(p)\}$ where $p = \sigma(A)$. Therefore, since $p \in A$ and $\hat{f}(A) = f(A)$, $f(p) = g(p)$.

1.8 THEOREM. Let X be a one-dimensional Peano continuum. Then, X is a dendrite if and only if whenever f is a map of X onto a continuum Y such that $\hat{f}$ is universal, f must be universal.

PROOF. Since dendrites are $C(X)$-selection continua ([2] or [50, p. 371]), half of the theorem follows from the preceding lemma. To prove the other half, assume the given condition about maps and suppose X contains a simple closed curve S. Then, since $\dim(X) = 1$, there is a retraction r from X onto S [19, p. 83]. It follows easily that $\hat{r}: C(X) \to C(S)$ is universal (since $C(S)$ is a 2-cell and $C(X)$ is acyclic [61], so $\hat{r}$ is AH-essential). But, clearly, r is not universal (as can be seen by following r with a small rotation of S, or see [17, p. 433]). Therefore, X does not contain a simple closed curve and, hence, is a dendrite [66, p. 88].

1.9 PROBLEM. Would the theorem above remain true if the condition that X is one-dimensional were omitted?

Regarding this problem, note the next theorem and the problem following it (a map is <u>monotone</u> provided that each of its point inverses is connected).

1.10 THEOREM. Let X be a continuum (not necessarily Peanian) such that whenever f is a map of X onto a Peano continuum Y such that $\hat{f}$ is universal, f must be universal. Then, every Peano continuum which is a monotone image of X has the fixed point property (in particular, X does if X is Peanian).

PROOF. Let f be a monotone map of X onto a Peano continuum Y. Then, by 2.3 of [45], $\hat{f}$ is universal. Hence, f is universal. Therefore, Y has the fixed point property [17].

1.11 PROBLEM. If X is a Peano continuum such that every monotone image of X has the fixed point property, must dim(X) $\leq$ 1? If the answer is yes, then the answer to the preceding problem is also yes (by the last two theorems).

Moving away from Peano continua, note the following result proved in [48, p. 226].

1.12 THEOREM. If f is any map from a continuum X onto an arc-like continuum Y, then $\hat{f}$: C(X) → C(Y) is universal.

This result was used in [48] to obtain as easy corollaries two well-known and different types of results: (1) If Y is an arc-like continuum, then C(Y) has the fixed point property [60]; (2) Arc-like continua are in Class (W) (Theorem 4 of [52]). Thus, the following problem is of interest.

1.13 PROBLEM. We say that a continuum Y is in <u>Class (Û)</u> if for every map f of any continuum X onto Y, $\hat{f}$: C(X) → C(Y) is universal. Determine an intrinsic characterization of Class (Û).

It is clear that Class (Û) is contained in Class (W). J. F. Davis has an example (unpublished) of a weakly chainable continuum in Class (W) which contains a simple triod. Hence, by the corollary to the following theorem, Class ($\hat{U}$) is not the same as Class (W).

1.14 THEOREM. Let X be a continuum such that $\dim[C(X)] = 2$. If f is a mapping of X onto a continuum Y such that $\hat{f}$: $C(X) \to C(Y)$ is universal, then Y is a-triodic.

PROOF. Suppose that Y contains a triod Z (a <u>triod</u> is a continuum Z containing a subcontinuum N such that Z-N is the union of three mutually separated sets). Then, C(Y) contains a 3-cell B [54]. Let r be a retraction of C(Y) onto B. It is readily seen that $r \circ \hat{f}$: $C(X) \to B$ is universal (since if g maps C(X) into B, then, $\hat{f}$ being universal, there exists $A \in C(X)$ such that $\hat{f}(A) = g(A)$ - thus, since $g(A) \in B$, $\hat{f}(A) \in B$ so $r \circ \hat{f}(A) = \hat{f}(A)$, hence $r \circ \hat{f}(A) = g(A)$). Thus, $r \circ \hat{f}$ is AH-essential and, therefore, $\dim[C(X)] \geq 3$.

1.15 COROLLARY. If Y is in Class ($\hat{U}$), then Y is a-triodic if Y is a continuous image of an arc-like continuum, a circle-like continuum, or a one-dimensional hereditarily indecomposable continuum X such that $H^1(X)$ is finitely generated.

PROOF. In each case, the previous theorem applies by Corollary 1 of [6] in the first two cases and by Theorem 1 of [56] in the third case.

We mention that continuous images of arc-like continua have been studied in [8], [24], and [36] (we remark that the first statement of the proof in the footnote in [36, p. 179] is false even in the special case considered - compare with Theorem 3 of [3]). Continuous images of certain circle-like continua have been investigated in [7] and [55].

II. FIXED POINT SETS

The <u>fixed</u> <u>point</u> <u>set</u> of a map f is written FPS(f) and is simply the set of all fixed points of f. The Brouwer fixed point theorem says that FPS(f) $\neq \phi$ for any self-map f of an n-cell. In Theorem 1 of [53] it was shown that any nonempty, closed subset of an n-cell is the FPS of some self-map of the n-cell. The proof is surprisingly simple and was extrapolated into the following result in [65, p. 554]:

2.1 THEOREM. Let (X,d) be a metric space, and let $p \in X$. If there is a homotopy h: $X \times [0,1] \rightarrow X$ such that

h(x,0) = x for all $x \in X$

h(x,t) $\neq$ x for all $x \in X - \{p\}$ and $t \neq 0$,

then for any closed subset K of X such that $p \in K$, there is a map f: $X \rightarrow X$ such that K = FPS(f).

PROOF. Assuming diam(X) $\leq$ 1, define f: $X \rightarrow X$ by letting f(x) = h(x,d(x,K)) for each $x \in X$. It is evident that f has the desired properties.

A space is said to have the <u>complete</u> <u>invariance</u> <u>property</u> [65], CIP, provided that every nonempty, closed subset of it is a FPS for some self-map of the space. In [65, p. 553], it was asked if every Peano continuum has CIP. This question was answered negatively in [31] with elegant examples which are even acyclic, (n+1)-dimensional and LC^{n-1} for each n = 1, 2, ... Later it was shown that all one-dimensional Peano continua have CIP [34]. In connection with this discussion, the reader should also see the examples in [32] and in the follow-up paper [33].

One direction which seems worthwhile for future study is to investigate fixed point sets for particular types of maps. Fixed point sets for homeomorphisms have been examined by a number of people (e.g.: see [30], [53], [58], [59], and the survey article in [57]). The following theorem is due to Y. Ohsuda and me [51] (a <u>free</u> <u>arc</u> $\overset{\frown}{pq}$ <u>in</u> <u>X</u> satisfies $\overset{\frown}{pq}$ - (p,q) is open in X):

2.2 THEOREM. Let X be an arcwise connected continuum with a free arc. Then, every nonempty subcontinuum of X is a FPS for a homeomorphism of X onto X if and only if X is a simple closed curve.

2.3 EXAMPLE. In relation to the theorem above, the continuum X drawn here has free arcs and every closed subset of it is a FPS for a homeomorphism

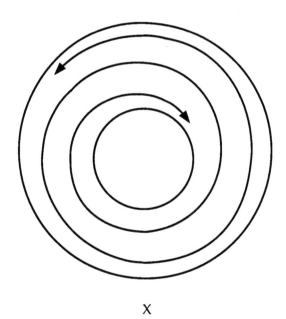

X

of it onto itself but X is not arcwise connected.

The theorem and example just given lead to the following question.

2.4 PROBLEM. Is the simple closed curve the only one-dimensional Peano continuum such that every nonempty, closed subset (or subcontinuum) of it is the FPS for a homeomorphism of the continuum onto itself? We note that every

nonempty, closed subset of a 2n-cell is the fixed point set of a homeomorphism

of the 2n-cell onto itself by Theorems 1 and 2 of [59] - this was observed but

proved incorrectly in [53].

If $\mathfrak{C}$ is a class of maps, then we say X has <u>CIP</u> <u>with</u> <u>respect</u> <u>to</u> $\mathfrak{C}$ provided

that every nonempty, closed subset of X is a FPS for some map in $\mathfrak{C}$. Some

early works can be thought of as being related to the study of fixed point

sets for particular types of maps (e.g., see [64, pp. 76-77]), but a system-

atic study when $\mathfrak{C}$ is, for example, the monotone maps (maps with connected

point inverses) has not been done. In particular:

2.5 PROBLEM. What continua have CIP with respect to monotone maps

(surjective or not, your choice)? What about confluent or even weakly

confluent maps?

For the following very specific problem, let B^n denote the closed unit

ball in Euclidean n-space and let $B_0^n = \{x \in B^n: \|x\| \leq 1/2\}$. The proof of

Theorem 2 of [53] says B_0^n is not the FPS of any homeomorphism of B^n onto B^n

when n is odd.

2.6 PROBLEM. Is B_0^n the FPS of a monotone map of B^n onto B^n when n is

odd? If the answer is yes, then the question becomes: Does B^n have CIP with

respect to monotone maps when n is odd (n even is taken care of by homeomor-

phisms [59])?

Another direction in the study of fixed point sets was begun in [10]

where fixed point sets for multivalued maps were investigated. More pre-

cisely: Let X be a continuum, let C(X) be the hyperspace of subcontinua of X

[44], and let F: X → C(X) be continuous. In this situation, a <u>fixed</u> <u>point</u> <u>of</u>

<u>F</u> is a point p ∈ X such that p ∈ F(p). The <u>multivalued</u> <u>fixed</u> <u>point</u> <u>set</u> <u>of</u> <u>F</u>,

written MFPS(F), is (p ∈ X: p ∈ F(p)). These ideas go back to fundamental work some of which is summarized in [64, pp. 74-76] (however, note that here we are using a continuous F, not a usc F; compare with [10, p. 86]). Now, a continuum X is said to have <u>MCIP</u> [10] provided that every nonempty, closed subset of it is a MFPS(F) for some map F: X → C(X).

Clearly, a continuum has MCIP if it has CIP. The question in [65] and its negative answer in [31] were the prime motivation for obtaining the following result in 2.2 of [10, p. 87]:

2.7 THEOREM. Every Peano continuum has MCIP.

The natural question now is: Can the theorem above be proved by using continuum-valued maps which are arbitrarily close to single-valued maps? In other words: Let us say that a continuum X has <u>ε-MCIP</u> provided that for each $\delta > 0$, every nonempty, closed subset of X is a MFPS for some map F: X → C(X) with diam[F(x)] < δ for all x ∈ X; then:

2.8 PROBLEM. Does every Peano continuum have ε-MCIP? An affirmative answer would improve 2.7 and be the best possible result in view of the fact that Peano continua do not have CIP [31].

2.9 PROBLEM. What continua have ε-MCIP?

The following result can be used to give some partial answers to 2.8.

2.10 THEOREM. Let X be a Peano continuum with metric d, and let p ∈ X. Assume that for each $\delta > 0$ there is a map G_δ: X → C(X) satisfying (1)-(3) below:

(1)   diam[$G_\delta(x)$] < δ for all x ∈ X;

(2)   x ∉ $G_\delta(x)$ for all x ≠ p;

(3) $d(x, G_\delta(x))$ < δ for all x ∈ X.

Then, for any closed subset A of X such that $p \in A$ and for any $\eta > 0$, there is

a map F: X → C(X) such that $\text{diam}[F(x)] < \eta$ for all $x \in X$ and A = MFPS(F).

PROOF. We may assume the metric d for X is convex ([1], [37]). Then,

letting $K_d(\epsilon, L)$ denote the closed ϵ-ball about a subcontinuum L of X (i.e.:

$K_d(\epsilon, L) = \{x \in X: d(x,y) \leq \epsilon \text{ for some } y \in L\}$), we see that the function

$$(\epsilon, L) \to K_d(\epsilon, L)$$

is continuous and maps $[0, +\infty) \times C(X)$ into C(X) (comp., [39, pp. 170-172]).

Now, letting A, η, and G_δ with $\delta = \eta/3$ be as in the theorem, and letting

φ: X → [0,1] be continuous such that $\varphi^{-1}(1) = A$, define F by

$$F(x) = K_d(\varphi(x) \cdot d(x, G_\delta(x)), G_\delta(x))$$

for each $x \in X$. We see that F has the desired properties.

As an illustration of the applicability of this theorem, we remark that

it can be used quite easily to prove that the cone over any Peano continuum

has ϵ-MCIP (even though such cones need not have CIP [33]).

III. DISCONNECTION NUMBERS

The following definitions are from [42]. Let X be a connected space. A

cardinal number $n \leq \aleph_0$ is called a _disconnection_ _number_ _for_ X provided that

whenever $A \subset X$ such that $|A| = n$, then X-A is not connected. We write

$D(X) \leq \aleph_0$ to mean there is a disconnection number for X. When $D(X) \leq \aleph_0$, we

let $D^s(X)$ denote the smallest disconnection number for X.

In [42], we proved the following theorem (a _graph_ is a one-dimensional,

compact, connected polyhedron or a point; $\chi(X)$ denotes the Euler characteris-

tic of a graph X (i.e., $\chi(X) = |\text{vertices}| - |\text{edges}|$), and P(X) denotes the set

of end points, or pendant nodes, of X).

3.1 THEOREM. Let X be a nondegenerate continuum. If $D(X) \leq \aleph_0$, then X is a graph and

(#) $D^s(X) = 2 - \chi(X) + |P(X)|$.

Conversely: If X is a graph, then $D(X) \leq \aleph_0$ and, thus, (#) holds.

One interesting consequence of the theorem is that if $\aleph_0$ is a disconnection number for a continuum X, then so is some integer.

I have recently discovered a paper of A. H. Stone [63] in which he proved a general result which implies the theorem above except that he does not obtain the formula in (#). (To understand the connection between Stone's Theorem 1 [63, p. 268] and the theorem above, the reader should note the statement in lines 15-16 of [63, p. 266] - the same fact was proved in the same way in 4.1 of [42] but for continua.) The general approaches in [42] and [63] are substantially different. Also, in [42] the formula in (#) was used to obtain a number of other results among which is the following one (6.3 of [42]).

3.2 THEOREM. Let X be a continuum. Then, $D^s(X) = 3$ if and only if X is one of the following five continua: an arc, a noose, a figure eight, a theta curve, a dumbbell.

As noted in [42], several known results follow easily from this theorem: $D^s(X) = 2$ where X is a continuum if and only if X is a simple closed curve [38, p. 342], and a continuum X is an arc if and only if X has exactly two non-cut points (see [23, p. 179]).

The results above lead to the following two problems from [42].

3.3 PROBLEM. For each n = 2, 3, ..., let D_n^s denote the number of topologically different graphs X such that $D^s(X) = n$. Find a formula which can be used to calculate D_n^s for each n or to determine upper bounds, lower bounds, or estimates concerning how fast $D_n^s \to \infty$ as $n \to \infty$. We have seen that $D_2^s = 1$ and $D_3^s = 5$. I suspect that $D_4^s = 26$; $D_4^s \geq 26$ and all X such that $D^s(X) = 4$ are planar (8.2 of [42]).

3.4 PROBLEM. Can $\aleph_0$ be the smallest disconnection number for a connected metric space? Note that by the comment following the first theorem above, $\aleph_0$ can not be the smallest disconnection number for a continuum.

The answer to 3.4 would seem to be "yes" by a parenthetical comment in (c) of [63, p. 274], but I do not know how to construct such a space.

For each n = 2, 3, ..., let $G(n) = \{X: D^s(X) = n\}$. We say that a graph U is <u>universal</u> <u>for</u> $G(n)$ provided that each X in $G(n)$ can be topologically embedded in U. We note the following proposition which we include mainly as a point of departure for the discussion following it.

3.5 PROPOSITION. For each n = 2, 3, ..., there is a universal graph U for $G(n)$ such that $D^s(U) \leq (n-1)D_n^s + 1$.

PROOF. For two graphs X and Y, let XvY denote the join of X and Y at $x \in X$ and $y \in Y$ (i.e., XvY is the quotient space obtained from the disjoint union of X and Y by identifying x with y). Obviously, XvY is a graph and it is easy to see that $\chi(XvY) = \chi(X) + \chi(Y) - 1$. Thus, by using 3.1 and taking cases depending on whether or not $x \in P(X)$ or $y \in P(Y)$, we see that

(1) $D^s(XvY) \leq D^s(X) + D^s(Y) - 1$.

Now, fix n. Let X_i, $1 \leq i \leq k = D_n^s$ be an indexing of the k topologically different members of $G(n)$. Let U be the graph obtained as the final result of the successive joins X_1vX_2, $(X_1vX_2)vX_3$, ..., $(X_1v\cdots vX_{k-1})vX_k = U$. Clearly, U is universal for $G(n)$. Finally, by using (1) k times, we obtain the desired inequality $D^s(U) \leq (n-1)k + 1$.

It is easy to improve this result by redoing its proof as follows. Start with a tree X_1 having n-1 end points p_1, ..., p_{n-1}. For each i such that $2 \leq i \leq n$, join X_i to X_1 at p_{i-1}. Then, join the remaining graphs X_j, $n+1 \leq j \leq D_n^s$, as before. The resulting universal graph U satisfies

$D^s(U) \leq (n-1)D_n^s - n + 2$.

However, this inequality is also far from being the best possible. Part of the reason is that for $j < D_n^s$, $X_1 \vee \cdots \vee X_j$ may already be universal and further joins may increase the smallest disconnection number. One does better by joining graphs some of which are not in G(n). For example, by joining a simple closed curve to a theta curve at a point of order two in the theta curve, we obtain a universal graph U for G(3) such that $D^s(U) = 4$.

This brings us to the following problem. We let g(n) = min{j: there is a universal graph U for G(n) such that $D^s(U) = j$}, for each n = 2, 3, ...

3.6 PROBLEM. Find g(n) or at least some interesting estimates for it. We note that g(2) = 2 and that g(3) = 4 (as has just been shown above).

Let K_n denote the complete graph on n vertices (i.e., any two vertices of K_n are joined by one and only one edge). We see that K_6 is universal for G(3) and that no K_n has this property for n < 6. Also, clearly, K_3 is universal for G(2). This brings us to the following problem.

3.7 PROBLEM. For $n \geq 4$, find the minimum m such that K_m is universal for G(n).

The last five problems in this section are of a different nature than the previous ones.

Recall that a map f: X → Y is said to be n-to-1 provided that $|f^{-1}(y)| = n$ for all $y \in Y$. There is a vast literature on n-to-1 maps and even the basic papers are too numerous to cite here. We remark that there are a number of interesting recent papers of Jo Heath and A. J. W. Hilton about

n-to-1 maps between graphs. No work has yet been done concerning the

following general problem (as is necessitated by the forthcoming discussion,

three more specific problems will be stated later).

3.8 PROBLEM. Obtain results about how n-to-1 maps between graphs affect

their smallest disconnection numbers.

We note immediately that even 3-to-1 maps between graphs may lower the

smallest disconnection number of the domain graph to 2 or may raise it an

arbitrarily large amount. The first fact can be seen by observing that if T

is a tree, then $D^s(T) = 1 + |P(T)|$ and there are trees T such that $|P(T)|$ is

as large as we wish and for which there is a 3-to-1 map of T onto S^1 (such

trees are characterized in [16]). The second fact can be seen as follows.

First, note that there are maps α and β of [0,1] onto [0,1] such that

$$|\alpha^{-1}(t)| = |\beta^{-1}(t)| = 3 \text{ for } 0 < t < 1$$
$$|\alpha^{-1}(0)| = 2, \ |\alpha^{-1}(1)| = 1, \text{ and } |\beta^{-1}(0)| = |\beta^{-1}(1)| = 1$$

(such maps are drawn below - α is essentially the map used in [15, p. 789]

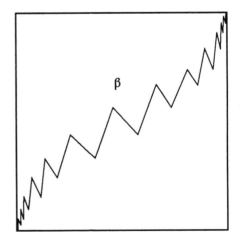

to give an example of a 3-to-1 map of $[0,1]$ onto S^1). Now, fix an integer

$n \geq 2$. Let X_n denote the graph drawn below and obtained by glueing n disjoint

simple closed curves C_1, ..., C_n to S^1 at n distinct points q_1, ..., q_n. Let Z

denote a circle in the plane such that $Z \cap X_n = \phi$, and let p_1, ..., p_{2n} denote

2n distinct points of Z. Assume as in the figures that each p_i precedes p_{i+1}

and each q_i precedes q_{i+1} in the clockwise direction. Let A_i denote the arc in

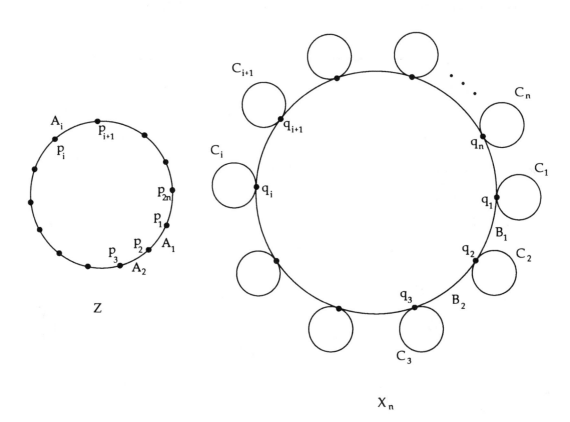

Z from p_i to p_{i+1} which contains no other p_j, and let B_i denote the arc in

$S^1 \subset X_n$ from q_i to q_{i+1} which contains no other q_j. Note that if $\alpha' = \gamma \circ \alpha$

where γ is the quotient map identifying 0 and 1, then α' is a 3-to-1 map of

[0,1] onto a simple closed curve and $\alpha'(0) = \alpha'(1)$. It is now easy to see how to use $\alpha': A_i \to C_i$ for i odd and $\beta: A_i \to B_i$ for i even to produce a 3-to-1 map f of Z onto X_n. Since $D^s(X_n) = n+2$, f raises $D^s(Z)$ by n. (We can also raise $D^s(Z)$ by one by obtaining a 3-to-1 map of Z onto a figure eight - assuming that $n \geq 2$ was only a convenience for the procedure used above.)

In view of these examples, we modify the previous problem with the following three more specific ones.

3.9 PROBLEM. Characterize those graphs X for which there is an n-to-1 map $(n \geq 2)$ onto another graph with smaller disconnection number than X. The dual problem with smaller replaced by larger is also of interest. Also, what can be said about the integers $n \geq 2$?

3.10 PROBLEM. For what graphs X and for what integers $n \geq 2$ is there an upper bound on the smallest disconnection numbers of all graphs which are an n-to-1 image of X?

3.11 PROBLEM. For what graphs X and Y such that $D^s(X) = D^s(Y)$ and for what integers $n \geq 2$ is there an n-to-1 map of X onto Y?

In working on 3.9-3.11, the reader should be aware of [9] and [14]. We also note that results concerning 3.11 may be able to be viewed as generalizations of the work in [35] where graphs X for which there is an n-to-1 map of X onto X for some $n \geq 2$, or for all odd n, are characterized.

For the problem in 3.13, we introduce a companion notion to that of a disconnection number.

A subset A of a continuum X is said to <u>weakly cut</u> X provided that X-A is not continuumwise connected. A cardinal number $n \leq \aleph_0$ is called a <u>cutting number</u> <u>for</u> X provided that whenever $A \subset X$ such that $|A| = n$, A weakly cuts X.

For example, the smallest cutting numbers for the three continua X, Y, and Z
pictured below are one, two, and three, respectively. Note the next theorem.

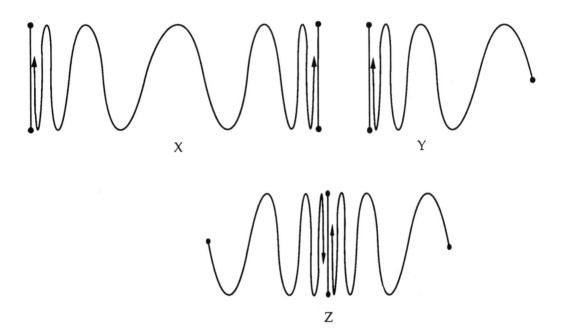

X   Y

Z

   3.12   THEOREM.   A (nondegenerate) continuum X which is irreducible about
some n points, n < ∞, has smallest cutting number ≤ n+1.

   PROOF.   Assume X is irreducible about the n points x_1, ..., x_n.   Let
A ⊂ X such that $|A|$ = n+1.   Let B_1, ..., B_n be mutually disjoint, nondegen-
erate subcontinua of X such that $x_i \in B_i$ and $B_i \cap A \subset \{x_i\}$ for each i (such
subcontinua exist by using Theorem 1 of [23, p. 172]).   Let $q_i \in B_i - \{x_i\}$ for
each i.   Now, suppose A does not weakly cut X.   Then (since $q_i \notin A$ for any i),
there is subcontinuum Q_i of X-A such that q_1, $q_i \in Q_i$ for each i.   Let

$$C = \left[\bigcup_{i=1}^{n} B_i \right] \cup \left[\bigcup_{i=1}^{n} Q_i \right].$$

Then, C is a continuum and, since $C \cap A \subset (x_1, \ldots, x_n)$ and $|A| = n+1$, some

point of A is not in C, so $C \neq X$. But, since $x_i \in C$ for each i, this contra-

dicts our assumption that X is irreducible about the points $x_1, \ldots, x_n$.

Therefore, A weakly cuts X.

This brief discussion of cutting numbers brings us to the following:

3.13 PROBLEM. Determine various classes of continua which have cutting

numbers and find formulas or conditions which give their smallest cutting

numbers.

Finally, it is appropriate to mention here that my doctoral student Gary

Seldomridge is in the process of finishing his dissertation on extended dis-

connection numbers. Roughly, this means he is concerned with removing n

disjoint nondegenerate subcontinua, or various types of subcontinua, to dis-

connect the continuum. In other words, points in the definition of disconnec-

tion number are replaced by various types of subcontinua.

IV. QUOTIENT MAPS

We note the following ideas from the literature. First, one-to-one

images of non-compact intervals of reals have been studied in [20], [25],

[40], [46], and [49], and confluent images of non-compact intervals have been

studied in [47]. Second, those continua such that every map of all continua

onto them has a particular property P have been characterized for some choices

of P (e.g., see [4], [11], [13], and [26]). Third, there are a number of

results which say that a map g has a certain property if the composition g∘f

has that property (a summary of such results is in the table in [29, pp. 48-

49]). The theorem below is from [43] and gives two characterizations of trees in the context of these ideas (a <u>tree</u> is a graph containing no simple closed curve or, equivalently, a one-dimensional, compact, connected, acyclic poly-hedron or a point; a <u>quotient map</u>, or <u>identification map</u>, is a map f from a space X onto a space Y such that $f^{-1}(U)$ is open in X if and only if U is open in Y).

4.1 THEOREM. Let I be a non-compact interval in the reals R^1, and let Y be a continuum which is a continuous image of I (i.e., Y is an arcwise connected continuum which is the union of countably many Peano continua). Then (1), (2), and (3) below are equivalent:

(1)   Every continuous function from I onto Y is a quotient map.

(2)   For every continuous function f from I onto Y and every function g from Y into R^1 such that g∘f is continuous, g is continuous.

(3)   Y is a tree.

We remark that (1) and (2) would be equivalent immediately by 3.1 of [5, p. 123] if we allowed the range of g in (2) to be any topological space.

In what follows, a <u>semi-continuum</u> is a metric space S such that any two points of S can be joined by a subcontinuum of S [23, p. 188]. Also, recall our agreement that a map means a continuous function.

The discussion and theorem above lead to the following notion. Let us say that a continuum Y is in <u>Class (Q)</u> provided that every map of every semi-continuum onto Y is a quotient map (the terminology is analogous to that for notions such as Class (W) and Class (P) [11]-[13]). The following theorem gives an intrinsic characterization of the continua in Class (Q) (recall - a continuum is said to be <u>finitely irreducible</u> provided that it is irreducible about a finite set).

4.2 THEOREM. For a continuum Y, (a)-(c) below are equivalent:

(a)   Y is in Class (Q).

(b)   Y is finitely irreducible.

(c)   Every map from any locally compact semi-continuum onto Y is a quotient map.

PROOF. The proof that (b) implies (a) is easy and is proved in the Proposition near the end of [43]. Also, it is obvious that (a) implies (c). Hence, it only remains to prove (c) implies (b).

Assume (b) is false. Then, by [28], there are subcontinua Y_i of Y, i = 1, 2, ..., such that $Y_i \subset Y_{i+1} \neq Y_i$ for each i and $\overset{\infty}{\underset{i=1}{\cup}} Y_i = Y$. Fix a point $p \in Y_1$. Let S be the subspace of $Y \times (0,1]$ defined by

$$S = \left[\overset{\infty}{\underset{i=1}{\cup}} (Y_i \times \{1/i\}) \right] \cup \{(p,t): 0 < t \leq 1\}.$$

(p,1)

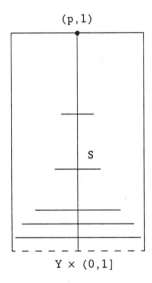

S

Y × (0,1]

Let f be the vertical projection from S to Y, i.e.,

f(y,t) = y for each (y,t) ∈ S.

It is evident that S is a locally compact semi-continuum, f is continuous, and, since $\overset{\infty}{\underset{i=1}{\cup}} Y_i = Y$, f maps S onto Y. We show that f is not a quotient map as follows. Since each Y_i is properly contained in Y_{i+1}, there exists a point $y_{i+1} \in Y_{i+1} - Y_i$ for each $i = 1, 2, \ldots$. We assume without loss of generality that $(y_i)_{i=2}^{\infty}$ converges to a point $y \in Y$. Let

$$Z = \begin{cases} (y_2, y_3, \ldots) & \text{, if } y \neq y_i \text{ for any } i \\ (y_2, y_3, \ldots) - (y_k) & \text{, if } y = y_k \text{ for some k.} \end{cases}$$

In either case, Z is not closed in Y and, since only finitely many points of $f^{-1}(Z)$ are in $Y_i \times (1/i)$ for any given i, $f^{-1}(Z)$ is closed in S. Hence, f is not a quotient map. Thus, (c) is false. This completes the proof that (c) implies (b) and, therefore, we have proved the theorem.

Note the following two corollaries, after which we shall state some open problems.

4.3 COROLLARY. An arcwise connected continuum is in Class (Q) if and only if it is a tree.

PROOF. An arcwise connected continuum is a tree if and only if it is finitely irreducible (the "if part" is an easy induction, and the "only if part" is due to the fact that a tree must be irreducible about its finitely many end points by 1.1 (ii) of [66, p. 88] and Theorem 4 of [23, p. 183]). Therefore, the corollary follows from the theorem above.

4.4 COROLLARY. If Y is a finitely irreducible continuum, then any point q in the boundary of any nondegenerate subcontinuum of Y weakly cuts Y (i.e., Y-(q) is not a semi-continuum).

PROOF. Suppose there exists q in the boundary of a nondegenerate subcontinuum K of Y such that Y-(q) is a semi-continuum. Fix $p \in K-(q)$. Then, letting S be the subspace of $Y \times [0,1]$ defined by

$$S = [(Y-(q)) \times (1)] \cup [K \times (0)] \cup ((p,t) : 0 \leq t \leq 1),$$

we see that S is a semi-continuum. Let f be the vertical projection of S onto

Y (f(y,t) = y for each (y,t) ∈ S). To see that f is not a quotient map, let

$(y_i)_{i=1}^{\infty}$ be a sequence in Y-K converging to q. Clearly, Z = {y_i: i = 1, 2, ...}

is not closed in Y but $f^{-1}(Z)$ is closed in S (since (q,1) ∉ S). Thus, f is

not a quotient map. Hence, by the theorem above, Y is not finitely irreduc-

ible. Therefore, we have proved 4.4.

   4.5 PROBLEM. Find a characterization of continua irreducible about a

countable, closed set which is similar to the equivalence of (a) and (b) in

4.2.

   4.6 PROBLEM. Give an inherent characterization of those continua Y

which have the property that every map of every semi-continuum S onto Y is a

uniform limit of quotient maps of S onto Y.

   4.7 PROBLEM. It would be intere ing to have more information about the

continua satisfying the condition in the conclusion of 4.4. Can this class of

continua be characterized in some appealing way? Must a continuum satisfying

the condition in the conclusion of 4.4 be irreducible about a countable,

closed set? The continuum drawn below shows that even a dendrite satisfying

this condition need not be finitely irreducible (thus, the converse of 4.4 is

false).

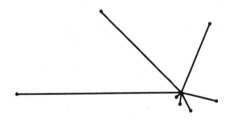

REFERENCES

1. R. H. Bing, Partitioning a set, Bull. Amer. Math. Soc., 55 (1949), 1101-1110.

2. C. E. Capel and W. L. Strother, Multi-valued functions and partial order, Portugaliae Math., 17 (1958), 41-47.

3. H. Cook and W. T. Ingram, Obtaining AR-like continua as inverse limits with only two bonding maps, Glasnik Math., 4 (1969), 309-311.

4. James Francis Davis, The equivalence of zero span and zero semispan, Proc. Amer. Math. Soc., 90 (1984), 133-138.

5. James Dugundji, Topology, Allyn and Bacon, Inc., Boston, Mass., 1966.

6. C. Eberhart and S. B. Nadler, Jr., The dimension of certain hyperspaces, Bull. Pol. Acad. Sci., 19 (1971), 1027-1034.

7. Lawrence Fearnley, Characterization of the continuous images of all pseudo-circles, Pac. J. Math., 23 (1967), 491-513.

8. _____, Characterizations of the continuous images of the pseudo-arc, Trans. Amer. Math. Soc., 111 (1964), 380-399.

9. Paul W. Gilbert, n-to-one mappings of linear graphs, Duke Math. J., 9 (1942), 475-486.

10. Jack T. Goodykoontz, Jr. and Sam B. Nadler, Jr., Fixed point sets of continuum-valued mappings, Fund. Math., 122 (1984), 85-103.

11. J. Grispolakis and E. D. Tymchatyn, Continua which admit only certain classes of onto mappings, Top. Proc., 3 (1978), 347-362.

12. _____ and _____, Continua which are images of weakly confluent mappings only, (II), Houston J. of Math., 6 (1980), 375-387.

13. _____ and _____, Weakly confluent mappings and the covering property of hyperspaces, Proc. Amer. Math. Soc., 74 (1979), 177-182.

14. O. G. Harrold, Jr., Exactly (k,1) transformations on connected linear graphs, Amer. J. Math., 62 (1940), 823-834.

15. _____, The non-existence of a certain type of continuous transformation, Duke Math. J., 5 (1939), 789-793.

16. Jo Heath and A. J. W. Hilton, Trees that admit 3-to-1 maps onto the circle, J. of Graph Theory, 14 (1990), 311-320.

17. W. Holsztyński, Universal mappings and fixed point theorems, Bull. Pol. Acad. Sci., 15 (1967), 433-438.

18. _____, Universality of the product mappings onto products of I^n and snake-like spaces, Fund. Math., 64 (1969), 147-155.

19. Witold Hurewicz and Henry Wallman, <u>Dimension Theory</u>, Princeton University Press, Princeton, N.J., 1948 (ninth printing, 1974).

20. F. Burton Jones, One-to-one continuous images of a line, Fund. Math., 67 (1970), 285-292.

21. Hisao Kato, The dimension of hyperspaces of certain 2-dimensional continua, Top. and its Appls., 28 (1988), 83-87.

22. J. Krasinkiewicz, On the hyperspaces of snake-like and circle-like continua, Fund. Math., 83 (1974), 155-164.

23. K. Kuratowski, <u>Topology</u>, <u>Vol</u>. II, Acad. Press, New York, N.Y., 1968.

24. A. Lelek, On weakly chainable continua, Fund. Math., 51 (1962), 271-282.

25. _____ and L. F. McAuley, On hereditarily locally connected spaces and one-to-one images of the line, Colloq. Math., 17 (1967), 319-324.

26. _____ and David R. Read, Compositions of confluent mappings and some other classes of functions, Colloq. Math., 29 (1974), 101-112.

27. O. W. Lokuciewski, On a theorem on fixed points, Усп. Мат. Наук, 12 3 (75), (1957), 171-172 (Russian).

28. T. Maćkowiak, A characterization of finitely irreducible continua, Colloq. Math., 56 (1988), 71-83.

29. _____, Continuous mappings on continua, Dissertationes Math., 158 (1979).

30. John R. Martin, Fixed point sets of homeomorphisms of metric products, Proc. Amer. Math. Soc., 103 (1988), 1293-1298.

31. _____, Fixed point sets of Peano continua, Pac. J. Math., 74 (1978), 163-166.

32. _____ and Sam B. Nadler, Jr., Examples and questions in the theory of fixed point sets, Can. J. Math., 31 (1979), 1017-1032.

33. _____, Lex G. Oversteegen, and E. D. Tymchatyn, Fixed point sets of products and cones, Pac. J. Math., 101 (1982), 133-139.

34. _____ and E. D. Tymchatyn, Fixed point sets of 1-dimensional Peano continua, Pac. J. Math., 89 (1980), 147-149.

35. S. Miklos, Exactly (n,1) mappings on graphs, Per. Math. Hungarica, 20 (1989), 35-39.

36. J. Mioduszewski, A functional conception of snake-like continua, Fund. Math., 51 (1962), 179-189.

37. E. E. Moise, Grille decomposition and convexification theorems for compact locally connected continua, Bull. Amer. Math. Soc., 55 (1949), 1111-1121.

38. Robert L. Moore, Concerning simple continuous curves, Trans. Amer. Math. Soc., 21 (1920), 333-347.

39. Sam B. Nadler, Jr., A characterization of locally connected continua by hyperspace retractions, Proc. Amer. Math. Soc., 67 (1977), 167-176.

40. _____, Continua which are a one-to-one continuous image of [0,∞), Fund. Math., 75 (1972), 123-133.

41. _____, Continuum Theory: An Introduction, Marcel Dekker, Inc., to appear (Spring, 1992).

42. _____, Continuum theory and graph theory: disconnection numbers, to appear J. of London Math. Soc.

43. _____, Functional characterizations of trees, to appear Glasnik Mat.

44. _____, Hyperspaces of Sets, Monographs and Textbooks in Pure and Applied Math., vol. 49, Marcel Dekker, Inc., New York, N.Y., 1978.

45. _____, Induced universal maps and some hyperspaces with the fixed point property, Proc. Amer. Math. Soc., 100 (1987), 749-754.

46. _____, On locally connected plane one-to-one continuous images of [0,∞), Colloq. Math., 37 (1977), 47-49.

47. _____, The metric confluent images of half-lines and lines, Fund. Math., 102 (1979), 183-194.

48. _____, Universal mappings and weakly confluent mappings, Fund. Math., 110 (1980), 221-235.

49. _____ and J. Quinn, Embeddability and structure properties of real curves, Memoirs of the Amer. Math. Soc., 125 (1972), 1-74.

50. _____, and L. E. Ward, Jr., Concerning continuous selections, Proc. Amer. Math. Soc., 25 (1970), 369-374.

51. Yutaka Ohsuda, Fixed point sets of homeomorphisms, Masters Thesis, West Virginia University, Morgantown, W.V., 1982; Sam B. Nadler, Jr., Director of Thesis.

52. David R. Read, Confluent and related mappings, Colloq. Math., 29 (1974), 233-239.

53. Herbert Robbins, Some complements to Brouwer's fixed point theorem, Israel J. Math., 5 (1967), 225-226.

54. J. T. Rogers, Jr., Dimension of hyperspaces, Bull. Pol. Acad. Sci., 20 (1972), 177-179.

55. _____, Pseudo-circles and universal circularly chainable continua, Ill. J. Math., 14 (1970), 222-237.

56. _____, Weakly confluent mappings and finitely-generated cohomology, Proc. Amer. Math. Soc., 78 (1980), 436-438.

57. Helga Schirmer, Fixed point sets of continuous self-maps, Fixed Point Theory Proc. (Sherbrooke, 1980), Lecture Notes in Math., vol. 866, Springer-Verlag, Berlin, 1981, pp. 417-428.

58. _____, Fixed point sets of homeomorphisms of compact surfaces, Israel J. Math., 10 (1971), 373-378.

59. _____, On fixed point sets of homeomorphisms of the n-ball, Israel J. Math., 7 (1969), 46-50.

60. J. Segal, A fixed point theorem for the hyperspace of a snake-like continuum, Fund. Math., 50 (1962), 237-248.

61. _____, Hyperspaces of the inverse limit space, Proc. Amer. Math. Soc., 10 (1959), 706-709.

62. S. Sirota, Spectral representation of spaces of closed subsets of bicompacta, Soviet Math. Dokl., 9 (1968), 997-1000.

63. A. H. Stone, Disconnectible spaces, Top. Conf. Arizona State Univ. 1967 (proceedings of, E. E. Grace, editor; published at A. S. U., 1968), 265-276.

64. T. Van Der Walt, Fixed And Almost Fixed Points, Math. Centre Tracts, Math. Centrum, Amsterdam, 1963.

65. L. E. Ward, Jr., Fixed point sets, Pac. J. Math., 47 (1973), 553-565.

66. Gordon Thomas Whyburn, Analytic Topology, Amer. Math. Soc. Colloq. Publ., Vol. 28, Amer. Math. Soc., Providence, R.I., 1942.

A Continuum Separated by Each of Its Nondegenerate Proper Subcontinua

SAM B. NADLER, JR.   West Virginia University, Morgantown, West Virginia

GARY A. SELDOMRIDGE   Potomac State College, Keyser, West Virginia

1.   **Introduction:** A fundamental result in the theory of continua is that every continuum contains a non-cut point [3,p 177]; that is, a point which does not separate the continuum. David Bellamy asked the first author if every nondegenerate continuum X contains a nondegenerate subcontinuum which does not separate X. The question was motivated by a similar question of Y. T. Leung who asked if every continuum X in the plane contains a nondegenerate subcontinuum which does not separate X. The question for continua in the plane is still unresolved, but we answer Bellamy's version of the question by giving an example of a Suslinean dendroid such that every nondegenerate proper subcontinuum of it separates it. Thus, even though every continuum has at least two non-cut points, our example shows that a continuum may fail to have any "fat non-cut points".

2.   **Construction of the Example:** In essence, the construction is similar to other constructions which start with a given continuum, systematically "replace" selected points with a

copy of the original continuum, and obtain the desired
continuum as an inverse limit. However, for our purposes it
is more convenient for verifying some of the properties to
describe the continuum geometrically.

Let X_0 be the unit interval as embedded in the plane
between $(0,0)$ and $(1,0)$. For each positive integer j, let
$S_{1,j}$ denote the closed interval from $(1,0)$ to $(\frac{1}{2},2^{-j})$,
$S_{2,j}$ the closed interval from $(\frac{1}{2},0)$ to $(0,2^{-j})$, $S_{3,j}$ the
closed interval from $(0,0)$ to $(\frac{1}{2},-2^{-j})$, and $S_{4,j}$ the closed
interval from $(\frac{1}{2},0)$ to $(1,-2^{-j})$ (See figure 1). Let $\mathcal{S}_1$
denote the set of all such $S_{i,j}$. Let $\mathcal{R}_1$ represent the
collection consisting of all members of $\mathcal{S}_1$ and the closed
intervals from $(0,0)$ to $(\frac{1}{2},0)$ and from $(\frac{1}{2},0)$ to $(1,0)$. Let
$X_1 = \cup\mathcal{R}_1$ and note that X_1 is a continuum. Let V_1 represent
the set of all points in X_1 which are endpoints of more than
one member of $\mathcal{S}_1$; that is the points $(0,0)$, $(\frac{1}{2},0)$, and $(1,0)$.

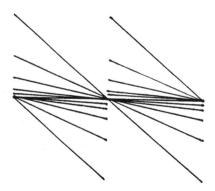

figure 1

For each fixed $S \in \mathcal{S}_1$, let φ_S be the minimum absolute value of the angle between S and T for any $T \in \mathcal{S}_1$ such that $T \neq S$ and $T \cap S \neq \phi$. Let $\vartheta_S = \frac{1}{4} \min\{\varphi_S, \text{Tan}^{-1}(1)\}$. If $S \in \mathcal{R}_1 - \mathcal{S}_1$, let $\vartheta_S = \frac{1}{4} \text{Tan}^{-1}(1)$. In either case, let $g_S : X_1 \longrightarrow \mathbb{R}^3$ be the homeomorphism which performs the following transformations in the order given:

2.1 Map any point $(x, y) \in X_1$ to $(s \cdot x, y)$ where s is the length of S.

2.2 Let C be the closed region in quadrants I and IV bounded by $y = x \tan (\vartheta_S)$ and $y = -x \tan (\vartheta_S)$. For any $S \in \mathcal{S}$, retract any point $x \in S - C$ to the "first point" of $S \cap C$ (See figure 2).

figure 2

2.3 If $S \in \mathcal{S}_1$, let p be the unique element of $S \cap V_1$; otherwise, let p be the point of $S \cap V_1$ with greater x-coordinate. Rotate, and translate the image in 2.2 until the interval on the x-axis maps onto S with the origin being mapped to p.

2.4   Rotate the plane containing the previous image about S
      so that the angle between the plane and the xy-plane is
      $\text{Tan}^{-1}(1)$ when viewed from the endpoint of S with
      greater x-coordinate.

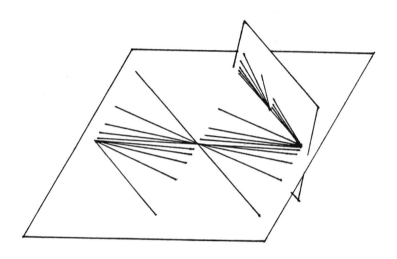

figure 3

Let $\mathscr{S}_2 = \{g_S(T):S \in \mathcal{R}_1 \text{ and } T \in \mathscr{S}_1\}$ and let
$\mathcal{R}_2 = \{g_S(T):S \in \mathcal{R}_1 \text{ and } T \in \mathcal{R}_1\}$.   Let $X_2 = \cup\mathcal{R}_2$.

We will define each $\mathscr{S}_n$ and $\mathcal{R}_n$ for $n > 1$ in a similar
manner.   For any $S \in \mathscr{S}_{n-1}$, let φ_S be as above and let
$\vartheta_S = \frac{1}{4} \min\{\varphi_S, \text{Tan}^{-1}(n) - \text{Tan}^{-1}(n - 1)\}$.   If $S \in \mathcal{R}_{n-1} - \mathscr{S}_{n-1}$,
let $\vartheta_S = \frac{1}{4} (\text{Tan}^{-1}(n) - \text{Tan}^{-1}(n - 1))\}$.   For any $S \in \mathcal{R}_{n-1}$, we
define $g_S:X_1 \rightarrow \mathbb{R}^3$ the same way as above except that the
angle used in 2.4 is $\text{Tan}^{-1}(n - 1)$ and V_n and $\mathscr{S}_n$ replace V_1
and $\mathscr{S}_1$ respectively in 2.3.   We then define

$$\mathscr{S}_n = \{g_s(T):S \in \mathscr{R}_{n-1} \text{ and } T \in \mathscr{S}_{n-1}\},$$

$$\mathscr{R}_n = \{g_s(T):S \in \mathscr{R}_{n-1} \text{ and } T \in \mathscr{R}_{n-1}\}, \text{ and}$$

$$X_n = \cup \mathscr{R}_n.$$

We note that for each n, X_n is a continuum with $\overline{X_n} \subset X_{n+1}$.

Let $X = \bigcup_{n=1}^{\infty} X_n$. Then X is easily seen to be a continuum.

3. **Verification of the Properties of the Example:** It is readily

seen that X is arcwise connected. We note that the continuum

was constructed in such a way that the Anderson – Choquet

Theorem applies (Theorem 1 of [1]) and, thus, $X = \varprojlim \{X_n, f_n\}$

where each f_n is the natural map of X_{n+1} onto X_n.

We will use the following notation in establishing

properties of X. For any $n \geq 1$, let V_n denote the set of all

points of X_n which are endpoints of more than one element of

$\mathscr{S}_n$. For each $n \geq 1$ and each $v \in V_n$, let $\mathscr{S}_n^v = \{S \in \mathscr{S}_n : v \in S\}$

and let $\mathscr{R}_n^v = \{S \in \mathscr{R}_n : v \in S\}$.

We establish the following:

3.1 Continuum X is a dendroid and as such is uniquely

arcwise connected.

Since X is an arcwise connected inverse limit of

dendroids, Theorem 4 of [6] implies that X is a dendroid. It

then follows immediately from [2] that X is uniquely arcwise
connected.

Since by construction, for any distinct S and T elements
of $\mathscr{S}_n^v$, with $p \in S - \{v\}$ and $q \in T - \{v\}$, there is an arc
joining p and q and passing through v, the following is an
immediate consequence of 3.1.

3.2   If for some n and some $v \in V_n$, S and T are distinct
      elements of $\mathscr{S}_n^v$, $p \in S - \{v\}$, $q \in T - \{v\}$, and K is a
      connected subset of X containing p and q, then $v \in K$.

Statement 3.3 is easily verified by recalling the method
of construction and 3.1 above.

3.3   For any n and for any u and v elements of V_n, there
      exists a unique arc ℓ in X_n from v to u such that
      $\ell \subset \cup\{\mathscr{R}_n - \mathscr{S}_n\}$ and that the cardinality of $V_n \cap \ell$ is
      finite.

We next show that:

3.4   If A is a compact subset of X such that X - A is
      connected and $v \in V_n \cap A$ for some n, then there exists
      at most one $S \in \mathscr{S}_n^v$ such that S is not contained in A.

Suppose that S and T are distinct elements of $\mathcal{S}_n^v$ such that neither is contained in A. Then there exists $p \in S - A$ and $q \in T - A$. Since $v \in A$, $p \in S - \{v\}$ and $q \in T - \{v\}$. Thus, $X - A$ is a connected set containing both p and q. Hence, by 3.2, $v \in X - A$. But this contradicts that $v \in A$ and 3.4 is proved.

In addition we show:

3.5 No subcontinuum of X is contained in $X - \bigcup_{n=1}^{\infty} X_n$.

Suppose that A is a subcontinuum of X with $A \subset X - \bigcup_{n=1}^{\infty} X_n$. Let p and q be distinct elements of A. By the method of construction, there exist distinct points s and $t \in \bigcup_{n=1}^{\infty} X_n$ such that there exist connected subsets A_1 and A_2 of X with $\{s, p\} \subset A_1$ and $\{t, q\} \subset A_2$. Since $X_n \subset X_{n+1}$ for each n, there exists N such that s and $t \in X_N$. Then since X_N is a dendroid, there exists an arc $A_3 \subset X_N$ such that $\{s, t\} \subset A_3$. It then easily follows that $A \cup A_1 \cup A_2 \cup A_3$ is a non-unicoherent subcontinuum of X. But since X is a dendroid, this is a contradiction and 3.5 is proved.

We will now show that every nondegenerate proper subcontinuum of X separates X. Let A be a nondegenerate proper subcontinuum of X and suppose that $X - A$ is a nondegenerate connected set. Since X is a dendroid, there exist arcs A_1 and A_2 contained in A and X - A respectively. Since by 3.5, $X - \bigcup_{n=1}^{\infty} X_n$ contains no arc, it is easily seen that there exist points x_1, x_2, y_1, and y_2 such that $\{x_i, y_i\} \subset A_i \cap \bigcup_{n=1}^{\infty} X_n$ for $i \leq 2$. Hence, there exists N such that $\{x_1, x_2, y_1, y_2\} \subset X_N$. Thus, for $i \leq 2$, the unique arc joining x_i and y_i in X_N must be contained in A_i. It then follows as a result of the method of construction that there exists a number n large enough so that $V_n \cap A_1 \neq \phi$ and $V_n \cap A_2 \neq \phi$. In particular, $V_n \cap A \neq \phi$ and $V_n \cap (X - A) \neq \phi$. Let $v \in V_n \cap A$ and $u \in V_n \cap (X - A)$. Let ℓ be the arc in $\bigcup\{\mathcal{R}_n - \mathcal{S}_n\}$ from v to u as guaranteed in 3.3. Let R be the arc in ℓ with one endpoint v and the other endpoint v_1 where v_1 is the first element of V_n in $\ell - \{v\}$ from v to u. By the method of construction, there exists a sequence $\{S_i\}_{i=1}^{\infty}$ of sets in $\mathcal{S}_n^v$ which converges to R. Since by 3.4 there exists at most one $S \in \{S_i\}_{i=1}^{\infty}$ such that S is not contained in A, we may assume without loss of generality that $S_i \subset A$ for each i. Then since A is compact, $R \subset A$ and in particular $v_1 \in A$. Since by 3.3, the cardinality of $V_n \cap \ell$ is finite, this process can be repeated until eventually u is shown to be an element of A. But this contradicts $u \in X - A$ and, thus, X - A cannot be a nondegenerate connected set. Since X - A cannot be degenerate, X - A is not connected.

We now show that X is Suslinean; i.e., every collection of mutually disjoint nondegenerate subcontinua of X is countable [4]. Suppose that $\mathcal{A}$ is a collection of mutually disjoint nondegenerate subcontinua of X. Let $A \in \mathcal{A}$. Since A is a nondegenerate connected subset of a dendroid, A contains an arc B. Let B_1 and B_2 be disjoint subarcs of B. By 3.5, there exists $p_1 \in B_1 \cap X_n$ for some n. Similarly, there exists $p_2 \in B_2 \cap X_m$ for some m. Let $N = \max\{n,m\}$. Since $X_i \subset X_{i+1}$ for every i, $\{p_1,p_2\} \subset X_N$. Let C be the unique arc in X_N joining p_1 and p_2. Since $X_N \subset X$, C is also the unique arc in X joining p_1 and p_2. Thus, $C \subset B$. Since $C \subset X_N$ and since X was constructed so that $\bigcup_{n=1}^{\infty} V_n$ is continuumwise dense in X_N, there exists $v \in \bigcup_{n=1}^{\infty} V_n \cap C$. Thus $v \in \bigcup_{n=1}^{\infty} V_n \cap A$. We have now implicitly defined a function $\varphi: \mathcal{A} \to \bigcup_{n=1}^{\infty} V_n$ such that $\varphi(A) \in A$ for each $A \in \mathcal{A}$. Therefore, φ is one-to-one and we conclude that the cardinality of $\mathcal{A}$ is less than or equal to the cardinality of $\bigcup_{n=1}^{\infty} V_n$. Since by construction, $\bigcup_{n=1}^{\infty} V_n$ is countable, $\mathcal{A}$ is countable.

Finally we note that X is nonplanar since it contains a copy of the figure below (where the line ℓ is assumed to intersect the two harmonic fans only in one of its endpoints).

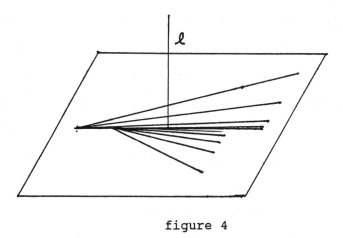

figure 4

4. Comments: A general result in [7] shows that any continuum
    which becomes disconnected upon the removal of any countably
    infinite collection of points must be a graph (the converse
    is obvious). This was proved by different methods in [5]
    with the author of [5] being unaware of [7] until recently.
    In [5], it was also shown that for a (connected) graph G, the
    smallest number n such that the graph becomes disconnected
    upon the removal of any n points, denoted $D^s(G)$, is given by
    the formula

$$D^s(G) = 2 - \chi(G) + |G^{[1]}|$$

    where $\chi(G)$ is the Euler characteristic of G and $|G^{[1]}|$ is the
    number of endpoints of G. In particular, when $D^s(G)$ exists,
    it is always finite. Also since every continuum contains a
    non-cut point, $D^s(G) \geq 2$. Let us say that a cardinal number
    n is a subcontinuum disconnection number of a continuum X
    provided that for any collection $\mathcal{A}$ of disjoint subcontinua of
    X with $|\mathcal{A}| = n$, $X - \bigcup \mathcal{A}$ is not connected. Our example shows

that the smallest subcontinuum disconnection number of a continuum can be 1 in contrast to $D^s(G) \geq 2$. We do not know, however, if the smallest subcontinuum disconnection number of a continuum can be $\aleph_o$.

REFERENCES

1. R. D. Anderson and Gustave Choquet, A plane continuum no two of whose nondegenerate subcontinua are homeomorphic: an application of inverse limits, Proc. Amer. Math. Soc., 10(1959), 347-353.

2. K. Borsuk, A theorem on fixed points, Bull. Pol. Acad. Sci., 2(1954), 17-20.

3. K. Kuratowski, Topology, Vol II, Academic Press, New York, N.Y., 1968.

4. A. Lelek, On the topology of curves II, Fund. Math., 70(1971), 131-138

5. Sam B. Nadler, Jr., Continuum theory and graph theory: disconnection numbers, to appear Journal of the London Math. Soc.

6. Sam B. Nadler, Jr., Multicoherence techniques applied to inverse limits, Trans. Amer. Math. Soc., 157(1971), 227-234.

7. A. H. Stone, Disconnectible spaces, Topology Conference Arizona State University, 1967, E. E. Grace, editor, 265-276.

Dense Embeddings into Cubes and Manifolds

J. NIKIEL   Texas A&M University, College Station, Texas

H. M. TUNCALI   Nipissing University, North Bay, Ontario, Canada

E. D. TYMCHATYN   University of Saskatchewan, Saskatoon, Saskatchewan, Canada

1. Introduction

All spaces considered here are separable and metric. A space X is said to be *nowhere locally compact* provided each compact subset of X is nowhere dense in X.

Our purpose is to show that, under natural assumptions, nowhere locally compact spaces admit dense embeddings into n-cubes and n-manifolds. For $n = \infty$ the results are already well-known. However, the main ideas of our proof in the case $n = \infty$ are easily adjusted to suit the finite dimensional case.

In 1978, J. C. Oxtoby and V. S. Prasad asked which spaces can be densely embedded into the Hilbert cube, [**13**, p. 496]. Concerning their question, D. W. Curtis proved that, in particular, each σ-compact and nowhere locally compact space is homeomorphic to a dense subset of the Hilbert cube, [**6**]. Curtis' result was improved in P. L. Bowers' papers [**3**] and [**4**]. In [**3**] it was shown that if $f : X \to Z$ is a continuous map of a σ-compact nowhere locally compact space X into a connected Fréchet manifold Z (e.g. $Z = \ell_2$) such that $f(X)$ is dense in Z, and U is an open covering of Z, then there exists an embedding $g : X \to Z$ such that $g(X)$ is dense in Z and g is U-close to f. In [**4**] the result of [**3**] was applied to give a proof that the assumption of σ-compactness of X is not essential in the result we recalled above. In [**3**] and [**4**], Bowers considered also finite dimensional spaces obtaining results about dense embeddings of n-dimensional nowhere locally compact spaces into spaces satisfying the discrete n-cells property.

In Theorem 1, we reprove a part of Bowers' result by showing that each nowhere locally compact space admits a dense embedding into the Hilbert space ℓ_2. The proof we give is rather elementary (and lengthy). It does not involve the discrete n-cells property and, therefore, can be modified to show in Theorem 3 that a nowhere locally compact space which admits an embedding into a finite dimensional cube $[0,1]^n$ is homeomorphic to a dense subset of each connected triangulable $(n + 1)$-dimensional manifold (e.g. $[0,1]^{n+1}$ or R^{n+1}). Incidentally, the latter result

This research was supported in part by NSERC grant number A5616 and by a grant from the U of S President's Fund. The first named author acknowledges support from an NSERC International Fellowship.

solves an old problem of B. Knaster concerning dense embeddings of composants of hereditarily indecomposable continua into Euclidean spaces (see [11, p. 63]; that was communicated to the authors by Prof. A. Lelek during the Spring Topology Conference 1991 in Sacramento).

The following concept will play an important role in our constructions and arguments. Let X be a space, $\mathbf{F}$ a collection of closed subsets of X and $A \subset X$. We shall say that A is *loosely embedded* in X with respect to $\mathbf{F}$ if, for each $F \in \mathbf{F}$, each $x \in \mathrm{Bd}_X(F) \cap A$ and each neighbourhood U of x in X, the set $\mathrm{Int}_X(F) \cap A \cap U$ is nonempty. Equivalently, A is loosely embedded in X with respect to $\mathbf{F}$ if $\mathrm{Cl}\big(\mathrm{Int}_X(F) \cap A\big) = F \cap A$ for each $F \in \mathbf{F}$.

2. Embeddings into the Hilbert cube and into related spaces

The Hilbert cube Q is the product space $[0,1]^\infty$ with its standard metric $d\big((x_n)_{n=1}^\infty, (y_n)_{n=1}^\infty\big) = \sum_{n=1}^\infty \frac{1}{2^n}|x_n - y_n|$. The pseudo-interior s of Q is the set $]0,1[^\infty$. Recall that both s and $Q - s$ are dense in Q. It is well-known that s is homeomorphic to the Hilbert space ℓ_2, see e.g. [1].

Throughout this section, if $U \subset Q$ then $\mathrm{Cl}(U)$, $\mathrm{Int}(U)$ and $\mathrm{Bd}(U)$ denote the closure, interior and boundary of U in Q, respectively.

Recall that a (separable metric) space M is said to be a *Q-manifold* (resp. a *Fréchet manifold*) if each point of X has a neighbourhood in X which is homeomorphic to an open subset of Q (resp. of ℓ_2).

A closed subset A of a space X is said to be a *Z-set* in X if for each open cover $\mathbf{C}$ of X there exists a continuous mapping $f_{\mathbf{C}} : X \to X - A$ such that id_X and $f_{\mathbf{C}}$ are *C-close*, i.e. for each $x \in X$ there exists $C \in \mathbf{C}$ such that $x, f_{\mathbf{C}}(x) \in C$.

A function $f : X \to Y$ is a *Z-embedding* if f is an embedding and $f(X)$ is a Z-set in Y. A continuous mapping $f : X \to Y$ is said to be *proper* if $f^{-1}(A)$ is compact for each compact subset A of Y.

Lemma 1 ([2], see also [5, Theorem 18.2]). *Let X be a locally compact separable metric space, A be a closed subset of X, M be a connected Q-manifold and $f : X \to M$ be a proper map such that $f|_A$ is a Z-embedding. Then for each open cover $\mathbf{C}$ of M there exists a Z-embedding $f_{\mathbf{C}} : X \to M$ such that $f_{\mathbf{C}}|_A = f|_A$ and f and $f_{\mathbf{C}}$ are C-close.*

Lemma 2 ([1], see also [5, Theorem 11.1]). *If $A, B \subset Q$ are Z-sets and $h : A \to B$ is a homeomorphism, then h can be extended to a homeomorphism $H : Q \to Q$ of Q.*

Let k be a positive integer and $\mathbf{U}$ a finite collection of subsets of Q which covers Q. We shall say that $\mathbf{U}$ is a *k-partition* of Q if each $U \in \mathbf{U}$ is of the form $U = [a_{i_1}^1, a_{i_1+1}^1] \times ... \times [a_{i_k}^k, a_{i_k+1}^k] \times [0,1]^\infty$, where a_i^j are points of $[0,1]$ for $j = 1, ..., k$ and $i = 0, ..., n_j$; $n_j > 1$ and $0 = a_0^j < a_1^j < ... < a_{n_j}^j = 1$ for $j = 1, ..., k$; and $i_j \in \{0, ..., n_j - 1\}$ for $j = 1, ..., k$.

Let $\mathbf{U}$ be a k-partition of Q. Clearly, $\mathrm{Int}(U) \cap \mathrm{Int}(U') = \emptyset$ where $U, U' \in \mathbf{U}$ and $U \neq U'$. If $k > 1$, then $\mathrm{Bd}(U)$ is connected for each $U \in \mathbf{U}$. A subcollection $\mathbf{V}$ of $\mathbf{U}$ is said to be a *chain* if $\mathrm{Int}(\bigcup \mathbf{V})$ is connected and $\mathbf{V}$ can be labelled as $\mathbf{V} = \{V_1, ..., V_m\}$ in such a way that the conditions $V_i \cap V_j \neq \emptyset$ and $|i - j| \leq 1$ are equivalent for all $i, j \in \{1, ..., m\}$. A subcollection $\mathbf{W}$ of $\mathbf{U}$ is *chain-connected* if for any $W, W' \in \mathbf{W}$ there exists $\mathbf{V} \subset \mathbf{W}$ such that $\mathbf{V}$ is a chain and $W, W' \in \mathbf{V}$.

Let $\mathbf{U}$ and $\mathbf{V}$ be k-partitions of Q and $\epsilon > 0$. We shall say that $\mathbf{U}$ and $\mathbf{V}$ are ϵ-*close* if there are integers $n_j > 1$ for $j = 1, ..., k$; and points $a_i^j, b_i^j \in [0,1]$ for $j = 1, ..., k$ and $i = 0, ..., n_j$ such that

(a) $0 = a_0^j < a_1^j < ... < a_{n_j}^j = 1$ and $0 = b_0^j < b_1^j < ... < b_{n_j}^j = 1$ for $j = 1, ..., k$,
(b) $a_{i-1}^j < b_i^j < a_{i+1}^j$ where $i = 1, ..., n_j - 1$ and $j = 1, ..., k$,
(c) $|a_i^j - b_i^j| < \epsilon$ for any $i = 0, ..., n_j$ and $j = 1, ..., k$,
(d) each $U \in \mathbf{U}$ is of the form $U = [a_{i_1}^1, a_{i_1+1}^1] \times ... \times [a_{i_k}^k, a_{i_k+1}^k] \times [0,1]^\infty$ for some $i_j \in \{0, ..., n_j - 1\}$ where $j = 1, ..., k$, and
(e) each $V \in \mathbf{V}$ is of the form $V = [b_{j_1}^1, b_{j_1+1}^1] \times ... \times [b_{i_k}^k, b_{i_k+1}^k] \times [0,1]^\infty$ for some $i_j \in \{0, ..., n_j - 1\}$ where $j = 1, ..., k$.

If $\mathbf{U}$ and $\mathbf{V}$ are a k-partition and an l-partition of Q respectively, then we shall say that $\mathbf{U}$ *refines* $\mathbf{V}$ provided for each $U \in \mathbf{U}$ there is $V \in \mathbf{V}$ such that $U \subset V$. If the latter condition is satisfied then obviously $k \geq l$.

Lemma 3 (see e.g. [10, §24, section VII, Theorem 3, p.265]). *Let X be a separable metric space and $\{A_t : t \in [0,1]\}$ a family of closed subsets of X such that $A_t \subset A_s$ if $t < s$. Then the equality $A_s = \text{Cl}(\bigcup_{t<s} A_t)$ holds for every $s \in [0,1]$ except for a countable set of indices.*

Finitely many subsequent applications of Lemma 3 give the following fact:

Lemma 4. *Let $X \subset Q$ and $\mathbf{U}$ and $\mathbf{V}$ be a k-partition and an l-partition of Q, respectively. Suppose that X is loosely embedded with respect to $\mathbf{V}$ and $\mathbf{U}$ refines $\mathbf{V}$. Then for each $\epsilon > 0$ there is a k-partition $\mathbf{U}_\epsilon$ of Q such that X is loosely embedded with respect to $\mathbf{U}_\epsilon$, $\mathbf{U}_\epsilon$ refines $\mathbf{V}$, and $\mathbf{U}_\epsilon$ and $\mathbf{U}$ are ϵ-close.*

<u>Proof.</u> There exist integers m_j for $j = 1, ..., k$ and points $a_i^j \in [0,1]$ for $j = 1, ..., k$ and $i = 0, ..., m_j$ such that $0 = a_0^j < a_1^j < \cdots < a_{m_j}^j = 1$, for $j = 1, ..., k$, and $\mathbf{U}$ is the collection of all sets U of the form $U = [a_{i_1}^1, a_{i_1+1}^1] \times \cdots \times [a_{i_k}^k, a_{i_k+1}^k] \times [0,1]^\infty$. Also, there exist integers n_j for $j = 1, ..., l$ and points $b_i^j \in [0,1]$ for $j = 1, ..., l$ and $i = 0, ..., n_j$ such that $0 = b_0^j < b_1^j < \cdots < b_{n_j}^j = 1$, for $j = 1, ..., l$, and $\mathbf{V}$ is the collection of all sets V of the form $V = [b_{i_1}^1, b_{i_1+1}^1] \times \cdots \times [b_{i_l}^l, b_{i_l+1}^l] \times [0,1]^\infty$. Since $\mathbf{U}$ refines $\mathbf{V}$, it follows that $l \leq k$ and $\{b_0^j, b_1^j, ..., b_{n_j}^j\} \subset \{a_0^j, a_1^j, ..., a_{m_j}^j\}$ for each $j = 1, ..., l$.

Suppose that $\epsilon > 0$. We may assume that $\epsilon < \frac{1}{2}(a_{i+1}^j - a_i^j)$ for each selection of j and i.

Let $t \in]0, \epsilon[$. Let $c_0^j(t) = a_0^j = 0$ and $c_{m_j}^j(t) = a_{m_j}^j = 1$ for $j = 1, ..., k$. Let $c_i^j(t) = a_i^j$ if $a_i^j \in \{b_0^j, b_1^j, ..., b_{n_j}^j\}$ for some $j \in \{1, ..., l\}$ and $i \in \{0, ..., m_j\}$. If $c_i^j(t)$ was not defined by the above rules for some $j \in \{1, ..., k\}$ and $i \in \{0, ..., m_j\}$, then let $c_i^j(t) = a_i^j + t$.

Let $\mathbf{W}_t$ denote the partition of Q into the sets W of the form $W = [c_{i_1}^1(t), c_{i_1+1}^1(t)] \times \cdots \times [c_{i_k}^k(t), c_{i_k+1}^k(t)] \times [0,1]^\infty$. Clearly, $\mathbf{W}_t$ refines $\mathbf{V}$, and $\mathbf{U}$ and $\mathbf{W}_t$ are ϵ-close. It suffices to show that X is loosely embedded with respect to $\mathbf{W}_t$ for some $t \in]0, \epsilon[$.

We consider the case when $k = l$ and $\{a_0^j, a_1^j, \ldots, a_{m_j}^j\} = \{b_0^j, b_1^j, \ldots, b_{n_j}^j\}$ for $j \geq 2$. In the general case, the proof can be obtained by induction and application of analogous methods.

Note that if $j \geq 2$, then $m_j = n_j$ and $a_i^j = b_i^j = c_i^j(t)$ for each $i \in \{0, 1, \ldots, m_j\}$. Let I denote the set of all $i \in \{1, \ldots, m_1 - 1\}$ such that $a_i^1 \notin \{b_0^1, \ldots, b_{n_1}^1\}$.

Let $i \in I$. For each $t \in [0, \epsilon]$ and each $(i_2, \ldots, i_l)$ such that $i_j \in \{0, 1, \ldots, n_j - 1\}$ for $j = 2, \ldots, l$, let
$$A_t(i, i_2, \ldots, i_l) = X \cap \left([0, a_i^1 + t] \times [a_{i_2}^2, a_{i_2+1}^2] \times \cdots \times [a_{i_l}^l, a_{i_l+1}^l] \times [0, 1]^\infty\right) \text{ and}$$
$$B_t(i, i_2, \ldots, i_l) = X \cap \left([a_i^1 + \epsilon - t, 1] \times [a_{i_2}^2, a_{i_2+1}^2] \times \cdots \times [a_{i_l}^l, a_{i_l+1}^l] \times [0, 1]^\infty\right). \text{ By}$$
Lemma 3, there are countable subsets T_i' and T_i'' of $[0, \epsilon]$ such that $A_s(i, i_2, \ldots, i_l) = \text{Cl}(\bigcup_{t < s} A_t(i, i_2, \ldots, i_l))$ when $s \in]0, \epsilon[- T_i'$,
and $B_s(i, i_2, \ldots, i_l) = \text{Cl}(\bigcup_{t < s} B_t(i, i_2, \ldots, i_l))$ when $s \in]0, \epsilon[- T_i''$.

Let $T = \bigcup_{i \in I}(T_i' \cup T_i'')$. Then T is countable. Let $t \in]0, \epsilon[- T$. It follows that X is loosely embedded with respect to $\mathbf{W}_t$. Now, it suffices to let $\mathbf{U}_\epsilon = \mathbf{W}_t$.

Lemma 5. *Let X be a locally compact space, M an open subset of Q, $\mathbf{U}$ a k-partition of Q, $V, V' \in \mathbf{U}$ be such that $V \neq V'$ and $\text{Int}(V \cup V')$ is a connected subset of M, $f : X \to M$ be an embedding such that $f(X)$ is a Z-set in M, and let C be an open subset of $\text{Int}(V \cup V')$ such that the set $A = f(X) \cap C \cap V \cap V'$ is compact. If D is a an open subset of Q such that $A \subset D$ and $\text{Cl}(D) \subset C$, then there exists an embedding $g : X \to M$ such that*

(i) *$g(X)$ is a Z-set in M,*

(ii) *$f(x) = g(x)$ if $x \in X$ and $f(x) \notin D$,*

(iii) *$g(x) \in D$ if $x \in X$ and $f(x) \in D$, and*

(iv) *$g(X) \cap D$ is loosely embedded in D with respect to $\{D \cap V, D \cap V'\}$.*

Proof. If $A = \emptyset$, we let $f = g$. In case when $A \neq \emptyset$, it suffices to employ Lemma 3 and a suitably chosen piecewise linear homeomorphism $G : Q \to Q$ such that $G(q) = q$ for every $q \notin D$ (see Fig.1).

THEOREM 1. *If X is a nowhere locally compact separable metric space, then X is homeomorphic to a dense subset of the pseudo-interior s of the Hilbert cube Q.*

Proof. Let Y be a compact metric space such that $X \subset Y$ and X is dense in Y. Obviously, Y has no isolated point. We let ρ denote a metric on Y.

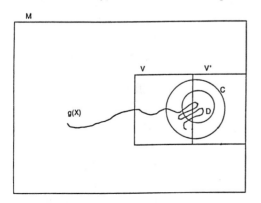 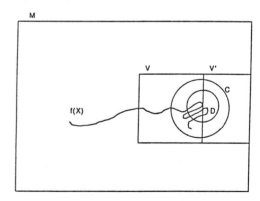

Figure 1.

For each integer $n > 1$ we let

$$S_n = (\{0,1\} \times [0,1]^\infty) \cup \bigcup_{i=1}^{n-1} ([0,1]^i \times \{0,1\} \times [0,1]^\infty).$$

Then each S_n is a compact nowhere dense subset of Q and $s = Q - \bigcup_{n=2}^\infty S_n$. First, we apply an induction to define for $n = 1, 2, \ldots$ the following objects:

$$A_n, Y_n, \mathbf{U}_n, g_n, \text{ and } P_z, \mathbf{V}_z \text{ and } W_z \text{ for } z \in A_n$$

such that for all positive integers n and l with $l < n$ the following conditions (1)–(24) are satisfied:

(1) $A_1 = \emptyset$ and $\mathbf{U}_1 = \{Q\}$.
(2) A_n is a finite subset of $Y - X$ and $A_n \cap A_l = \emptyset$.
(3) $Y_n = Y - \bigcup_{i=1}^n A_i$.
(4) $\mathbf{U}_{n+1}$ is a k_{n+1}-partition of Q for some integer $k_{n+1} > 1$.
(5) mesh $\mathbf{U}_{n+1} \leq \frac{1}{n+1}$.
(6) $\mathbf{U}_{n+1}$ refines $\mathbf{U}_n$.
(7) $g_n : Y_n \to s \subset Q$ is an embedding.
(8) $g_n(Y_n) \cap \text{Int}(U)$ is a non-empty Z-set in $\text{Int}(U)$ for each $U \in \mathbf{U}_n$.
(9) $g_n(Y_n)$ is loosely embedded in Q with respect to $\mathbf{U}_{n+1}$.
(10) If $z \in A_n$, then P_z is a closed neighbourhood of z in Y.
(11) If $z, z' \in A_n$ and $z \neq z'$, then $P_z \cap P_{z'} = \emptyset$.
(12) If $z \in A_n$, then $P_z \cap A_l = \emptyset$.
(13) If $z \in A_n$, then $W_z \in \mathbf{U}_n$ and $W_z \cap g_{n-1}(Y_{n-1}) = \emptyset$.
(14) If $n > 1$, $W \in \mathbf{U}_n$ and $W \cap g_{n-1}(Y_{n-1}) = \emptyset$, then there is exactly one $z \in A_n$ such that $W = W_z$.
(15) $\mathbf{V}_z$ is a chain in $\mathbf{U}_{n+1}$ if $z \in A_n$.
(16) If $z \in A_n$, then the first link of $\mathbf{V}_z$ contains $g_{n-1}(P_z \cap Y_{n-1})$ and the last link of $\mathbf{V}_z$ is contained in $\text{Int}(W_z)$.
(17) If $z \in A_n$, then diam $g_{n-1}^{-1}(\bigcup \mathbf{V}_z) \leq \frac{1}{n}$.
(18) If $z \in A_n$, then there is $U \in \mathbf{U}_{n-1}$ such that $\bigcup \mathbf{V}_z \subset \text{Int}(U) - (S_n \cup \bigcup_{y \in A_{n-1}} \bigcup \mathbf{V}_y)$.
(19) If $z, z' \in A_n$ and $z \neq z'$, then $\bigcup \mathbf{V}_z \cap \bigcup \mathbf{V}_{z'} = \emptyset$.
(20) If $U \in \mathbf{U}_n$, then the collection $\{V \in \mathbf{U}_{n+1} : V \subset U \text{ and } V \not\subset \bigcup_{z \in A_n} \mathbf{V}_z\}$ is chain-connected and $(\text{Int}(U) - \bigcup_{z \in A_n} \bigcup \mathbf{V}_z) \cap g_n(Y_n) \neq \emptyset$.
(21) If $z \in A_n$, then $g_n(P_z \cap Y_n) \subset \text{Int}(W_z \cup \bigcup \mathbf{V}_z) - \bigcup_{z' \in A_n, z' \neq z} \bigcup \mathbf{V}_{z'}$ and $\text{Cl}(g_n(P_z \cap Y_n)) = g_n(P_z \cap Y_n) \cup \text{Cl}(\text{Bd}(W_z) - \bigcup_{z' \in A_n} \bigcup \mathbf{V}_{z'}) \cup (W_z \cap S_n)$.
(22) If $z \in A_n$ and $V \in \mathbf{V}_z$, then $V \cap g_n(P_z \cap Y_n) \neq \emptyset$ and $g_n^{-1}(V)$ is compact.
(23) If $n > 1$ and $x \in Y_n - \bigcup_{z \in A_n} P_z$, then $g_n(x) = g_{n-1}(x)$.
(24) If $n > 1$ and $x \in Y_n$ is such that $g_{n-1}(x) \in \bigcup_{y \in A_{n-1}} \bigcup \mathbf{V}_y$, then $g_n(x) = g_{n-1}(x)$.

Note that the conditions (1)–(24) are not independent, for example (2) can be derived from (10), (13), (14) and (12), and (24) can be obtained from (18), (16) and (23). The second part of the condition (21) should be highlighted, since it contains the major idea of avoiding larger and larger portions of Q by the maps g_n.

We let $A_1 = \emptyset$, $Y_1 = Y$ and $\mathbf{U}_1 = \{Q\}$. Since Y is a compact metric space, there exists a Z-embedding $g_1 : Y_1 \to s \subset Q$ (see e.g. Lemmas 1 and 2). By Lemma 3, there exists a 2-partition $\mathbf{U}_2$ of Q such that mesh $\mathbf{U}_2 \leq \frac{1}{2}$ and $g_1(Y_1)$ is loosely embedded with respect to $\mathbf{U}_2$.

Suppose that for some positive integer m, we have already defined

$$A_1, ..., A_m; \quad Y_1, ..., Y_m; \quad \mathbf{U}_1, ..., \mathbf{U}_m, \mathbf{U}_{m+1}; \quad g_1, ..., g_m$$

$$\text{and } P_z, \mathbf{V}_z \text{ and } W_z \text{ for all } z \in A_1 \cup ... \cup A_m$$

such that the conditions (1)–(24) are satisfied whenever $1 \leq n \leq m$ and $1 \leq l < n$.

For each $U \in \mathbf{U}_m$ let $\mathbf{U}_{m+1}^U = \{V \in \mathbf{U}_{m+1} : V \subset U\}$, $\mathbf{S}^U = \{V \in \mathbf{U}_{m+1}^U : V \in \bigcup_{z \in A_m} \mathbf{V}_z\}$ and $\mathbf{T}^U = \{V \in \mathbf{U}_{m+1}^U : V \cap g_m(Y_m) = \emptyset\}$. By the inductive assumption (22) for $n = m$, $\mathbf{S}^U \cap \mathbf{T}^U = \emptyset$, and by (20), $U - \bigcup \mathbf{S}^U$ and $\text{Int}(U - \bigcup \mathbf{S}^U)$ are connected sets.

Let $U \in \mathbf{U}_m$. For each $T \in \mathbf{T}^U$ let B_T be an arc in $\text{Int}(U - \bigcup \mathbf{S}^U)$ from $\text{Int}(T)$ to $g_m(Y_m)$ such that $B_T \cap B_{T'} = \emptyset$ for $T \neq T' \in \mathbf{T}^U$, there exists $p_T \in Y_m - X$ such that $B_T \cap g_m(Y_m) = \{g_m(p_T)\}$, $B_T - \{g_m(p_T)\}$ is piecewise linear, B_T is loosely embedded in Q with respect to $\mathbf{U}_{m+1}$ and $g_m(p_T) \in \text{Int}(W)$ for some $W \in \mathbf{U}_{m+1}$.

Let $\mathbf{U}_{m+2}$ be a k_{m+2}-partition of Q for some integer k_{m+2} such that $k_{m+2} > k_{m+1}$ and mesh $\mathbf{U}_{m+2} \leq \frac{1}{m+2}$. For $U \in \mathbf{U}_m$ and $T \in \mathbf{T}^U$ let $\mathbf{E}^T = \{V \in \mathbf{U}_{m+2} : V \cap B_T \neq \emptyset\}$. We may assume that the following conditions are satisfied when $U \in \mathbf{U}_m$ and $T, T' \in \mathbf{T}^U$ with $T \neq T'$:

(27) $\mathbf{E}^T$ is a chain with initial link E_I^T and last link E_L^T,

(28) E_I^T is the only element of $\mathbf{E}^T$ which meets $g_m(Y_m)$, $g_m(p_T) \in \text{Int}(E_I^T)$, $E_L^T \subset \text{Int}(T)$ and $B_T \subset \text{Int}(\bigcup \mathbf{E}^T)$,

(29) $\bigcup \mathbf{E}^T \subset \text{Int}(U) - (S_{m+1} \cup \bigcup_{y \in A_m} \bigcup \mathbf{V}_y)$, and

(30) $\bigcup \mathbf{E}^T \cap \bigcup \mathbf{E}^{T'} = \emptyset$.

We may also suppose (by taking a refinement of $\mathbf{U}_{m+2}$ if necessary) that

(31) $W - \bigcup_{T \in \mathbf{T}^U} \bigcup \mathbf{E}^T$ and $\text{Bd}(W) - (\bigcup_{T \in \mathbf{T}^U} \bigcup \mathbf{E}^T \cup S_{m+1})$ are connected sets for all $W \in \mathbf{U}_{m+1}^U$ and $U \in \mathbf{U}_m$, and

(32) $\text{diam } g_m^{-1}(E_I^T) < \frac{1}{m+1}$ for each $T \in \mathbf{T}^U$, $U \in \mathbf{U}_m$,

because $E_I^T \subset \text{Int}(U)$ and so $g_m^{-1}(E_I^T)$ is compact, by (8). Since $g_m(Y_m)$ has no isolated point, we may suppose (by taking again a refinement of $\mathbf{U}_{m+2}$ if necessary) that

(33) $\text{Int}(V) \cap g_m(Y_m) - \bigcup_{T \in \mathbf{T}^U, \, U \in \mathbf{U}_m} \bigcup \mathbf{E}^T \neq \emptyset$ for each $V \in \mathbf{U}_{m+1} - \bigcup_{U \in \mathbf{U}_m} \mathbf{T}^U$.

For $U \in \mathbf{U}_m$ and $T \in \mathbf{T}^U$, let $P_T = g_m^{-1}(E_I^T)$ and

$$M_T = \text{Int}(T \cup \bigcup \mathbf{E}^T) - \Big(\bigcup_{T' \in \mathbf{T}^U - \{T\}} \bigcup \mathbf{E}^{T'} \cup (\text{Bd}(\bigcup \mathbf{E}^T) - E_L^T) \cup S_{m+1} \Big).$$

Then, by (31), M_T is a connected open subset of Q and there is a homeomorphic copy J_T of $[0, 1)$ in M_T such that

(34) $B_T \subset J_T$ and $g_m(p_T)$ is the end-point of J_T,

and

(35) $\mathrm{Cl}(J_T) = J_T \cup \mathrm{Cl}\big(\mathrm{Bd}(T) - \bigcup_{T' \in \mathbf{T}^U} \bigcup \mathbf{E}^{T'}\big) \cup (T \cap S_{m+1}).$

Notice that J_T is a closed subset of M_T and if $V \in \mathbf{E}^T$ then $\mathrm{Int}(V) \cap J_T \neq \emptyset$. We may suppose that J_T is loosely embedded in Q with respect to $\mathbf{U}_{m+2}$. By the last part of (28), if $q \in J_T \cap \mathrm{Bd}(V) \cap B_T$ for some $V \in \mathbf{E}^T$, then there is a unique $V' \in \mathbf{E}^T - \{V\}$ such that $q \in V \cap V'$ and $q \in \mathrm{Int}(V \cup V')$. We may also suppose that if $q \in \mathrm{Bd}(V) \cap J_T - B_T$ for some $V \in \mathbf{U}_{m+2}$, then there is a unique $V' \in \mathbf{V}_{m+2} - \{V\}$ such that $q \in V \cap V'$. Then $V \cup V'$ is a neighbourhood of q in Q.

There exists a collection $\mathbf{C}^T = \{C_1, C_2, \dots\}$ of connected open subsets of Q such that

(36) $\mathbf{C}^T$ irreducibly covers J_T, $\mathrm{Cl}(C_i) \subset M_T$ for each i, and $C_i \cap C_j \neq \emptyset$ if and only if $|i - j| \leq 1$,

(37) $g_m(p_T) \in C_1 \subset \mathrm{Cl}(C_1) \subset \mathrm{Int}(E_I^T)$,

(38) for each i there exist $V, V' \in \mathbf{U}_{m+2}$ such that $\mathrm{Cl}(C_i) \subset \mathrm{Int}(V \cup V')$,

(39) if $V \in \mathbf{U}_{m+2}$ is such that $V \cap J_T \neq \emptyset$, then there exists i such that $\mathrm{Cl}(C_i) \subset V$,

(40) $\mathrm{card}\big(J_T \cap \mathrm{Bd}(V) \cap C_i\big) \leq 1$ if $V \in \mathbf{U}_{m+2}$ and $i \geq 1$,

(41) $\mathrm{diam}\, C_i \leq \frac{1}{i}$ for each i;

– see Fig. 2.

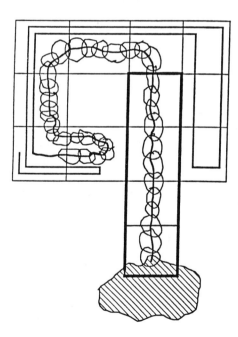

Figure 2.

Since Y_m has no isolated point, there is a continuous function
$$f_T : \mathrm{Cl}(P_T - \{p_T\}) \to J_T \cup E_I^T \subset \big(E_I^T \cup \bigcup \mathbf{C}^T\big)$$ such that

(42) $f_T(x) = g_m(x)$ if $x \notin g_m^{-1}(C_1)$,

(43) $J_T \subset f_T(P_T - \{p_T\}) \subset E_I^T \cup \bigcup \mathbf{C}^T$,
and

(44) for each $t \in J_T - B_T$, $f_T^{-1}(J_T^t) \cup \{p_T\}$ is a neighbourhood of p_T in Y_m, where J_T^t denotes the component of $J_T - \{t\}$ which misses B_T.

Recall that, by (8), $g_m(Y_m) \cap \mathrm{Int}(U)$ is a Z-set in $\mathrm{Int}(U)$. Hence, by Lemma 1, there is a Z-embedding $g_T : \mathrm{Cl}(P_T) - \{p_T\} \to s \cap \left(E_I^T \cup \bigcup \mathbf{C}^T\right)$ such that

(45) $g_T(x) = f_T(x) = g_m(x)$ if $x \notin g_m^{-1}(C_1)$,
and

(46) f_T and g_T are $\mathbf{C}^T$-close on $g_m^{-1}(C_1)$.

By Lemma 5 (infinitely many independent applications) and the properties of J_T and $\mathbf{C}^T$, we may assume that

(47) $g_T(P_T - \{p_T\})$ is loosely embedded in Q with respect to $\mathbf{U}_{m+2}$.

By (36) (39) and (46), it follows that $V \cap g_T(P_T - \{p_T\}) \neq \emptyset$ for each $V \in \mathbf{E}^T$, because $V \cap J_T \neq \emptyset$. Moreover, (41), (46) and (35) imply that

$$\mathrm{Cl}\big(g_T(P_T - \{p_T\})\big) = g_T(P_T - \{p_T\}) \cup \mathrm{Cl}(\mathrm{Bd}(T) - \bigcup_{T' \in \mathbf{T}^U} \bigcup \mathbf{E}^{T'}) \cup (T \cap S_{m+1}).$$

In fact, $g_T(P_T - \{p_T\}) \subset E_I^T \cup \bigcup \mathbf{C}^T$.

Now, let $A_{m+1} = \{p_T : T \in \mathbf{T}^U \text{ for some } U \in \mathbf{U}_m\}$ and $Y_{m+1} = Y_m - A_{m+1}$. Let $g_{m+1} : Y_{m+1} \to Q$ be defined by

$$g_{m+1}(x) = \begin{cases} g_m(x) & \text{if } x \in Y_m - \bigcup\{\mathrm{Int}(P^T) : T \in \mathbf{T}^U \text{ for some } U \in \mathbf{U}_m\} \\ g_T(x) & \text{if } x \in P_T - \{p_T\} \text{ for some } T \in \mathbf{T}^U \text{ and } U \in \mathbf{U}_m . \end{cases}$$

Recall that if $T, T' \in \mathbf{T}^U$ for some $U \in \mathbf{U}_m$ and $T \neq T'$, then $\bigcup \mathbf{E}^T \cap \bigcup \mathbf{E}^{T'} = \emptyset$. Hence, $P_T \cap P_{T'} = \emptyset$, and so g_{m+1} is well-defined. A straightforward argument shows that g_{m+1} is an embedding. Since a union of finitely many Z-sets is a Z-set again—see e.g. [5, Theorem 3.1(3)], it follows that $g_{m+1}(Y_{m+1}) \cap \mathrm{Int}(V) \cap s$ is a Z-set in $\mathrm{Int}(V) \cap s$ for each $V \in \mathbf{U}_{m+1}$.

Suppose that $z \in A_{m+1}$ and let T be such that $z = p_T$. Then T is unique. We let $W_z = T, P_z = P_T$ and $\mathbf{V}_z = \mathbf{E}^T$. It is routine to verify that the conditions (1)–(24) are satisfied for $n = m + 1$. This finishes the inductive construction.

We emphasize that the inductive construction given above is not going to be involved in proofs below. Those proofs will make use of the inductive conditions (1) – (24) only.

Let $Y_\infty = \bigcap_{n=1}^\infty Y_n$. By (23), the first part of (21), (18) and (16), it follows that if $x \in Y_\infty, U \in \mathbf{U}_n$ and $g_n(x) \in U$, then $g_{n+1}(x) \in U$. An inductive application of (6) gives the following

(48) if $x \in Y_\infty, U \in \mathbf{U}_n$ for some positive integer n and $g_n(x) \in U$, then $g_{n+i}(x) \in U$ for $i = 1, 2, \dots$.

Note that the above mentioned conditions imply also that

(49) if $x \in Y_\infty$, $U \in \mathbf{U}_n$ and $g_n(x) \in \mathrm{Bd}(U)$, then $g_{n+i}(x) = g_n(x)$ for $i = 1, 2, \dots$

and

(50) if $x \in Y_\infty, U \in \mathbf{U}_n$ and $g_n(x) \in \text{Int}(U)$, then $g_{n+i}(x) \in \text{Int}(U)$ for $i = 1, 2, \ldots$.

Let $x \in Y_\infty$. For each positive integer n, let $K_x^n = \bigcup \{U \in \mathbf{U}_n : g_n(x) \in U\}$. By (6) and (48), $K_x^1 \supset K_x^2 \supset \ldots$. By (5), the set $\bigcap_{n=1}^\infty K_x^n$ consists of a single point $g(x)$. Thus $g : Y_\infty \to Q$ is a well-defined function. By (5), the sequence $g_n|Y_\infty$ uniformly converges to g, and so g is continuous. By (48), we have

(51) if $x \in Y_\infty, U \in \mathbf{U}_n$ and $g_n(x) \in U$, then $g(x) \in U$.

First, we shall prove the following powerful strenghtening of (50):

(52) if $x \in Y_\infty, U \in \mathbf{U}_n$ and $g_n(x) \in \text{Int}(U)$, then $g(x) \in \text{Int}(U)$.

Let $x \in Y_\infty$ and $U \in \mathbf{U}_n$ be such that $g_n(x) \in \text{Int}(U)$. If $x \notin \bigcup_{z \in A_{n+i}} P_z$ for all $i = 1, 2, \ldots$, then (23) implies that $g_n(x) = g_{n+1}(x) = \ldots = g(x) \in \text{Int}(U)$. Hence, we may assume that k is the first positive integer such that $x \in P_z$ for some $z \in A_{n+k}$. Then $g_n(x) = \ldots = g_{n+k-1}(x)$. Let $U_{n+k-1} \in \mathbf{U}_{n+k-1}$ be such that $g_{n+k-1}(x) \in U_{n+k-1}$. By (6), $U_{n+k-1} \subset U$. By (18) and (16), $W_z \subset U_{n+k-1}$ and $\bigcup \mathbf{V}_z \subset \text{Int}(U_{n+k-1})$. Moreover, the first part of (21) implies that

(53) $g_{n+k}(x) \in \text{Int}(W_z \cup \bigcup \mathbf{V}_z) - \bigcup_{z' \in A_{n+k}, z' \neq z} \bigcup \mathbf{V}_{z'}$.

Suppose that $g_{n+k}(x) \in V$ for some $V \in \mathbf{V}_z$. By (24), $g_{n+k+1}(x) = g_{n+k}(x)$. Since $V \in \mathbf{U}_{n+k+1}$ – by (15), it follows that $g(x) \in V \subset \text{Int}(U)$ – see (51).

Now suppose that

(54) $g_{n+k}(x) \in \text{Int}(W_z) - \bigcup \mathbf{V}_z$.

By (23), if $x \notin \bigcup_{y \in A_{n+k+j}} P_y$ for all $j = 1, 2, \ldots$, then $g_{n+k}(x) = g_{n+k+1}(x) = \ldots = g(x)$ and $g(x) \in \text{Int}(W_z) \subset \text{Int}(U)$. Hence, we may assume that l is the first positive integer such that $x \in P_y$ for some $y \in A_{n+k+l}$. Then $g_{n+k}(x) = \ldots = g_{n+k+l-1}(x)$. Let $U_{n+k+l-1} \in \mathbf{U}_{n+k+l-1}$ be such that $g_{n+k+l-1}(x) \in U_{n+k+l-1}$.

By (6), $U_{n+k+l-1} \subset W_z \subset U_{n+k-1} \subset U$. As above, we can conclude that $W_y \subset U_{n+k+l-1}$, $\bigcup \mathbf{V}_y \subset \text{Int}(U_{n+k+l-1})$ and $g_{n+k+l}(x) \in \text{Int}(W_y \cup \bigcup \mathbf{V}_y)$. Moreover, if $g_{n+k+l}(x) \in V'$ for some $V' \in \mathbf{V}_y$, then $g(x) \in V' \subset \text{Int}(U)$. Hence, we may assume that $g_{n+k+l}(x) \in \text{Int}(W_y)$. If $W_y \subset \text{Int}(U)$, then $g(x) \in W_y \subset \text{Int}(U)$ by (51).

Suppose that $W_y \cap \text{Bd}(U) \neq \emptyset$. Since $W_y \subset W_z \subset U$, it follows that

(55) $W_y \cap \text{Bd}(W_z) \neq \emptyset$.

For $j = 0, 1, \ldots, l$, let $U_{n+k+j} \in \mathbf{U}_{n+k+j}$ be such that $W_y \subset U_{n+k+j}$. Then $W_z = U_{n+k}$, $U_{n+k+l-1}$ is the same as above and $W_y = U_{n+k+l}$. By (53) and (54), $U_{n+k+1} \not\subset \bigcup_{z' \in A_{n+k}} \mathbf{V}_{z'}$. Now (55) and a simple geometric argument imply that

(56) $U_{n+k+j} \cap \text{Bd}(W_z) - \bigcup_{z' \in A_{n+k}} \bigcup \mathbf{V}_{z'} \neq \emptyset$ for $j = 1, 2, \ldots, l$.

Let $P = g_{n+k}^{-1}(W_y)$. By the first part of (21), (13) and (23), it follows that $P \subset P_z$. By (7), P is a closed subset of Y_{n+k}. By (3), (11) and (12), P is a closed subset of $P_z - \{z\}$. By (56) for $j = l$ and the last part of (21), it follows that $P \neq \emptyset$ and $z \in \text{Cl}_Y(P)$. By induction and (20), $W_y \cap g_{n+k+j}(Y_{n+k+j}) \neq \emptyset$ for each $j > 0$. Therefore, $W_y \cap g_{n+k+l-1}(Y_{n+k+l-1}) \neq \emptyset$. This is a contradiction. So $W_y \cap \mathrm{Bd}(U) = \emptyset$. The proof of (52) is complete.

By (2), $X \subset Y_\infty$. Hence (8) and (51) imply that $g(X)$ meets each $U \in \mathbf{U}_n$ for $n = 1, 2, \ldots$. Therefore $g(X)$ is dense in Q. It remains to show that g is one-to-one, $g^{-1} : g(Y_\infty) \to Y_\infty$ is continuous and $g(Y_\infty) \subset s$.

Let x and x' be distinct points of Y_∞. We show that $g(x) \neq g(x')$. Let m be an integer such that $\rho(x, x') \geq \frac{2}{m}$. By (7), $g_m(x) \neq g_m(x')$. If $x, x' \notin \bigcup_{j=1}^\infty \bigcup_{z \in A_{m+j}} P_z$ then, by (23), $g(x) = g_m(x)$ and $g(x') = g_m(x')$, and so $g(x) \neq g(x')$. Hence, we may assume that there is a positive integer k such that $x \in P_z$ for some $z \in A_{m+k}$. We show that then

(57) $\quad g(x) \in \left(\mathrm{Int}(W_z) \cup \bigcup \mathbf{V}_z\right) - \bigcup_{y \in A_{m+k}, y \neq z} \bigcup \mathbf{V}_y$.

By (16), $g_{m+k-1}(x) \in \bigcup \mathbf{V}_z$ and, by (21), $g_{m+k}(x) \in \mathrm{Int}(W_z \cup \bigcup \mathbf{V}_z) - \bigcup_{y \in A_{m+k}, y \neq z} \bigcup \mathbf{V}_y$. If $g_{m+k}(x) \in \bigcup \mathbf{V}_z$, then $g_{m+k+1}(x) = g_{m+k}(x)$ by (24). By (15), $\mathbf{V}_z \subset \mathbf{U}_{m+k+1}$. Hence, by (51), $g(x) \in \bigcup \mathbf{V}_z$ and it suffices to recall (19) in order to get (57). Suppose that $g_{m+k}(x) \notin \bigcup \mathbf{V}_z$. Since, by (15), $\bigcup \mathbf{V}_z$ is closed in Q, it follows that $g_{m+k}(x) \in \mathrm{Int}(W_z) - \bigcup_{y \in A_{m+k}} \mathbf{V}_y$. First, we prove that

(58) $\quad g_{m+k+1}(x) \in \mathrm{Int}(W_z) - \bigcup_{y \in A_{m+k}} \mathbf{V}_y$.

If $x \notin \bigcup_{s \in A_{m+k+1}} P_s$ then, by (23), $g_{m+k+1}(x) = g_{m+k}(x)$, and so (58) holds. Suppose that $x \in P_s$ for some $s \in A_{m+k+1}$. By (16), $g_{m+k}(x) \in g_{m+k}(P_s) \subset \bigcup \mathbf{V}_s$. Hence, (18) and the fact that $g_{m+k}(x) \in W_z \in \mathbf{U}_{m+k}$ imply that $\bigcup \mathbf{V}_s \subset \mathrm{Int}(W_z) - \bigcup_{y \in A_{m+k}} \bigcup \mathbf{V}_y$. Therefore, by the last part of (16), $W_s \in \mathbf{U}_{m+k+1}$ is such that $W_s \subset W_z \in \mathbf{U}_{m+k}$ and $W_s \notin \bigcup_{y \in A_{m+k}} \mathbf{V}_y \subset \mathbf{U}_{m+k+1}$. Now, by (21), $g_{m+k+1}(x) \in g_{m+k+1}(P_s) \subset \mathrm{Int}(W_s \cup \bigcup \mathbf{V}_s) \subset \mathrm{Int}(W_z) - \bigcup_{y \in A_{m+k}} \bigcup \mathbf{V}_y$. This finishes the proof of (58).

Let $U \in \mathbf{U}_{m+k+1}$ be such that $g_{m+k+1}(x) \in U$. By (58), $U \notin \bigcup_{y \in A_{m+k}} \mathbf{V}_y$. If $g_{m+k+1}(x) \in \mathrm{Bd}(U)$ then, by (49), $g(x) = g_{m+k+1}(x)$. Thus $g(x) \in \mathrm{Int}(W_z) - \bigcup_{y \in A_{m+k}} \bigcup \mathbf{V}_y$. If $g_{m+k+1}(x) \in \mathrm{Int}(U)$ then, by (52), $g(x) \in \mathrm{Int}(U) \subset \mathrm{Int}(W_z) - \bigcup_{y \in A_{m+k}} \bigcup \mathbf{V}_y$. This completes the proof of (57).

Now, let us establish the properties of $g(x')$. First, suppose that $x' \in P_{z'}$ for some $z' \in A_{m+k}$. Since $\rho(x, x') \geq \frac{2}{m} > \frac{2}{m+k}$, it follows by (17) and (16) that $z \neq z'$ – because $z \in P_z$ and $z' \in P_{z'}$, by (10). We have the following condition analogous to (57):

(59) $\quad g(x') \in \left(\mathrm{Int}(W_{z'}) \cup \bigcup \mathbf{V}_{z'}\right) - \bigcup_{y \in A_{m+k}, y \neq z'} \bigcup \mathbf{V}_y$.

Since $z \neq z'$, it follows that $\bigcup \mathbf{V}_z \cap \bigcup \mathbf{V}_{z'} = \emptyset$ – by (19), and $W_z \neq W_{z'}$ whence $\mathrm{Int}(W_z) \cap \mathrm{Int}(W_{z'}) = \emptyset$ – by (14). Therefore (57) and (59) imply that $g(x) \neq g(x')$.

Now, suppose that $x' \notin \bigcup_{y \in A_{m+k}} P_y$. By (23), $g_{m+k-1}(x') = g_{m+k}(x')$. Let $U \in \mathbf{U}_{m+k}$ be such that $g_{m+k}(x') \in U$. Consider the case when $g_{m+k}(x') \neq g_{m+k+1}(x')$. Then $x' \in P_s$ for some $s \in A_{m+k+1}$ by (23). An argument analogous to the proof of (57) shows that, in particular,

(60) $\quad g(x') \in \mathrm{Int}(W_s) \cup \bigcup \mathbf{V}_s$.

By (16) and (18), $\bigcup V_s \subset \text{Int}(U) - \bigcup_{y \in A_{m+k}} \bigcup V_y$ – because $g_{m+k}(x') \in U \in U_{m+k}$. Moreover, by (16), $W_s \subset U$. By (17), (16) and the fact that $\rho(x, x') > \frac{2}{m+k}$, it follows that $g_{m+k-1}(x') \notin \bigcup V_z$. Thus $g_{m+k}(x') \notin \bigcup V_z$. By (13), $g_{m+k}(x') \notin W_z$. Hence $W_z \neq U$. Furthermore, by (22) and (13), $W_s \notin \mathbf{V}_z$. It follows that $(\text{Int}(W_z) \cup \bigcup V_z) \cap (\text{Int}(W_s) \cup \bigcup V_s) = \emptyset$. By (57) and (60), $g(x) \neq g(x')$.

Finally, consider the case when $g_{m+k}(x') = g_{m+k+1}(x')$. Let $V \in U_{m+k+1}$ be such that $g_{m+k+1}(x') \in V$. By (17), (16) and the inequality $\rho(x, x') > \frac{2}{m+k}$, we have that $g_{m+k-1}(x') \notin \bigcup V_z \cup W_z$. Since $g_{m+k-1}(x') = g_{m+k+1}(x')$ and $\mathbf{V}_z \subset U_{m+k+1}$, it follows that $V \notin \mathbf{V}_z$. Moreover, by (13), V is not a subset of W_z. By (49), if $g_{m+k+1}(x') \in \text{Bd}(V)$, then $g(x') = g_{m+k+1}(x') \notin \bigcup V_z \cup \text{Int}(W_z)$. By (52), if $g_{m+k+1}(x') \in \text{Int}(V)$, then $g(x') \in \text{Int}(V)$. By (57), it follows that $g(x) \neq g(x')$. This finishes the proof that g is one-to-one.

Continuity of g^{-1} is equivalent to the following condition:

(61) for any $x, x_1, x_2, \ldots \in Y_\infty$, if $\lim g(x_n) = g(x)$ then $\lim x_n = x$.

Therefore, in order to prove that g^{-1} is continuous, it suffices to show that

(62) for any $x, x_1, x_2, \ldots \in Y_\infty$, if $\lim x_n \neq x$ then $\lim g(x_n) \neq g(x)$.

Let $x, x_1, x_2, \ldots \in Y_\infty$ be such that $\lim x_n \neq x$. If there is $x' \in Y_\infty$ such that $\lim x_n = x'$, then $x \neq x'$, and so $g(x) \neq g(x')$. Moreover, the continuity of g implies that $\lim g(x_n) = g(x')$. Thus we may assume that the sequence does not converge in Y_∞. Since $Y_\infty \subset Y$ and Y is compact, we have two cases to consider.

Case 1. There exist points $x', x'' \in Y_\infty$ such that $x' \neq x''$ and x' and x'' are limits of subsequences of $\{x_n\}$. Then $g(x') \neq g(x'')$ and, by continuity of g, $g(x')$ and $g(x'')$ are limits of subsequences of $\{g(x_n)\}$. Hence, $\lim g(x_n)$ does not exist.

Case 2. There exists a point $z \in Y - Y_\infty$ which is the limit of a subsequence of $\{x_n\}$. We are going to show that $\lim g(x_n)$ does not exist in the set $g(Y_\infty)$. We may assume that $z = \lim x_n$. By (3) and (2) there is the unique integer $m > 1$ such that $z \in A_m$. By (10), we may assume that $x_1, x_2, \ldots \in P_z$. By the first part of (21), $g_n(P_z \cap Y_n) \subset \text{Int}(W_z \cup \bigcup V_z) - \bigcup_{y \in A_m, y \neq z} \bigcup V_y$. By (3), (12) and (11), $z \in P_z \subset Y_m \cup \{z\}$ and $z \notin Y_m$. Moreover, by the last part of (22), $g_m^{-1}(V)$ is a compact subset of Y_m for all $V \in \mathbf{V}_z$. Since $\mathbf{V}_z$ is finite — by (15), it follows that $g_m^{-1}(\bigcup V_z)$ is a compact subset of Y_m. Let $L_0 = P_z - g_m^{-1}(\bigcup V_z)$. By the above remarks, L_0 is a neighbourhood of z in Y. Hence, there is an integer n_0 such that $x_n \in L_0$ provided $n > n_0$.

Let $C = \text{Cl}(\text{Bd}(W_z) - \bigcup_{y \in A_m} \bigcup V_y)$. For each positive integer i let

$$C_i = \bigcup \{V \in U_{m+i} : V \subset \text{Cl}(W_z - \bigcup_{y \in A_m} \bigcup V_y) \text{ and } V \cap C \neq \emptyset\}.$$

Then $C_1 \supset C_2 \supset \ldots$ and $C = \bigcap_{i=1}^\infty C_i$. By (21), $\text{Cl}(g_m(L_0 - \{z\})) = g_m(L_0 - \{z\}) \cup C$. For each positive integer i let $L_i = \{z\} \cup g_m^{-1}(C_i) - \bigcup \{P_y : y \in A_{m+1} \cup \ldots \cup A_{m+i}\}$. Since the sets P_y are compact, $P_y \subset Y - \{z\}$ and $A_{m+1} \cup \ldots \cup A_{m+i}$ is finite, it follows that each L_i is a closed neighbourhood of z in Y. Hence there are positive integers $n_1, n_2, \ldots$ such that $n_0 < n_1 < n_2 < \ldots$ and, for any $i = 0, 1, \ldots$ and $n > n_i$,

$g_m(x_n) \in C_i$. By (23), if $n > n_i$ then $g_m(x_n) = g_{m+1}(x_n) = \ldots = g_{m+i}(x_n) \in C_i$. Since C_i is a union of members of $\mathbf{U}_{m+i}$, (51) implies that $g(x_n) \in C_i$ provided $n > n_i$. Therefore if $\lim g(x_n) = c$ exists in Q, then $c \in C$.

Now, it remains to show that $C \cap g(Y_\infty) = \emptyset$. Suppose that $x \in Y_\infty$. By (7), $g_m(x) \notin C$ (because if $g_m(x) \in C$ then we get a contradiction with (21)). Let $W \in \mathbf{U}_m$ be such that $g_m(x) \in W$. If $g_m(x) \in \mathrm{Bd}(W)$, then $g_m(x) = g(x)$ – by (49), and if $g_m(x) \in \mathrm{Int}(W)$ then $g(x) \in \mathrm{Int}(W)$ – by (52). In any case, $g(x) \notin C$ — because $C \subset \mathrm{Bd}(W_z)$ and $W_z \in \mathbf{U}_m$.

Finally, let us prove that $g(Y_\infty) \subset s$. Suppose that $x \in Y_\infty$ is such that $g(x) \in Q - s$. Let $k > 1$ be such that $g(x) \in S_k$ and $U \in \mathbf{U}_k$ be such that $g_k(x) \in U$. By (48) and (51), $g_{k+1}(x), g_{k+2}(x), \ldots, g(x) \in U$. Since $g_k(x) \in s$ (by (7)) and $g(x) \notin s$, it follows that there is $n > k$ such that $g_k(x) = g_{k+1}(x) = \cdots = g_{n-1}(x)$ and $g_k(x) \neq g_n(x)$. By (23), there is $z \in A_n$ such that $x \in P_z$. Then $g_n(x) \in W_z \cup \bigcup V_z$, by the first part of (21). Recall that $S_k \subset S_n$ and $S_n \cap \bigcup V_z = \emptyset$, by (18). Moreover, if $g_n(x) \in \bigcup V_z$, then (24) implies that $g_{n+1}(x) = g_n(x)$. Since $V_z \subset \mathbf{U}_{n+1}$, we have $g(x) \in \bigcup V_z$, by (51). This contradicts the facts that $g(x) \in S_k \subset S_n$ and $S_n \cap \bigcup V_z = \emptyset$. Hence, $g_n(x) \in W_z \in \mathbf{U}_n$. By (51), $g(x) \in W_z$. By the second part of (21), $g(x) \in \mathrm{Cl}\big(g_n(P_z \cap Y_n)\big)$. Since $P_z \cap Y_n = P_z - \{z\}$ (by (11) and (12)) and $g(x) \notin g_n(Y_n) \subset s$, it follows that there are points $y_1, y_2, \cdots \in P_z \cap Y_\infty$ such that $\lim_{i\to\infty} y_i = z$ and $\lim_{i\to\infty} g_n(y_i) = g(x)$. It is not difficult to see that $\lim_{i\to\infty} g(y_i) = g(x)$. Therefore, g^{-1} is not continuous, a contradiction. The proof of Theorem 1 is complete.

Remark 1. Theorem 1 reproves a part of Bowers' result, [4], with the use of quite different methods. We remark here that our construction can be slightly modified to prove that:

If X is a nowhere locally compact separable metric space, f is a continuous map of X onto a dense subset of the Hilbert cube Q, and $\mathbf{W}$ is an open covering of Q, then there is an embedding $g : X \to s \subset Q$ such that $g(X)$ is dense in Q, and f and g are $\mathbf{W}$-close.

The latter fact is a special case of a result of Bowers, who showed that in the case when $f : X \to s$ and $\mathbf{W}$ is an open covering of s, then there is a dense embedding $g : X \to s$ which is $\mathbf{W}$-close to f. Our proof can not be strenghtened to get that result. The reason is that Q is compact while s is not even locally compact.

Let us sketch an argument proving the above mentioned fact. One may assume that $\mathbf{W}$ is a finite open covering of Q. Hence, for some $k_1 > 1$, there is a k_1-partition $\mathbf{U}_1$ of Q such that $\mathbf{U}_1$ refines an open covering $\mathbf{W}'$ of Q which is a shrinking of $\mathbf{W}$.

By [7, 3.6.E, p. 234], there is a compact metric space Y such that $X \subset Y$, X is dense in Y and $f : X \to Q$ admits a continuous extension $F : Y \to Q$. One can apply Lemma 4 and [5, Theorem 8.1] to get a Z-embedding $g_1 : Y \to s \subset Q$ such that g_1 is $\mathbf{W}'$ close to F, $g_1(Y)$ is loosely embedded in Q with respect to $\mathbf{U}_1$, and $U \cap g_1(Y) \neq \emptyset$ for each $U \in \mathbf{U}_1$. Let $A_1 = \emptyset$. It is easy to construct A_n, Y_n, $\mathbf{U}_n$, g_n and P_z, V_z and W_z, for $z \in A_n$ and $n = 1, 2, \ldots$, such that the conditions (2) – (24) are satisfied and g_n is $\mathbf{W}'$-close to $F|_{Y_n}$. Let $g : Y_\infty \to Q$ be defined as in the proof of Theorem 1. Then $X \subset Y_\infty$, g is an embedding, $g(Y_\infty) \subset s$, $g(Y_\infty)$ and $g(X)$ are dense in Q, and $g|_X$ is $\mathbf{W}$-close to f.

Remark 2. One can modify the proof of Theorem 1 to get the following useful fact:

Let X be a nowhere locally compact separable metric space, C a compact subset of X and B be a subset of s Which is a Z-set in Q. If $h : C \to B$ is an embedding, then there exists an embedding $H : X \to s$ such that $H(X)$ is dense in s, $H(x) = h(x)$ for each $x \in C$, and $H(X) \cap B = H(C)$.

In fact, let Y be a compact metric space such that $X \subset Y$ and X is dense in Y. Clearly, C is a closed and nowhere dense subset of Y. Let ρ be a metric on Y such that $\text{diam}_\rho(Y) \leq 1$.

Let $L = \{(0,0)\} \times [0,1]^\infty$ and $K = [0,1] \times \{0\} \times [0,1]^\infty$. Then $L \subset K \subset Q$. By Lemma 2, there is a homeomorphism $\phi : Q \to Q$ such that $\phi(B) \subset L$ and $\phi(B)$ is a Z-set in L. By Lemma 1, there is an embedding $f : Y \to L$ such that $f(x) = \phi(h(x))$ for all $x \in C$. In coordinates, $f(x) = (0, 0, f^3(x), f^4(x), ...)$ for each $x \in Y$, where $f^i : Y \to [0,1]$ for each i. Let $F : Y \to Q$ be defined by $F(x) = (\rho(x, C), 0, f^3(x), f^4(x), ...)$ for each $x \in Y$. Then F is an embedding which extends $\phi \circ h$. Since C is closed, $L \cap F(Y) = F(C)$. Since $F(Y) \subset K$, it follows that $F(Y)$ is a Z-set in Q. By [5, Theorem 11.1], there is a homeomorphism $\psi : Q \to Q$ such that $\psi(K) \subset s$ and $\psi(\phi(x)) = x$ for all $x \in B$. Let $g_1 = \psi \circ F : Y \to Q$. Then g_1 is an embedding, $g_1(Y) \subset \psi(K) \subset s$, $g_1(Y)$ is a Z-set in Q, $g_1|_C = h$ and $g_1(Y) \cap B = g_1(C)$.

For each set $A \subset Q$, let $\text{I}(A) = \text{Int}(A) - B$ and $\text{B}(A) = \text{Bd}(A) \cup (A \cap B)$. Then $\text{I}(A)$ is open in Q, and if A is closed then $\text{B}(A)$ is closed, $A = \text{I}(A) \cup \text{B}(A)$ and $\text{I}(A) \cap \text{B}(A) = \emptyset$.

Now, observe that the inductive construction given in the proof of Theorem 1 can be modified in such a way that we get for $n = 1, 2, ...$ the following objects: $A_n, Y_n, \mathbf{U}_n, g_n$, and $P_z, \mathbf{V}_z$ and W_z for $z \in A_n$ such that g_1 is as defined above and analogues of the conditions (1)–(24) are satisfied after replacing everywhere $\text{Int}(\cdot)$ and $\text{Bd}(\cdot)$ by $\text{I}(\cdot)$ and $\text{B}(\cdot)$, respectively. In fact, since C is nowhere dense in Y, it is nowhere dense in each Y_n. Therefore, for each n, each $x \in C$ and each neighbourhood S of $g_n(x) = F(x)$ in Q, the set $g_n(Y_n) \cap \text{I}(S)$ is nonempty (this strenghtens the property (9) at points of C).

Define the map $g : Y_\infty \to Q$ as previously. An analogous proof shows that g is an embedding and $g(X)$ is dense in Q. Since $h(x) = F(x) = g_1(x) = g_2(x) = ...$ for $x \in C$, g extends h. Note that the property (52) can be restated as follows:

(52') if $x \in Y_\infty, U \in \mathbf{U}_n$ and $g_n(x) \in \text{I}(U)$, then $g(x) \in \text{I}(U)$.

Note that if $x \in Y_\infty - C$, then $g_1(x) \in \text{I}(Q) = Q - B$. Recall that $Q \in \mathbf{U}_1$. By (52'), if $x \in Y_\infty - C$ then $g(x) \in \text{I}(Q)$, and so $g(Y_\infty) \cap B = g(C)$. Now, it suffices to let $H = g|_X$.

Remark 3. With the use of the result of Remark 2 it is possible to get the following strenghtening of Theorem 1 which was obtained by quite different methods in [4].

If X is a nowhere locally compact separable metric space and M is a connected Fréchet manifold (or a connected Q-manifold), then X is homeomorphic to a dense subset of M.

Before sketching our proof we shall recall certain useful facts. It is well-known that each Fréchet manifold M (resp. each Q-manifold N) can be triangulated, i.e., there exists a locally compact polyhedron P such that M is homeomorphic to $P \times s$, see [9] (resp. N is homeomorphic to $P \times Q$, see [5, Theorem 37.2]). Moreover, if R is a locally compact polyhedron, then $R \times s$ is a Fréchet manifold, [15, Theorem 3], and $R \times Q$ is a Q-manifold, [14] or [5, Theorem 28.1]. Recall also that $[0,1]^n \times s$ is homeomorphic to s (see e.g. [12, Exercise 6.6.2, p. 299]).

Let X be a nowhere locally compact separable metric space and P be a locally compact and connected polyhedron. Since s is dense in Q it suffices to show that X is homeomorphic to a dense subset of $P \times s$. Let $\mathcal{K}$ be a triangulation of P, $\mathcal{L} = \{\sigma \in \mathcal{K} : \sigma$ is not a face of any $\sigma' \in \mathcal{K} - \{\sigma\}\}$, and for each positive integer n let P_n denote the n-dimensional skeleton of P with respect to $\mathcal{K}$ (then each P_n is connected).

Let Y' be a metric compactification of X. If $\mathcal{K}$ is finite, set $Y = Y'$, and otherwise select a point $y_0 \in Y' - X$ and let $Y = Y' - \{y_0\}$.

Let $\mathbf{U} = \{\sigma \times s : \sigma \in \mathcal{L}\}$. It suffices to construct a Z-embedding $f : Y \to P \times s$ such that $f(Y)$ is loosely embedded in $P \times s$ with respect to $\mathbf{U}$, and $f^{-1}(U)$ is a non-empty compact subset of Y for each $U \in \mathbf{U}$. Then the result of Remark 2 can be employed to get a dense embedding $h : X \to P \times s$ such that h coincides with f on each set $\partial(\sigma) \times s$ for $\sigma \in \mathcal{L}$.

The auxiliary Z-embedding f is constructed as $f = \lim f_n$, where $f_n : Y \to P_n \times s$, for $n = 1, 2, \ldots$, and the sequence $\{f_n\}$ is locally eventually constant. The construction of f_1 is the most complicated part of that argument.

3.   Finite dimensional case

Now, we shall prove some analogues of Theorem 1 for the case of embedding of nowhere locally compact spaces into finite dimensional cubes and manifolds. We start by adapting certain definitions from Section 2 to the current case. Let k be a positive integer.

We shall say that a (finite) collection $\mathbf{U}$ of subsets of $[0,1]^k$ is a *partition* of $[0,1]^k$ provided $\mathbf{U}$ covers $[0,1]^k$ and each $U \in \mathbf{U}$ is of the form $U = [a_{i_1}^1, a_{i_1+1}^1] \times \ldots \times [a_{i_k}^k, a_{i_k+1}^k]$, where a_i^j are points of $[0,1]$ for $j = 1, ..., k$ and $i = 0, ..., n_j$; $n_j > 1$ and $0 = a_0^j < a_1^j < \ldots < a_{n_j}^j = 1$ for $j = 1, ...k$; and $i_j \in \{0, ..., n_j - 1\}$ for $j = 1, ..., k$. For a given partition $\mathbf{U}$ of $[0,1]^k$, *chains* in $\mathbf{U}$ and *chain-connected* subcollections of $\mathbf{U}$ are defined exactly as in Section 2. We stress here that, again, if $\mathbf{V}$ is a chain then $V, V' \in \mathbf{V}$ and $V \cap V' \neq \emptyset$ imply that $\text{Int}(V \cup V')$ is connected.

If M is a k-manifold, $\partial(M)$ will denote its combinatorial boundary. We also let $\partial_k = \partial([0,1]^k)$. By a k-dimensional simplex we mean a space homeomorphic to $[0,1]^k$.

A closed subset P of a k-manifold M is said to be *accessible* in M provided, for each $x \in P$, there is an arc I_x in M such that $P \cap I_x = \{x\}$. Obviously, an accessible subset of M is nowhere dense in M, and a closed subset of an accessible set is accessible again.

The following result has a proof very similar to the construction employed in the proof of Theorem 1 (and Remark 2).

Theorem 2. *Let X be a nowhere locally compact space and $k \geq 3$ be an integer. If there exists an embedding $f : X \to [0,1]^k$ such that $\mathrm{Cl}(f(X))$ is an accessible subset of $[0,1]^k$, then there exists a homeomorphism $g : X \to g(X)$ of X onto a dense subset $g(X)$ of $[0,1]^k$.*

If, moreover, $f^{-1}(\partial_k)$ contains no non-void open subset of X, then g can be assumed to satisfy the following additional property: $f^{-1}(\partial_k) = g^{-1}(\partial_k)$ and $f(x) = g(x)$ for all $x \in f^{-1}(\partial_k)$.

Proof. Without loss of generality, one may assume that $f^{-1}(\partial_k)$ contains no non-void open subset of X (e.g. by selecting f such that $f(X) \subset [\frac{1}{3}, \frac{2}{3}]^k$). We are going to prove the expanded version of Theorem 2.

Since f is an embedding, there exists a compact space Y such that $X \subset Y$ and f admits an extension $g_1 : Y \to \mathrm{Cl}(f(X))$ such that g_1 is a homeomorphism of Y onto $\mathrm{Cl}(f(X))$. We let ρ be a metric on Y (e.g. such that g_1 is an isometry).

For each subset U of $[0,1]^k$ we let $\mathrm{I}(U) = \mathrm{Int}(U) - \partial_k$ and $\mathrm{B}(U) = \mathrm{Bd}(U) \cup (U \cap \partial_k)$.

It is easy to apply an induction and define for $n = 1, 2, \dots$ the following objects:

$$A_n, Y_n, \mathbf{U}_n, g_n, \text{ and } P_z, \mathbf{V}_z \text{ and } W_z \text{ for } z \in A_n$$

such that for all positive integers n and l with $l < n$ the following conditions (1_k)–(24_k) are satisfied:

(1_k) $A_1 = \emptyset$, $\mathbf{U}_1 = \{[0,1]^k\}$ and $g_1 = f$.

(2_k) A_n is a finite subset of $Y - X$ and $A_n \cap A_l = \emptyset$.

(3_k) $Y_n = Y - \bigcup_{i=1}^n A_i$.

(4_k) $\mathbf{U}_{n+1}$ is a partition of $[0,1]^k$.

(5_k) mesh $\mathbf{U}_{n+1} \leq \frac{1}{n+1}$.

(6_k) $\mathbf{U}_{n+1}$ refines $\mathbf{U}_n$.

(7_k) $g_n : Y_n \to]0,1[^k \cup g_1(Y_1) \subset [0,1]^k$ is an embedding.

(8_k) $g_n(Y_n) \cap \mathrm{I}(U)$ is a non-empty accessible subset of $\mathrm{I}(U)$ for each $U \in \mathbf{U}_n$.

(9_k) $g_n(Y_n)$ is loosely embedded in $[0,1]^k$ with respect to $\mathbf{U}_{n+1}$, and if $U \in \mathbf{U}_{n+1}$ and $x \in U \cap \partial_k \cap g_1(Y_1)$ then there are points $x_1, x_2, \dots \in g_1(Y_1) \cap \mathrm{I}(U)$ such that $x = \lim_{i \to \infty} x_i$.

(10_k) If $z \in A_n$, then P_z is a closed neighbourhood of z in Y.

(11_k) If $z, z' \in A_n$ and $z \neq z'$, then $P_z \cap P_{z'} = \emptyset$.

(12_k) If $z \in A_n$, then $P_z \cap A_l = \emptyset$.

(13_k) If $z \in A_n$, then $W_z \in \mathbf{U}_n$ and $W_z \cap g_{n-1}(Y_{n-1}) = \emptyset$.

(14_k) If $n > 1$, $W \in \mathbf{U}_n$ and $W \cap g_{n-1}(Y_{n-1}) = \emptyset$, then there is exactly one $z \in A_n$ such that $W = W_z$.

(15_k) $\mathbf{V}_z$ is a chain in $\mathbf{U}_{n+1}$ if $z \in A_n$.

(16_k) If $z \in A_n$, then the first link of $\mathbf{V}_z$ contains $g_{n-1}(P_z \cap Y_{n-1})$ and the last link of $\mathbf{V}_z$ is contained in $\mathrm{I}(W_z)$.

(17_k) If $z \in A_n$, then diam $g_{n-1}^{-1}(\bigcup \mathbf{V}_z) \leq \frac{1}{n}$.

(18_k) If $z \in A_n$, then there is $U \in \mathbf{U}_{n-1}$ such that $\bigcup \mathbf{V}_z \subset \mathrm{I}(U) - \bigcup_{y \in A_{n-1}} \bigcup \mathbf{V}_y$.

(19_k) If $z, z' \in A_n$ and $z \neq z'$, then $\bigcup \mathbf{V}_z \cap \bigcup \mathbf{V}_{z'} = \emptyset$.

(20_k) If $U \in \mathbf{U}_n$, then the collection $\{V \in \mathbf{U}_{n+1} : V \subset U \text{ and } V \not\subset \bigcup_{z \in A_n} \mathbf{V}_z\}$ is chain-connected and $\left(\mathrm{I}(U) - \bigcup_{z \in A_n} \bigcup \mathbf{V}_z\right) \cap g_n(Y_n) \neq \emptyset$.

(21_k) If $z \in A_n$, then $g_n(P_z \cap Y_n) \subset I(W_z \cup \bigcup V_z) - \bigcup_{z' \in A_n, z' \neq z} \bigcup V_{z'}$ and
$Cl(g_n(P_z \cap Y_n)) = g_n(P_z \cap Y_n) \cup Cl(B(W_z) - \bigcup_{z' \in A_n} \bigcup V_{z'})$.

(22_k) If $z \in A_n$ and $V \in \mathbf{V}_z$, then $V \cap g_n(P_z \cap Y_n) \neq \emptyset$ and $g_n^{-1}(V)$ is compact.

(23_k) If $n > 1$ and $x \in Y_n - \bigcup_{z \in A_n} P_z$, then $g_n(x) = g_{n-1}(x)$.

(24_k) If $n > 1$ and $x \in Y_n$ is such that $g_{n-1}(x) \in \bigcup_{y \in A_{n-1}} \bigcup \mathbf{V}_y$, then $g_n(x) = g_{n-1}(x)$.

The conditions above are analogous to the conditions (1)–(24) in the proof of Theorem 1. And the inductive construction is quite analogous to the one given in the proof of Theorem 1. When g_m is given, g_{m+1} alters it only on sets $P_z - \{z\}$, $z \in A_{m+1}$, and, roughly speaking, the alteration can be described as a piecewise linear stretching of $g_m(P_z - \{z\})$ along a selected ray in $]0,1[^k$ that misses $g_m(Y_m - \{z\})$. Of course, in order to assure that (9_k) holds, one has to apply Lemma 3 rather frequently. We remark also that the initial hypothesis $k \geq 3$ is necessary because of the first part of (20_k) and (24_k), and that the assumption that $f^{-1}(\partial_k)$ is nowhere dense in X imply that $g_1^{-1}(Y_1)$ is nowhere dense in Y, while the latter fact makes it possible to have the second part of (9_k).

Once the inductive construction is carried over, it suffices to let $Y_\infty = \bigcap_{n=1}^\infty Y_n$ and define $g' : Y_\infty \to [0,1]^k$ by $g'(x) = \lim_{i \to \infty} g_i(x)$, and let $g = g'|_X$. In fact, the argument following the inductive construction in the proof of Theorem 1, requires slight modifications only in order to show that $g'(Y_\infty)$ is dense in $[0,1]^k$. Furthermore, (18_k) and (23_k) imply that $g_1(x) = g_2(x) = \cdots = g'(x)$ for each $x \in Y_\infty$ such that $g_1(x) \in \partial_k$, and (7_k) and (21_k) together with the fact that g' is an embedding prove that $g'(Y_\infty) \cap \partial_k \subset g_1(Y_1)$.

Theorem 2 is a key to dense embedding results in the finite dimensional case. We are going to describe the two most significant applications.

Theorem 3. *Let X be a nowhere locally compact space and M be a connected and triangulable k-manifold for some positive integer k. If X admits an embedding into $[0,1]^{k-1}$, then X is homeomorphic to a dense subset of M.*

Proof. If $k = 2$, then X is 0-dimensional and one may assume that it is a dense subset of the space P of all irrational real numbers. A straightforward argument gives then a dense embedding of P into M.

Suppose that $k \geq 3$.

Let $\mathcal{K}$ be a triangulation of M. Then $\mathbf{U} = \{\sigma \in \mathcal{K} : \dim \sigma = k\}$ is what can be called a partition of M. Let $\sigma_0 \in \mathbf{U}$ and $Z \subset \sigma_0 - \partial(\sigma_0)$ be a linearly embedded copy of $[0,1]^{k-1}$. Also, let $f : X \to Z$ be an embedding. Since M is connected, one can easily find a countable and closed subset C of $Cl(f(X)) - f(X)$ and a deformation of $Z - C$ such that the resulting space Z' is a piecewise linear closed sub-manifold of M such that Z' meets no $(k-2)$-simplex from $\mathcal{K}$ and $Z' \cap \sigma \neq \emptyset$ for each $\sigma \in \mathbf{U}$. Obviously, $Z' \cap \sigma$ is accessible in σ for each $\sigma \in \mathbf{U}$. Since $\mathcal{K}$ is a triangulation of a manifold M, one can employ Lemma 3 to find an embedding $f' : X \to Z'$ such that $f'(X)$ is loosely embedded in M with respect to $\mathbf{U}$, and $f'(X) \cap \sigma \neq \emptyset$ for each $\sigma \in \mathbf{U}$.

Let $\sigma \in \mathbf{U}$ and $X_\sigma = (f')^{-1}(\sigma)$. Then $f'|_{X_\sigma} : X_\sigma \to \sigma$ embeds X_σ into an accessible subset $Z' \cap \sigma$ of σ. Since $f'(X)$ is loosely embedded in M with respect to $\mathbf{U}$, X_σ is nowhere locally compact and $(f')^{-1}(\partial(\sigma))$ contains no non-void open subset of X. By Theorem 2, there is an embedding $g_\sigma : X_\sigma \to \sigma$ such that $g_\sigma(X_\sigma)$ is dense in σ, $(f')^{-1}(\partial(\sigma)) = g_\sigma^{-1}(\partial(\sigma))$ and $f'(x) = g_\sigma(x)$ for each $x \in (f')^{-1}(\partial(\sigma))$.

Let $g : X \to M$ be defined by $g(x) = g_\sigma(x)$ for $x \in X$. Then g is a well-defined embedding and $g(X)$ is dense in M.

Let Z be a $(k-2)$-dimensional closed subset of $[0,1]^k$ for some $k \geq 2$. Then Z does not locally separate $[0,1]^k$ (see [8, Theorem 1.8.13]). Hence, it is easy to see that Z is accessible in $[0,1]^k$.

The most important $(k-2)$-dimensional closed subset of $[0,1]^k$ is the Menger space M_{k-2}^k – see [8, pp. 121-122 and p. 129]. A 65 years old and still unsettled conjecture due to K.Menger states that each compact $(k-2)$-dimensional space that embeds into $[0,1]^k$ can be embedded into M_{k-2}^k. Of course, it is also very interesting to know whether the Menger conjecture is true without the assumption on compactness of the space under consideration.

Since M_{k-2}^k is accessible in $[0,1]^k$ one can follow the proof of Theorem 3 and get the following Theorem 4.

Theorem 4. *Let X be a nowhere locally compact space and L be a connected and triangulable k-manifold for some integer $k \geq 3$. If X admits an embedding into the Menger space M_{k-2}^k, then X is homeomorphic to a dense subset of L.*

Corollary. *Let X be a nowhere locally compact separable metric space which is at most k-dimensional and M be a connected and triangulable $(2k+1)$-manifold, for some non-negative integer k. Then X is homeomorphic to a dense subset of M.*

Proof. If $k = 0$ then X can be embedded into $[0,1]$ and it suffices to apply Theorem 3. Suppose that $k \geq 1$. It is well-known that each k-dimensional separable metric space can be embedded into the Menger space M_k^{2k+1} (see e.g. [8, Theorem 1.11.7]). Obviously, $M_k^{2k+1} \subset M_{2k-1}^{2k+1}$. Now, it suffices to apply Theorem 4.

We recall here that in [4, Corollary 4] it was shown that a nowhere locally compact separable metric space which is at most k-dimensional embeds densely in the Menger space M_k^{2k+1}.

Problem. We do not know whether the assumption that M is a triangulable k-manifold in Theorem 3 (and also Theorem 4) can be replaced by the hypothesis that M is a connected k-dimensional polyhedron which can not be locally separated by a $(k-2)$-dimensional subset. Under the more general hypothesis, Lemma 3 does not provide the proof that $f'(X)$ is loosely embedded in M.

Example. Let C denote the Cantor set and let $Y = C \times [0,1]^{k-1}$ for some $k \geq 2$. It is easy to find a countable dense subset A of Y such that, for each $c \in C$, the set $(\{c\} \times [0,1]^{k-1}) \cap A$ contains at most one point. Let $X = Y - A$.

Obviously, X is a nowhere locally compact space which can be embedded into $[0,1]^k$. It is not difficult (but rather tedious) to prove that X does not admit a dense embedding into $[0,1]^k$.

REFERENCES

1. R.D.Anderson, Hilbert space is homeomorphic to the countable infinite product of lines, Bull. Amer. Math. Soc. 72 (1966), 515-519.
2. R.D.Anderson and T.A.Chapman, Extending homeomorphisms to Hilbert cube manifolds, Pacific J. Math. 38 (1971), 281-293.
3. P.L.Bowers, Dense embeddings of sigma-compact, nowhere locally compact spaces, Proc. Amer. Math. Soc. 95 (1985), 123-130.
4. P.L.Bowers, Dense embeddings of nowhere locally compact separable metric spaces, Topology Appl. 26 (1987), 1-12.
5. T.A.Chapman, Lectures on Hilbert cube manifolds, Regional Conference Series in Mathematics, no.28, Amer. Math. Soc., Providence, R.I., 1976.
6. D.W.Curtis, Sigma-compact, nowhere locally compact metric spaces can be densely imbedded in Hilbert space, Topology Appl. 16 (1983), 253-257.
7. R.Engelking, General Topology, PWN – Polish Scientific Publishers, 1977.
8. R.Engelking, Dimension Theory, North Holland and PWN – Polish Scientific Publishers, 1978.
9. D.W.Henderson, Open subsets of Hilbert space, Compositio Math. 21 (1969), 312-318.
10. K.Kuratowski, Topology, vol.I, Academic Press and PWN – Polish Scientific Publishers, 1966.
11. A.Lelek, Sur l'unicohérence, les homéomorphies locales et les continus irréductibles, Fund. Math. 45 (1957), 51-63.
12. J. Van Mill, Infinite Dimensional Topology, North Holland, 1989.
13. J.C.Oxtoby and V.S.Prasad, Homeomorphic measures in the Hilbert cube, Pacific J. Math. 77 (1978), 483-497.
14. J.E.West, Mapping cylinders of Hilbert cube factors, General Topology Appl. 1 (1971), 111-125.
15. J.E.West, Products of complexes and Fréchet spaces which are manifolds, Trans. Amer. Math. Soc. 166 (1972), 317-337.

An Example Concerning Disconnection Numbers

ROBERT PIERCE   West Virginia University, Morgantown, West Virginia

Abstract: An example is given of a connected subset, M, of Euclidean 3-space such that M is separated by any countably infinite subset but M is not separated by some n-point subset for each positive integer n.

Let X be a connected topological space such that every countably infinite subset of X separates X. In [1], the (smallest) disconnection number, $D^s(X)$, of X is defined to be the smallest cardinal number κ such that X becomes disconnected upon the removal of any κ points.

The author has asked in [1] whether there exists a connected metric space X such that $D^s(X) = \aleph_0$. The following example answers this question affirmatively.

For a subset A of R^3 (Euclidean 3-space) we let $|A|$ denote the cardinality of A and we let $\bar{A}$ denote the closure of A. We let $N = \{1, 2, \ldots\}$.

Let $M = \{(0,0,n) : n \in N\} \subset R^3$, and let $\Phi = \{F_1, F_2, \ldots\}$ be a partition of M such that $|F_j| = j$ for each $j \in N$. Let $\Pi = \{P_1, P_2, \ldots\}$ be the family of all those two-point subsets of M which intersect two members of Φ. For each $k \in N$ let φ_k be a homeomorphism from $[0,\infty)$ into the plane $y = kx$ such that $\varphi_k([0,\infty)) \cap M = \phi$ and $\overline{\varphi_k([0,\infty))} \cap M = P_k$. Let I_k denote the open arc $\varphi_k((0,\infty))$ for each $k \in N$. Define X to be $M \cup I_1 \cup I_2 \cup I_3 \cup \ldots$. Then for $F_1 = \{(0,0,n_1)\}$ we have:

$$n_1 \neq n \in N \Rightarrow \exists\, k \in N \ni: ((0,0,n_1),(0,0,n)) = P_k = \bar{I}_k \cap M. \qquad (1)$$

From (1) we see that, since each $\bar{I}_k$ contains some point $(0,0,n)$, X is connected. Moreover if $k \in N$ then $\bar{I}_k$ meets two members of Φ, so it also follows
from (1) that F_j does not disconnect X for $1 < j < \infty$. Note also that
$|F_j| = j$.

Now suppose that $B \subset X$ with $|B| = \aleph_0$. If B intersects some I_k then B
disconnects X because the open arc I_k is open in X. On the other hand, if
$B \subset M$ then no F_k contains B and hence, as in (1), B contains some P_k. But P_k
is the common boundary in X of the open sets I_k and X - $\bar{I}_k$, so B again disconnects X. It follows from what has been shown that $D^s(X) = \aleph_0$.

We note that this example does not show whether there is such an example
in the plane.

REFERENCE

[1]   Sam B. Nadler, Jr., Continuum theory and graph theory: disconnection
       numbers, J. London Math. Soc., to appear.

Indecomposable Continua, Prime Ends, and Julia Sets

JAMES T. ROGERS, JR.   Tulane University, New Orleans, Louisiana

The Julia set J of a quadratic polynomial is either a Cantor set or a one-dimensional continuum (i.e., a compact connected metric space) in $\mathbf{C}$. We are interested in the possibility that J may be an exotic continuum. For this paper, exotic means that the boundary of some component of the Fatou set is indecomposable. M. Herman [H] has constructed a C^∞-diffeomorphism of S^2 containing a local Siegel disk whose boundary is a pseudocircle; whether this behavior can be duplicated by an analytic function on $\hat{\mathbf{C}}$ is still an open question.

The link between such exotic Julia sets and the dynamics of the Fatou set is Carathéodory's notion of prime end. It is striking that N. E. Rutt wrote a paper [Ru] over fifty years ago setting up the precise relationship between indecomposable continua and prime ends. The next three sections of this paper are a salute to Rutt's work. The theorems and examples relating prime ends to indecomposable continua are his, though the proofs are new.

In the fifth section of this paper, we give a number of necessary and sufficient conditions for a Siegel disk G of a quadratic polynomial to have an indecomposable continuum as boundary. The best one for verifying or refuting the existence of indecomposable boundaries is that the impression of some prime end of G have nonempty interior with respect to the boundary of G.

This research was partially supported by a COR grant from Tulane University.

In the last section we announce some joint work with John Mayer giving necessary and sufficient conditions for the Julia set itself to be indecomposable. The similarities are striking, though the proofs are quite different.

Beside Rutt, others who have developed the theory of prime ends and indecomposable continua are M. Charpentier [C] and S. D. Iliadis [I].

The author thanks John Mayer for his comments on a preliminary version of this paper.

2. Prime ends and eating pie. Let $G \subset \widehat{\mathbf{C}}$ be a simply connected domain whose boundary is a nondegenerate continuum. Let $\mathbf{D}$ be the open unit disk in $\mathbf{C}$, let $\varphi : \mathbf{D} \to G$ be a conformal homeomorphism, and let $\varphi(0) = w_0$.

We assume the reader is familiar with the theory of prime ends as expounded, for instance, in [CL]. We identify a point in $S^1 = \partial \mathbf{D}$ with the prime end η of G corresponding to that point. We also identify $\partial \mathbf{D} = S^1$ with $\mathbf{R}/\mathbf{Z}$; hence η is a real number satisfying $0 \le \eta < 1$.

Associated with a sequence $\{q_n\}$ of cross cuts representing the prime end η is the sequence $\{e_n\}$ of domains, where e_n is the complementary domain of $q_n \cup \partial G$ containing q_{n+1}. The *impression* of the prime end η is the continuum

$$I(\eta) = \bigcap_{n=1}^{\infty} \overline{e}_n .$$

The *principal set* $P(\eta)$ of η is the subcontinuum

$$P(\eta) = \{x \in \partial G : q_n \to x, \text{ for some sequence } \{q_n\}$$

of cross cuts defining $\eta\}$.

Let $\eta \in S^1 = \partial \mathbf{D}$. Let $[0, \eta)$ denote the radial ray of $\mathbf{D}$ determined by η. Let $R_\eta = \varphi([0, \eta))$ be the image of this ray under the Riemann mapping function φ. The principal set $P(\eta)$ can be defined alternatively as the continuum

$$P_\eta = \overline{R}_\eta - R_\eta .$$

The impression $I(\eta)$ is defined by

$$I(\eta) = \{w \in \widehat{C} : \text{there is some ray } S \text{ in } D,$$

$$\text{perhaps not radial, from } 0 \text{ to } \eta \text{ such}$$

$$\text{that } w \in \overline{\varphi(S)} - \varphi(S)\} \ .$$

Some crosscuts determining prime ends are shown in Figure 1, while the images of some radial rays are shown in Figure 2.

Assume $S^1 = \partial D$ is oriented in the counterclockwise direction. If η_1 and η_1 are points of ∂D, let $[\eta_1, \eta_2]$ denote the oriented arc beginning at η_1 and ending at η_2. Define

$$I(\eta_1, \eta_2) = \cup\{I(\eta) : \eta_1 \le \eta \le \eta_2\} \ .$$

Theorem 2.1 (Pie Eating Theorem). $I(\eta_1, \eta_2)$ is a continuum.

Proof. Consider the "piece of pie without crust" in D defined by

$$P = \{z \in C : z = re^{2\pi i \eta}, \ \text{where } 0 \le r < 1 \text{ and } \eta_1 \le \eta \le \eta_2\} \ .$$

Then $\varphi(P)$ is a piece of pie without crust in G, and $\overline{\varphi(P)} - \varphi(P)$ is the crust of this piece of pie. Note that

$$I(\eta_1, \eta_2) = \left[\overline{\varphi(P)} - \varphi(P)\right] \cup I(\eta_1) \cup I(\eta_2) \ .$$

To show $I(\eta_1, \eta_2)$ is a continuum, it suffices to show the crust $\overline{\varphi(P)} - \varphi(P)$ is a continuum. This is easy if one eats the piece of pie one bite at a time, i.e., if the bite B_n is defined by

$$B_n = \left\{z \in C : z = re^{2\pi i \eta}, 1 - \frac{1}{n-1} \le r < 1 - \frac{1}{n}, \eta_1 \le \eta \le \eta_2\right\}, \quad n > 1,$$

then

$$\overline{\varphi(P)} - \varphi(P) = \bigcap_{n>1} \overline{\varphi(P - (B_2 \cup \ldots \cup B_n))} \ ,$$

the intersection of a nested sequence of continua, and hence a continuum. This completes the proof.

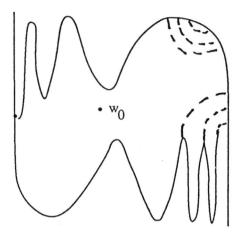

Figure 1

A continuum is *indecomposable* if it is not the union of two of its proper subcontinua. We conclude this section with a proof of the following theorem of Rutt.

Theorem 2.2. If ∂G is an indecomposable continuum, then $I(\eta) = \partial G$, for some prime end η.

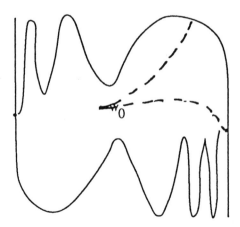

Figure 2

Proof. Note that

$$\partial G = \cup \{I(\eta) : \eta \in \partial D\} \ .$$

If η_1 and η_2 are diametrically opposite points of ∂D, then

$$\partial G = I(\eta_1, \eta_2) \cup I(\eta_2, \eta_1) \ .$$

Since ∂G is indecomposable, either $I(\eta_1, \eta_2)$ or $I(\eta_2, \eta_1)$ is equal to ∂G. Let us assume $I(\eta_1, \eta_2) = \partial G$. Let η_3 bisect $[\eta_1, \eta_2]$. It follows from indecomposability that either $I(\eta_1, \eta_3) = \partial G$ or $I(\eta_3, \eta_2) = \partial G$.

Continuing in the same manner, we find that there is a point η common to a decreasing sequence of closed intervals of S^1 such that $I(\eta) = \partial G$.

3. **Indecomposable continua and prime ends.** Recall that a continuum is *indecomposable* if it is not the union of two of its proper subcontinua. The closure of the unstable manifold of any periodic point of the Smale horseshoe is an example of an indecomposable continuum occurring in dynamical systems [Ba]. Continuum theorists know this example as the Knaster buckethandle K; it is pictured in Figure 3. The buckethandle is a nonseparating plane continuum with all proper nondegenerate subcontinua arcs.

According to Theorem 2.2, if $G = \widehat{C} - K$, then there must be a prime end η of G such that the impression $I(\eta)$ of η is equal to $\partial G = K$. There is exactly one such prime end η for G; it has the origin as its principal point. Two dotted crosscuts of η are indicated in Figure 3.

Let K_1 be the reflection of K through the y-axis. The continuum $K \cup K_1$ is pictured in Figure 4. There is a prime end η of $\widehat{C} - (K \cup K_1)$ having the origin as its principal point (two dotted crosscuts are indicated in Figure 4) such that the impression $I(\eta)$ of η is $K \cup K_1$. Hence the converse of Theorem 2.2 is not true. This is all that can happen, however, since Rutt has proved the following theorem.

Theorem 3.1. If there is a prime end η of G whose impression is equal to ∂G, then ∂G is an indecomposable continuum or the union of two indecomposable continua.

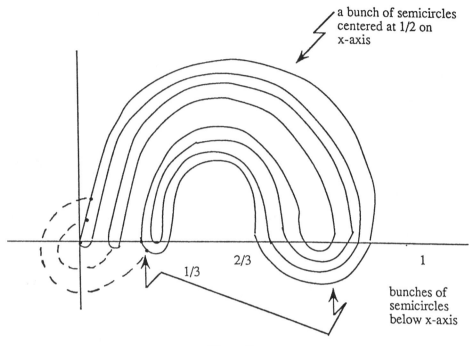

a bunch of semicircles
centered at 1/2 on
x-axis

1/3

2/3

1

bunches of
semicircles
below x-axis

Figure 3

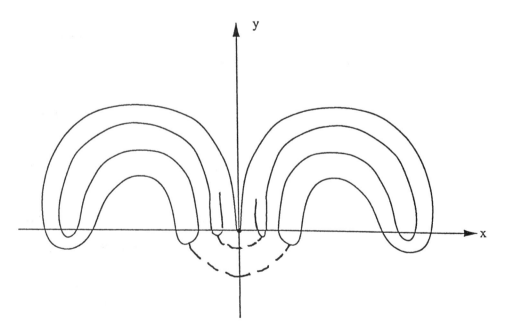

y

x

Figure 4

The converse of Theorem 3.1 is also not true, as the following example shows. Let K_2 be the reflection of K through the origin. The continuum $K \cup K_2$ is pictured in Figure 5. Even though $K \cup K_2$ is homeomorphic to $K \cup K_1$, the domain $\widehat{\mathbf{C}} - (K \cup K_2)$ has two prime ends whose impressions have nonempty interior with respect to $K \cup K_2$, but no prime end whose impression is equal to $K \cup K_2$.

A handy way to verify that the buckethandle is indecomposable is to use the following fact.

Theorem 3.2. If every proper nondegenerate subcontinuum of the continuum X is an arc, then either X is homeomorphic to $[0,1]$ or S^1, or X is indecomposable.

4. **Prime ends and composants.** Let X be an indecomposable continuum contained in the complex plane $\mathbf{C}$. The union of all proper subcontinua containing a point x of X is called a *composant* of X. In the case that each proper nondegenerate subcontinuum of X is an arc, the composants of X are precisely the arc components of X. The collection of composants of X forms a partition of X into disjoint sets. The continuum X contains uncountably many composants, and each composant is a dense F_σ in X [HY, p. 140].

Let G be a complementary domain of X. A point x of X is called an *accessible point* or *accessible from* G if there exists an arc A in $\overline{G}$ such that $A \cap X = \{x\}$. The composant of X containing x is called an *accessible composant* of X. The union of the accessible composants of X is a first category subset of X [Maz]. The points of X accessible from G are precisely the principal points of prime ends of the first or second kind [CL].

Kuratowski [Ku] has generalized this notion of accessible composant in the following manner. A composant C of X is called a K-*composant* provided there exist a continuum D contained in the composant C and a continuum L contained in $\overline{G}$ such that L intersects the complement of X, and $L \cap X = D$. Kuratowski [Ku] proved that the union of the K composants of X is also a first category subset of X. In particular, there exist composants

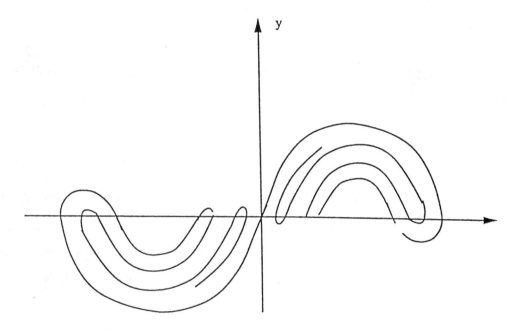

Figure 5

of X that are not K-composants.

The notion of a K-composant is quite apt for studying the prime end structure of indecomposable continua. Since the impression of a prime end of G is the limit set of a ray contained in G, we have the following theorem.

Theorem 4.1. If the boundary of the simply connected domain G is a nondegenerate indecomposable continuum X, then the impression of any prime end of G is either X or a subcontinuum of a K-composant of X.

Since not all composants of X are K-composants, we immediately have another proof of Theorem 2.2.

Corollary 4.2. For some prime end η of G, $I(\eta) = X$.

J. Krasinkiewicz [K1] introduced the notion of an internal composant of X. A composant of X is *internal* if every subcontinuum intersecting both X and the complement of X intersects all composants of X. Krasinkiewicz [K2] proved that X always contains internal composants, and in [K2], he proves the following theorem.

Theorem 4.3. Let X be a nonseparating plane continuum. If Y is an indecomposable subcontinuum of the boundary of X and C is an internal composant of Y, then any subcontinuum of X intersecting both C and the complement of Y contains Y.

This leads to an even stronger theorem, Theorem 4 of [Ru].

Theorem 4.4. If the boundary of the simply connected domain G contains a nondegenerate indecomposable subcontinuum D, then there is a prime end η such that $I(\eta)$ contains D.

Another theorem along these lines can be derived from [Ru, Lemma 4] or [Hag, Theorem 2].

Theorem 4.5. If the boundary of a simply connected domain contains an indecomposable continuum D and a continuum A such that A contains a nonempty open subset of D, then A contains D.

5.   Siegel disks of quadratics. Consider the one-parameter family $\{f_\lambda : \lambda \in \mathbf{C}\}$ of quadratic polynomials, where f_λ is defined by

$$f_\lambda(z) = \lambda z + z^2 \ .$$

Any quadratic polynomial can be conjugated to a unique f_λ, and so the study of the complex dynamics of quadratic functions can be reduced to the study of the dynamics of the family $\{f_\lambda\}$.

A point z_0 is *periodic* if $f_\lambda^n(z_0) = z_0$, for some positive integer n. The number $\beta = (f_\lambda^n)'(z_0)$ is the *eigenvalue* of z_0.

A periodic point is

(1) *superattracting* if $\beta = 0$

(2) *attracting* if $0 < |\beta| < 1$,

(3) *neutral* if $|\beta| = 1$, and

(4) *repelling* if $|\beta| > 1$.

The *Julia set* J_λ is the closure of the set of repelling periodic points of f_λ. The *Fatou set* F_λ is the complement of the Julia set.

The Julia set J_λ is either a Cantor set or a one-dimensional continuum. In what follows, we shall restrict ourselves to values of λ such that J_λ is a continuum.

If J_λ is a locally connected continuum, then it is reasonably well understood. This is the case, for instance, when the fixed point 0 is either attracting or superattracting. The neutral case is more interesting.

Note that λ is the eigenvalue of the neutral fixed point 0. Write $\lambda = e^{2\pi i\theta}$, $0 \leq \theta < 1$. If θ is rational, then 0 is called a *parabolic point* and J_λ is locally connected [M].

If θ is irrational, then there are two possibilities and each occurs. We say that f_λ is a *Siegel polynomial* if there is a domain G containing 0 and a conformal homeomorphism $\varphi : D \to G$ satisfying $\varphi(0) = 0$ and $f_\lambda \circ \varphi = \varphi \circ R_\theta$, where R_θ is the rotation through the angle θ. In other words, $f_\lambda \mid G$ is analytically conjugate to a rotation through the angle $2\pi\theta$. The domain G is called a *Siegel disk*.

In 1942, Siegel showed that this happens when θ is *Diophantine*: there exist $\varepsilon > 0$ and a natural number n such that

$$\left| \theta - \frac{p}{q} \right| > \frac{\varepsilon}{q^n}, \quad \text{for all rationals } \frac{p}{q} .$$

If θ is irrational and f_λ is not a Siegel polynomial, then f_λ is said to be a *Cremer polynomial* and 0 is said to be a *Cremer point*. Cremer polynomials exist [Cr].

The following information is available about Cremer points [M, p. 100].

Theorem 5.1. For all Cremer points, J_λ does not separate **C** and is not locally connected.

Question 5.2. What else can you say about J_λ for a Cremer point? for a Siegel disk?

Question 5.3. If G is a maximal Siegel disk for f_λ, then G is a component of F_λ and $\partial G \subset J_\lambda$. Must ∂G be a simple closed curve?

Question 5.4. Can the boundary of a Siegel disk be an indecomposable continuum?

Here it is appropriate to mention the example of M. Herman [H] of a C^∞-diffeomorphism $f : S^2 \to S^2$ with an invariant pseudocircle K such that one of the complementary domains of K is a "Siegel disk." The Herman example is adapted from Handel's example [Han]).

In [R2], we prove the following theorem.

Theorem 5.5 If G is a Siegel disk of the quadratic polynomial f_λ, then the following are equivalent:

(1) ∂G is an indecomposable continuum,

(2) $I(\zeta) = \partial G$, for some prime end ζ of G,

(3) $I(\zeta) = \partial G$, for all prime ends ζ of G,

(4) ∂G is the union of a countable number of impressions of prime ends of G,

(5) ∂G is the union of a countable number of indecomposable continua,

(6) The impression of some prime end of G has nonempty interior in ∂G, and

(7) Some indecomposable continuum of ∂G has nonempty interior in ∂G.

Remark. The theorem holds for any polynomial (see [R2]).

6. Julia sets of quadratics. Consider the one-parameter family $\{P_c : c \in \mathbf{C}\}$ of quadratic polynomials, where

$$P_c(z) = z^2 + c \ .$$

Any quadratic polynomial can be conjugated to a unique P_c, and this family is the customary one when we deal with the filled-in Julia set.

The *filled-in Julia set* K_c of P_c is the union of the Julia set J_c of P_c and the bounded complementary domains of J_c, if any. For a Cremer point $K_c = J_c$. In general, $J_c = \partial K_c = \partial(\widehat{\mathbf{C}} - K_c)$.

In case J_c is connected, let U denote the simply connected domain $\widehat{\mathbf{C}} - K_c$. There is a conformal homeomorphism $\varphi : \widehat{C} - \overline{D} \to U$ such that $\varphi(\infty) = \infty$ and $P_c \circ \varphi = \varphi \circ h$, where $h(z) = z^2$ [M].

John Mayer and the author [MR] have proved the following theorem relating the indecomposability of J_c to the prime end structure of U uniformized by φ.

Theorem 6.1. If the continuum J is the Julia set of some quadratic polynomial, then the following are equivalent:

(1) J is indecomposable,

(2) J contains an indecomposable subcontinuum with interior in J,

(3) The impression of some prime end of U has interior in J,

(4) The impression of some prime end of U is J,

(5) J is the union of a countable number of impressions of prime ends of U,

(6) J is the union of a countable number of indecomposable continua, and

(7) The impression of each prime end of U is J.

Remark. Theorem 6.1 is valid for any polynomial if we omit condition (7).

References

[Ba] M. Barge, *Horse shoe maps and inverse limits*, Pacific J. Math. **121** (1986), 29-39.

[B] B. Bielefeld, editor, *Conformal Dynamics Problem List*, Preprint #1990/1, Institute for Mathematical Sciences, SUNY–Stony Brook.

[BGM] B. L. Brechner, M. D. Guay, and J. C. Mayer, *Rotational dynamics on cofrontiers*, Contemporary Mathematics **117**, Amer. Math. Soc. Providence, RI, 1991, 39-48.

[C] M. Charpentier, *Sur certaines courbes fermées et leur bouts premiers*, 303–307.

[CL] E. F. Collingwood and A. J. Lohwater, "Theory of Cluster sets," Cambridge Tracts in Math. and Math. Physics **56**, Cambridge University Press, Cambridge, 1966.

[Cr] H. Cremer, *Zum Zentrum problem*, Math. Ann. **98** (1928), 151–163.

[DH1] A. Douady and J. H. Hubbard, *Étude dynamique des polynômes complexes (première partie)*, Publications Mathematiques D'Orsay **2** (1984), 1–75.

[DH2] _____, *Étude dynamique des polynômes complexes (deuxième partie)*, Publications Mathematiques D'Orsay **4** (1985), 1–154.

[Hag] C. L. Hagopian, *A fixed point theorem for plane continua*, Bull. Amer. Math. Soc. **77** (1971), 351–354.

[Han] M. Handel, *A pathological area preserving C^∞ diffeomorphism of the plane*, Proc. Amer. Math. Soc. **86** (1982), 163–168.

[H] M. R. Herman, *Construction of some curious diffeomorphisms of the Riemann Sphere*, J. London Math. Soc. **34** (1986), 375–384.

[HY] J. Hocking and G. Young, "Topology," Addison-Wesley Publishing Co., Reading, Mass., 1961.

[I] S. D. Iladis, *An investigation of plane continua via Carathéodory prime ends* , Soviet Math. Dokl. **13** (1972), 828–832.

[K1] J. Krasinkiewicz, *On the composants of indecomposable plane continua*, Bull. Pol. Acad. Sci. **20** (1972), 935–940.

[K2] —————————, *On internal composants of indecomposable plane continua*, Fund. Math. **84** (1974), 255–263.

[Ku] K. Kuratowski, *Sur une condition qui cáractérise les continus indécomposables*, Fund. Math. **14** (1929), 116–117.

[Maz] S. Mazurkiewicz, Sur les point accessibles les continus indecomposables, Fund. Math **14** (1929), 107–115.

[MO] J. C. Mayer and L. G. Oversteegen, *Denjoy meets rotation on an indecomposable cofrontier*, in preparation.

[MR] J. C. Mayer and J. T. Rogers, Jr., *Indecomposable continua and the Julia sets of polynomials*, Proc. Amer. Math. Soc.

[M]   J. Milnor, *Dynamics in one complex variable: introductory lectures*, Preprint #1990/5, Institute for Mathematical Sciences, SUNY-Stony Brook.

[R1] J. T. Rogers, Jr., *Intrinsic rotations of simply connected regions and their boundaries*, Complex Variable Theory Appl. (to appear).

[R2] —————————, *Singularities in the boundaries of local Siegel disks*, preprint.

[R3] —————————, *Rotations of simply connected regions and circle-like continua*, Contemporary Mathematics, **117**, Amer. Math. Soc. Providence, RI, 1991, 139–148.

[Ru] N. E. Rutt, *Prime ends and indecomposability*, Bull. Amer. Math. Soc. **41** (1935), 265–273.

Homeomorphisms on Cofrontiers with Unique Rotation Number

MARK H. TURPIN   Boston University, Boston, Massachusetts

ABSTRACT. On a class of homeomorphisms and cofrontiers we state some results relating the rate of convergence of the rotation number to the number theoretic properties of the rotation number and the differentiability of the homeomorphism. We offer some questions for homeomorphisms with unique rotation number on cofrontiers.

1. PRELIMINARIES

A *continuum* is a compact, connected set. A *subcontinuum* is subset of a continuum which is itself a continuum. A *cofrontier* is a planar continuum whose complement is the union of two connected open sets, and the common boundary of each of these open sets is the cofrontier.

Without loss of generality, let the origin be in the bounded component of the complement of the cofrontier. Let z be a point in the cofrontier and let $\Theta(z)$ be the usual angle assigning function modulo 2π. Given a homeomorphism f of a cofrontier the rotation number $\rho(z)$ is defined as a limit,

$$\rho(z) = \lim_{i \to \infty} \frac{\Theta(f^i(z)) - \Theta(z)}{i}.$$

when the limit exists. In general, the rotation number of a point need not exist. The rotation number measures the average angular progress of a point under iteration.

2. HISTORY

In 1932 G. D. Birkhoff [Bk] constructed an annulus homeomorphism with a strange attracting set in the interior. He noticed that the attracting set had interesting properties. In particular, the set was compact, connected, and the common boundary of two regions the closure of whose union was the annulus, yet the set was not homeomorphic to a circle. M. Charpentier [Ch], a student of Birkhoff's, showed this set to be an indecomposable continuum. That is, there was no way to write the set as the union of two proper subcontinua. The dynamics on Birkhoff's attractor are known to be quite interesting. The points in the set have a range of rotation numbers, in fact as shown by Le Calvez [LC] and Casdagli [Ca], for each number in this range there is a periodic or quasi-periodic orbit with this rotation number.

In 1948 R. H. Bing [Bi1] constructed a hereditarily indecomposable continuum, that is an indecomposable continuum for which every compact connected subset is indecomposable. The construction, being independent of any dynamical system, allows the continuum, called the pseudoarc, to be more easily studied and applied to other settings. In 1951 Bing [Bi2] gives an example of a hereditarily indecomposable continuum which, as subset of the plane, divides the plane into two open connected sets. This example, called the pseudocircle, was studied and given topological characterizations by L. Fearnly [Fe] and independently by J. T. Rogers. This work paved the way for M. Handel [Ha] who in 1982, using techniques of Fathi and Herman [FH], constructed a pseudocircle as an infinite intersection of annuli. On each annulus he defined a rotation, the resulting homeomorphism being the infinite composition of these rotations. He showed that the example could be extended to an annulus map, either area preserving or with the pseudocircle as an attractor. Also, the map restricted to the pseudocircle is not semi-conjugate to rotation on a circle, despite the fact that the rotation number is the same well-defined value for each point. This example indicates that the types of invariant sets possible for annulus homeomorphisms can be strange, even under the seemingly restrictive area-preserving assumption.

A class of examples derived from the Handel example has been studied by B.L. Brechner, M.D. Guay, and J.C. Mayer [BGM] . They relate the prime end structure of the cofrontier to the parameters of the construction, the indecomposability of the cofrontier, recurrence and other dynamical properties of the homeomorphism, and some number theoretic properties of the rotation number. We have also considered a class of examples derived from the Handel example, and have results relating the rate of convergence of the rotation number to the parameters of the construction, the number theoretic properties of the rotation number, and the differentiability of the extended homeomorphism [T] . We will state some of these results in this paper and pose some questions about the general case of a homeomorphism on a cofrontier with unique rotation number.

General results about the multiple rotation number case on a cofrontier are given by Barge and Gillette [BG1] and [BG2], and among the others who have considered similar examples is M. Herman [He] .

3. RESULTS

The class of cofrontiers considered is generated by taking an infinite intersection of nested annuli. Each annulus is embedded in the previous annulus in a 'wiggly' fashion, and the thickness of the annuli tend uniformly to zero (See figure 1).

The homeomorphism is generated by taking an infinite composition of rotations on each of the annuli and restricting to the cofrontier.

Proposition 1. *For any example in this class, the rotation number $\rho = \rho(z)$ exists for all points z in the cofrontier and is independent of the point.*

The rate of convergence of the rotation number is shown to influence the number theory of the rotation number and the differentiability of the homeomorphism. By

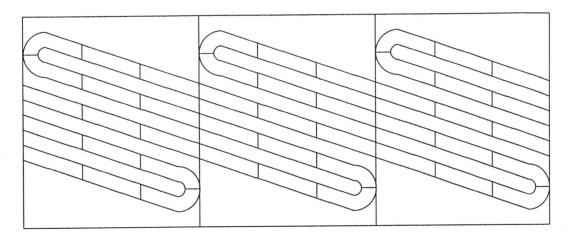

FIGURE 1. Embedding A_{n+1} in A_n

rate of convergence we mean the following. We know that for any z the quantity

$$\frac{\Theta(f^i(z)) - \Theta(z)}{i} - \rho = \frac{\Theta(f^i(z)) - \Theta(z) - i\rho}{i}$$

tends to zero as i tends to infinity. If there exists a point z and a sequence $\{i_n\}$ such that for any $K > 0$

$$\frac{\Theta(f^{i_n}(z)) - \Theta(z) - i_n\rho}{i_n} > K\frac{i_n^p}{i_n}$$

holds for infinitely many n then we say that the rotation number converges no faster than $\frac{i^p}{i}$. We note that if the rotation number converges, this statement will not be made with p equal to one.

Theorem 1. *If the rotation number converges no faster than $\frac{i^p}{i}$, $0 < p < 1$, then there exists a sequence of rationals, $\left\{\dfrac{\alpha_n}{\beta_n}\right\}$, converging to ρ and a constant K_p , depending only on p , such that $0 < \rho - \dfrac{\alpha_n}{\beta_n} < \dfrac{K_p}{\beta_n^{\frac{2-p}{1-p}}}$.*

As the rotation number converges more slowly, that is, as p can be made closer to one, the rotation number becomes more 'well approximable by rationals'. One can ask whether such a relationship exists in general.

Question 1. *Given a homeomorphism of a cofrontier with unique rotation number, do slower rates of convergence of the rotation number necessitate rotation numbers which are more 'well approximable'?*

Each member of the class of examples studied is extendable to a homeomorphism of the plane. The differentiability of the extension is limited by the rate of convergence of the rotation number on the cofrontier.

Theorem 2. *If the rotation number convergence is no faster than* $\dfrac{i^p}{i}$, $0 < p < 1$,
then the homeomorphism is no more than $\dfrac{1-p}{p}$ *times differentiable.*

The slower the rate of convergence of the rotation number, the less differentiable the homeomorphism may be. We note that it is not known how 'sharp' the theorem is. That is we do not have examples with any prescribed differentiability. Nonetheless the question can be posed in general.

Question 2. *Given a homeomorphism of a cofrontier with unique rotation number, extendable to the plane, do slower rates of convergence of the rotation number necessitate less differentiability of the homeomorphism?*

References

[BG1]   M. Barge and R.M. Gillette, *Indecomposability and Dynamics of Invariant Plane Separating Continua*, Contemporary Mathematics (to appear).

[BG2]   M. Barge and R.M. Gillette, *Rotation and Periodicity in Plane Separating Continua*, Ergodic Theory and Dynamical Systems (to appear)

[Bi1]   R.H. Bing, *A homogeneous indecomposable plane continuum*, Duke Mathematical Journal 15 (1948), 729-742 .

[Bi2]   R.H. Bing, *Concerning hereditarily indecomposable continua*, Pacific Journal of Math. 1 (1951), 43-51.

[Bk]   G.D. Birkhoff, *Sur quelques courbes fermées remarquables*, Bulletin de la Societe Mathematique de France 60 (1932), 1.

[BGM]   B.L. Brechner, M.D. Guay, and J.C. Mayer, *Rotational Dynamics on Cofrontiers*, (to appear) .

[Ca]   M. Casdagli, *Periodic orbits for dissipative twist maps*, Ergodic Theory and Dynamical Systems 7 (1987), 165-173.

[Ch]   M. Charpentier, *Sur quelques propriétés des courbes de M. Birkhoff*, Bulletin de la Societe Mathematique de France 62 (1934), 193.

[FH]   A. Fathi and M. Herman, *Existence de difféomorphismes minimaux*, Astérisque 49 (1977), 37-59.

[Fe]   L. Fearnley, *The pseudocircle is unique*, Transactions of the AMS 149 (1970), 45-64.

[Fr]   J. Franks, *Recurrence and fixed points of surface homeomorphisms*, Ergodic Theory and Dynamical Systems 8* (1988), 99-107.

[Ha]   M. Handel, *A pathological area-preserving diffeomorphism of the plane*, Proceedings of the American Mathematical Society 86 (1982), 163-168.

[He]   M. Herman, *Construction of some curious diffeomorphisms of the Riemann sphere*, Journal of the London Mathematical Society 34 (1986), 375-384.

[LC]   P. Le Calvez, *Propriétés des attracteurs de Birkhoff*, Ergodic Theory and Dynamical Systems 8 (1988), 241-310.

[T]   M. Turpin, *Rotation number properties on a class of cofrontiers*, preprint.

Self-Homeomorphic Star Figures

WLODZIMIERZ J. CHARATONIK   University of Wroclaw, Wroclaw,
Poland, and McNeese State University, Lake Charles, Louisiana

ANNE DILKS DYE and JAMES F. REED   McNeese State University,
Lake Charles, Louisiana

1. Introduction

The work done in sections 1, 2, and 3 of this paper was included [2]. Since part of the talk given at the Lafayette Conference on Continua Theory and Dynamical Systems was taken from that paper, we include for completeness a summary of the theorems and examples contained in it. Proofs and details are in the original paper. In Section 4 we include some new work which was in the talk in Lafayette, but not in the original paper.

Some basic definitions, which will be used throughout the paper, are necessary. By a continuum, we mean a compact connected metric space. A neighborhood of a point x in a topological space X is a set N satisfying x ε int(N). A point x ε X is called a cut point of X [4, III, p. 41] if X \ {x} is not connected. A point x ε X is called a local cut point of X provided there is a connected neighborhood N of x, such that x is a cut point of N. Note that a local cut point is a point of local connectedness of the space X. We will, however, mainly use the notion of a local cut-point in locally connected spaces anyway. For connected sets, this definition of a cut point is equivalent to the definition of a separating point in [4, III, 8, p. 58], and the definition of a local cut point is equivalent to that of a local separating point in [4, III, 9, p. 61]. For a given space X the set of local cut points of X is denoted by LC(X).

2. Definitions and Basic Properties

In this section we recall definitions of four types of self-homeomorphic spaces and we discuss interrelations between them and some of their basic properties.

2.1 Definition: A topological space X is called self-homeomorphic if for any open set U $\subseteq$ X there is a set V $\subseteq$ U such that V is homeomorphic to X.

2.2 Definition: A topological space X is called strongly self-homeomorphic if for any open set U $\subseteq$ X there is a set V with nonempty interior such that V $\subseteq$ U, and V is homeomorphic to X.

2.3 Definition: A topological space X is called pointwise self-homeomorphic at a point x ε X if for any neighborhood U of x, there is a set V such that x ε V $\subseteq$ U and V is homeomorphic to X. The space X is called pointwise self-homeomorphic if it is pointwise self-homeomorphic at each point.

    2.4 Definition: A topological space X is called strongly pointwise self-homeomorphic at a point x ε X if for any neighborhood U of x, there is a neighborhood V of x such that V $\subseteq$ U and V is homeomorphic to X. The space X is called strongly pointwise self-homeomorphic if it is strongly pointwise self-homeomorphic at each of its points.

    Recall that metric spaces X and Y are called *similar* if there is a surjection f: X $\to$ Y such that there is a positive constant c satisfying d(f(x), f(y)) = c·d(x, y). Such a map is called a similarity. If we replace the condition that V is homeomorphic to X in each of the above definitions with the condition that V is similar to X, we get definitions of self-similar spaces. Thus every (strongly, pointwise or strongly pointwise) self-similar space is (strongly, pointwise or strongly pointwise) self-homeomorphic, but not conversely. Of course we could consider any class of mappings besides homeomorphisms or similarities. For example, if we use affine transformations we get analogous definitions of self-affine spaces. However, homeomorphisms and similarities seem to be the most interesting ones.

    An arc, and *n*-dimensional cube and a one point union of two *n*-dimensional cubes with the identification point of the boundary of both of the *n*-dimensional cubes are examples of strongly pointwise self-homeomorphic continua.

    We note some basic properties of self-homeomorphic spaces.

    2.5 Theorem: The following diagram of implications applies to the above definitions.

$$2.2 \quad \to \quad 2.1$$
$$\uparrow \qquad\qquad \uparrow$$
$$2.4 \quad \to \quad 2.3$$

    In [2], one can find examples showing that none of the implications in the above diagram can be reversed, and that there is no relationship between 2.2 and 2.3.

3. Methods of Constructing Self-homeomorphic spaces.

    In this section we recall a method of constructing self-homeomorphic spaces. The method comes from [1, 3.7, pp. 80-85] and its details and proofs of the theorems can be found in [2]. Let us recall some common notations. For a given compact metric space X, a nonempty subset A of X, and a positive number r, we put N(A, r) = {x ε X: there exists a ε A with d(a, x) < r}.

    For two nonempty compact subsets A and B of X we define the Hausdorff distance dist(A, B)

$= \inf\{r > 0 : A \subseteq N(B, r) \text{ and } B \subseteq N(A, r)\}$, and we denote by 2^X the hyperspace of all nonempty compact subsets of X equipped with the Hausdorff distance.

3.1 Definition: Let (X, d) be a metric space. A map $f: X \rightarrow X$ is called contractive if there is a constant $0 \leq s < 1$ such that $d(f(x), f(y)) \leq s \cdot d(x, y)$ for every x, y ε X. Any such number s is called a contractivity factor for f.

For any given compact metric space X and a set of contractive mappings $\{f_i : i \varepsilon I\}$ having common contractivity factor $s < 1$, let $F: 2^X \rightarrow 2^X$ be defined by $F(A) = cl(\cup\{f_i(A): i \varepsilon I\})$. Then F is contractive with contractivity factor s (see [2, Theorem 3.3]) and therefore according to the Banach Contraction Mapping Theorem (see e.g.[1, Theorem 1, p. 76]), there is exactly one fixed set under F, i.e. a set A ε 2^X such that $F(A) = A$. Such a set A can be obtained as the limit Lim $F^n(B)$ for any B ε 2^X. Then we will write $A = F(X, \{f_i : i \varepsilon I\})$.

Now we present some conditions on mappings f_i under which the set $F(X, \{f_i : i \varepsilon I\})$ is self-homeomorphic.

3.2 Theorem. If the mappings $f_i : X \rightarrow X$ for i ε I are embeddings, then $F(X, \{f_i : i \varepsilon I\})$ is self-homeomorphic.

3.3 Theorem. If the mappings $f_i : X \rightarrow X$ for i ε I are embeddings and satisfy int $f_i(X) \neq \emptyset$ and int $f_i(X) \cap$ int $f_j(X) = \emptyset$ for i,j ε I with i$\neq$ j, then $F(X, \{f_i : i \varepsilon I\})$ is strongly self-homeomorphic.

3.4 Theorem. If the mappings $f_i : X \rightarrow X$ for i ε I are embeddings and the set I is finite, then $F(X, \{f_i : i \varepsilon I\})$ is pointwise self-homeomorphic.

3.5 Theorem. Under the assumptions of Theorem 3.4 the set is either a locally connected continuum, or is not connected.

4. A Family of Self-homeomorphic Plane Continua

In this section we will define a countable family of pointwise self-homeomorphic and strongly self-homeomorphic plane continua. Each member S_n, for $n \geq 2$, of this family will be located in a regular polygonal disk with vertices at $v_k = (1+a_n)\exp \frac{2k\pi i}{n}$, where a_n is a scale factor defined in Theorem 4.7. Also, each S_n will be defined as the only fixed set under some natural mapping. We will also investigate topological properties of S_n's.

Denote by P_n, for $n > 2$, the polygonal disk with its vertices $v_0, v_1, \ldots v_{n-1}$, as defined above. For k ε $\{0, 1, 2, 3, \ldots, n-1\}$, let f_k be the contractive mapping from P_n into P_n with contractivity factor a_n chosen in such a way that

(1) $f_i(P_n) \cap f_{i+1}(P_n) \neq \emptyset$ and   (2) int $f_i(P_n) \cap$ int $f_j(P_n) = \emptyset$ for $i \neq j$

and such that f_k maps P_n onto a similar copy, of P_n centered at v_k as defined above. See Figure 1. We note that the image of P_n is "scaled down" by a factor of a_n (as defined in Theorem 4.7). Condition (2) above, along with Theorem 3.3 give that the fixed set under this set of mappings is strongly self-homeomorphic. Condition (1) above gives that $\bigcup_{k=0}^{n-1} f_k(P_n) = F(P_n)$ is connected. By the remarks following definition 3.1, $S_n = F(P_n, \{f_i : i \ \varepsilon \ \{0, 1, \ldots, n-1\}\}) = \lim_{m \to \infty} F^m(P_n)$ and hence S_n is connected. Therefore, by Theorems 3.4 and 3.5 we have the following.

4.1 Proposition. For any $n > 2$, the set S_n is a strongly self-homeomorphic, pointwiese self-homeomorphic, locally connected continuum.

Observe that S_2 is the segment with vertices $v_0 = 1$ and $v_1 = -1$, S_3 is the Sierpiński Triangle (see Figure 2) and S_4 is the solid square with vertices ± 1, $\pm i$. Observe also that $a_2 = a_3 = a_4 = \frac{1}{2}$. Pictures of S_3, S_5, S_6, and S_8 are in Figure 2.

Now we will prove that S_m is homeomorphic to S_n if and only if m and n are multiples of 4 with $m, n > 4$. We start with a lemma.

4.2 Lemma. For each $n \geq 2$ the intersection $f_i(P_n) \cap f_{i+1}(P_n)$ is a common edge of the polygons $f_i(P_n)$ and $f_{i+1}(P_n)$ if n is a multiple of 4, and it is a common vertex of these polygons otherwise.

Proof. Because of the definition of S_n and the convexity of P_n, the intersection $f_i(P_n) \cap f_{i+1}(P_n)$ is either an edge or a vertex of both $f_i(P_n)$ and $f_{i+1}(P_n)$. If it is an edge, then it is perpendicular to $f_i(v_i v_{i+1})$. Thus $f_i(P_n)$, and hence P_n has two perpendicular edges. This can happen only in the case when n is a multiple of 4.

4.3 Theorem. If $n > 4$ is a multiple of 4, then S_n is homeomorphic to the Sierpiński Universal Plane Curve.

Proof. For any such n, by 4.2, $f_i(P_n) \cap f_{i+1}(P_n)$ is a common edge. In the limit, $\text{Lim } F^n(P_n)$, this creates a Cantor set of points and hence the continuum S_n will contain no local cut points. According to the Whyburn Characterization Theorem (see [5, Theorem 3, p. 322]) every locally connected, one-dimensional, plane continuum with no local cut points is homeomorphic to the Sierpiński Universal Plane Curve.

Now we are going to prove that if one of the numbers m or n is not a multiple of 4, then S_m is not homoeomophic to S_n. To this aim we recall the notion of the structure of the set of local cut-points of a continuum as defined in [2, Definition 5.1].

4.4 Definition. We say that two continua X and Y have the same structure of the sets of local

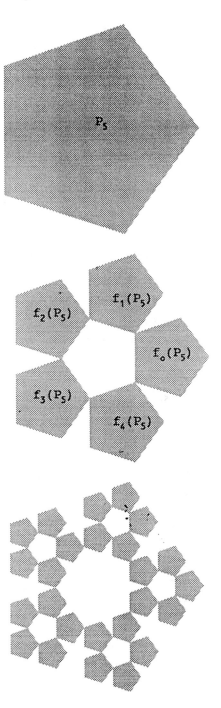

Figure 1

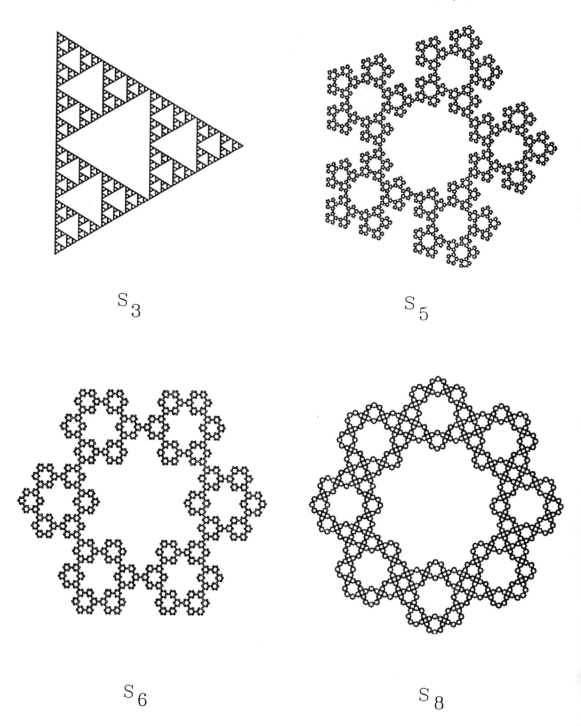

S_3

S_5

S_6

S_8

Figure 2

cut points if there is a homeomorphism α: LC(X) $\rightarrow$ LC(Y) such that a set A$\subseteq$ LC(X) disconnects X if and only if α(A) disconnects Y and the number of components of X\A is the same as the number of components of Y\α(A).

In general the structure of the set of local cut points does not distinguish self-homeomorphic continua.   In [2, Example 5.2] an example is shown of two nonhomeomorphic strongly self-homeomorphic and pointwise self-homeomorphic plane curves U and H having the same structure of the sets of local cut points.   However, we will show that for the family $\{S_n : n \varepsilon$ N with n$\geq$ 2 and n $\neq$ 4$\}$ the structures of local cut points determines homeomorphic members of the family.

4.5 Proposition.   Let m, n $\geq$ 2 be two distinct natural numbers.   If one of them is not a multiple of 4, then S_m and S_n do not have the same structure of the sets of local cut points.

Proof.   Observe that S_2 is the only member of the family $\{S_n : n \; \varepsilon$ N with n$\geq$ 2$\}$ having (global) cut points, and that S_n has no local cut points if and only if n is a multiple of 4.

Fix a number n$\geq$ 3 which is not a multiple of 4, and let p ε S_n be any local cut point.   We will show that there are exactly n$-$1 points q_1, q_2, ... q_{n-1} of S_n such that each of the sets $\{p, q_j\}$ for j ε $\{1,...,n-1\}$ disconnects S_n.   Note that this property depends on the structure of the sets of local cut points only, and it distinguishes S_n's if n is not a multiple of 4.

Indeed, if p is a local cut point of S_n, then there exists a natural number k such that p is the only point of the intersection

$$f_{i_k}(f_{i_{k-1}}(\ldots f_{i_1}(P_n)\ldots)) \cap f_{(i_k+1)\bmod n}(f_{i_{k-1}}(\ldots f_{i_1}(P_n)\ldots)) \; ,$$

for some i_1, ... i_k ε $\{0,...,n-1\}$.   For each i ε $\{0,...,n-1\}\backslash\{i_k\}$ the intersection

$$f_i(f_{i_{k-1}}(\ldots f_{i_1}(P_n)\ldots)) \cap f_{(i+1)\bmod n}(f_{i_{k-1}}(\ldots f_{i_1}(P_n)\ldots))$$

is a one point set.   Denote the points of the intersections by q_1, ...,q_{n-1} and note that each two element set $\{p, q_j\}$ for j ε $\{1,...,n-1\}$ disconnects S_n as required.   This finishes the proof.

The following corollary is a consequence of Theorem 4.3 and Proposition 4.5.

4.6 Corollary.   Two continua S_m and S_n for m, n $\geq$ 2 are homeomorphic if and only if either m = n or m, n > 4 and both m and n are multiples of 4.

We end the paper by calculating the scale a_n of the similarities f_0, f_1, ...,f_{n-1} needed to construct the continuum S_n, as in the definitions of S_n (see the beginnning of this section).

4.7 Theorem.   For every n $\geq$ 2 let m be the nearest integer to $\frac{n+2}{4}$ if n is not a multiple of 4 and let $m = \frac{n}{4}$ if n is a multiple of 4. Then the scale a_n is expressed by

$$(4.7.1) \qquad a_n = \frac{\sin\frac{\pi}{n}}{2\sin\frac{m\pi}{n}\cos\frac{(m-1)\pi}{n}}.$$

Proof. Let p be the point of intersection $f_0(P_n) \cap f_1(P_n)$. Let m and j be integers such that $p \; \varepsilon \; f_m(f_0(P_n)) \cap f_j(f_0(P_n))$, i.e.,

$$(4.7.2) \qquad p = f_m(f_0(v_m)) \text{ and } p = f_j(f_0(v_j)),$$

where v_k is the k-th vertex of P_n, for $k \; \varepsilon \; \{1, \dots, n-1\}$ (see the beginning of this section).

Because of the symmetry of S_n with respect to the line $y = \tan\frac{\pi}{n}$ we have $m+j = n+1$. Observe that among all directions of vectors $\overline{f_0(0)f_0(v_k)}$ the direction of $\overline{f_0(0)f_0(v_m)}$ is the closest to the direction of the vector $\overline{f_0(0)f_1(0)}$. Calculating the directions we conclude that among the numbers $\frac{2\pi k}{n}$ the number $\frac{2\pi m}{n}$ is the closest to $\frac{(n+2)\pi}{2n}$, and hence m is the closest integer to $\frac{n+2}{4}$. Further, by (4.7.2) we have $1 + a_n(v_m - 1) = v_1 + a_n(v_j - v_1)$, and so $a_n(1 - v_1 + v_j - v_m) = 1 - v_1$. Passing to the real parts of both sides we get

$$a_n(1 - \cos\frac{2\pi}{n} + \cos\frac{2j\pi}{n} - \cos\frac{2m\pi}{n}) = 1 - \cos\frac{2\pi}{n}.$$

After some elementary calculations we obtain (4.7.1).

References

1. M. Barnsley, (1988) Fractals Everywhere, Academic Press, Inc., New York, New York.

2. W.J. Charatonik and A. Dilks, *On self-homeomorphic spaces*, Topology Appl. (to appear).

3. K. Kuratowski, (1968) Topology, Vol. II, Academic Press, Inc., New York, New York.

4. G.T. Whyburn, (1942) Analytic Topology, Amer. Math. Soc. Colloq. Publ., vol. 28, Amer. Math. Soc., Providence, R. I.

5. G.T. Whyburn, *Topological Characterizations of the Sierpiński Curve*, Fund. Math. 45 (1958), 320-324.

Index

A

accessible, 1, 17, 190

accessible composant, 269

accessible in M, 256

accessible point, 269

accessible rotation number, 1

accessible with respect to
the domain, 63

accessible with respect to the
prime ends, 63

accessing arc, 17

admissible vertex map, 111

affine topological contraction,
122

affine transformation, 121

AH-essential, 202

Alexander's Isotopy Lemma, 44

almost periodic 62, 63, 79, 80

almost periodic point, 90

α -chain transitive, 39

α -invariant, 39

arc of accessibility to b, 63

arclike, 174

arcwise connected, 235

area preserving homeomorphism,
43, 44

attracted, 159

attracting, 272

attracting neighborhood, 45

attracting set, 17, 103

attractor, 45, 114

B

basic repetitive unit, 67, 68

basin of attraction, 159

bonding maps, 102

Brouwer Translation Arc Lemma, 48

C

Cantor set, 263

chain, 244

chainable, 93

chain-connected, 244, 256

chains in U, 256

chain recurrent, 17

chain-transitive, 15, 17, 36, 39

chaos, 89

chaotic attractor, 1

chaotic in X, 94

chaotic map, 94

chaotic set, 94

CIP, 211

circlelike, 174

circle of prime ends, 16

Class(Q), 222

Class(S), 204

Class($\widehat{U}$), 207

cofrontier, 52, 60, 81, 183,
   184, 277

compact semigroup, 114

complete invariance property, 209

composant, 186, 269

conjugacy, 102

conjugate mappings, 101

constant type, 78

continuum, 60, 100, 201, 277

contraction, 113

Cremer point, 273

Cremer polynomial, 273

cross cuts, 18, 264

C-set, 204

cutting number for X, 219

C(X), 202

C(X)-selection continuum, 206

D

$D^s(G)$, 240

decomposable continuum, 93

δ -chain, 35

δ -pseudo orbit, 35, 93

dendroid, 231

Denjoy, 18, 52, 183

Denjoy map, 104

derivative of f, 121

Diophantine, 273

disconnection number, 213, 261

E

elongated horseshoe map, 105

ϵ -chain, 17

ϵ -MCIP, 212

ϵ -shadowing, 93

equicontinuous, 62, 115

equivalent, 17, 60

ergodic, 95

exotic continuum, 263

F

Fatou, 263, 272

filled-in Julia set, 274

finitely irreducible, 222

fixed point, 211

fixed point set, 209

forced damped pendulum, 159

fractional part, 68

Fréchet manifold, 244

free arc, 210

free homeomorphism, 45

free modification, 45

G

generalized rotation number, 61, 62

global attractor, 154

graph, 213

H

Hawaiian disk, 54

H-map, 103

Hénon, 15, 21

Hénon attractor, 22

Hénon family of maps, 5

hereditarily indecomposable continuum, 93, 278

horseshoe map, 103

hyperbolic arcs, 47

hyperbolic IFS, 114

I

identification map, 222

IFS, 114

impression, 63, 264

Ind(h,C), 46

indecomposable continuum 51, 60, 93, 159, 263, 265, 267, 277

induced homeomorphism, 60, 61

inessential fixed point, 45

inessential neighborhood, 45

internal composant, 271

invariant set, 130

inverse limit, 102, 232

inverse limit space, 17, 89

inverse sequence, 102

J

Julia set, 263, 272

K

K-composant, 269

k-partition, 244

kernel, 130

kernel of the IFS, 115

Knaster buckethandle, 267

Knaster continuum, 99

Knaster U-continuum, 191

L

Lake-of-Wada channels, 190

length of g, 120

lift, 35

linear topological contraction,
119

links, 66

local Siegal disk, 51

loosely embedded in X, 244

Lozi map, 84

M

map, 202

MCIP, 212

metamorphis, 4

MFPS(F), 212

minimal homeomorphism, 62, 80

minimal ideal, 130

minimally invariant, 184

minimal on Λ , 184

minimal under f_α , 79

monotone, 207

N

neutral, 272

non-cut point, 231

nowhere locally compact, 243

O

ω -chaotic, 95

ω -limit point, 90

ω -scrambled, 94

orientation preserving
homeomorphism,
15, 18, 43, 44

P

parabolic point, 272

ϕ conjugate of A, 135

Poincaré return map, 159, 160

positive topological entropy,
89, 93

prime end, 15, 63, 183, 263

prime end compactification, 60

prime end homeomorphism, 18

prime end rotation number, 16,
60, 61

prime end uniformization, 184

principal continuum, 63

principal saddles, 5

principal set, 264

pseudoarc, 56, 93, 278

pseudocircle, 278

pseudo-orbit, 36

pseudorotation, 61, 71, 81,
183, 184

pseudo-rotation interval, 39

pseudo-rotation set, 39

pseudo-shear, 83

Q

Q-manifold, 244

quasi-Lebesgue measure, 49

quotient map, 222

R

ray, 190

ray-composant, 186

recurrent, 62, 80, 81, 184

recurrent point, 90

refines, 245

repelling, 272

repelling neighborhood, 45

repellor, 45

reversible, 83

right shift map, 102

rotary homoclinic tangency, 6

rotation interval, 36

rotation number, 16, 36, 52, 60, 277

rotation sequence, 36

rotation set, 62

rotation shadowing property, 35, 36

S

selection, 206

semiconjugacy, 102

semiconjugate mappings, 101

semi-continuum, 222

sensitive dependence on initial conditions, 94

simple dense canals, 190

shadowing property, 89, 93

shear map, 37

shift homeomorphism, 17

shift map, 89

Siegel disk, 183, 263, 272, 273

Siegel Disk Problem, 51, 52

Siegel polynomial, 273

simple n-odd, 205

singular values, 119

Smale horeshoe map, 99, 267

snakelike, 93

spans, 66, 68

spectral raduis, 118

spiral map, 37

stable manifold, 175

stable value, 202

standard pseudorotation, 71

strange attractor, 15

strong topological contraction, 113

strongly mixing, 95

subcontinuum, 277

subcontinuum disconnection number,
  240
superattracting, 272
Suslinean, 231, 238

       T

thumb disk, 54
topological entropy, 89
topologically equivalent, 113
topologically transitive, 93, 94
tree, 222
triod, 208
2^X, 202

       U

uniformization, 184

universal, 202, 203
universal covering space, 35
universal for G(n), 215
unstable manifold, 175

       W

Wada property, 158, 160
weakly cut, 219
weakly mixing, 95
weak topological contraction,
  113

       Z

Z-embedding, 244
Z-semigroup, 115